T0138096

Advances in Intelligent Systems and Computing

Volume 677

Series editor

Janusz Kacprzyk, Polish Academy of Sciences, Warsaw, Poland
e-mail: kacprzyk@ibspan.waw.pl

About this Series

The series "Advances in Intelligent Systems and Computing" contains publications on theory, applications, and design methods of Intelligent Systems and Intelligent Computing. Virtually all disciplines such as engineering, natural sciences, computer and information science, ICT, economics, business, e-commerce, environment, healthcare, life science are covered. The list of topics spans all the areas of modern intelligent systems and computing.

The publications within "Advances in Intelligent Systems and Computing" are primarily textbooks and proceedings of important conferences, symposia and congresses. They cover significant recent developments in the field, both of a foundational and applicable character. An important characteristic feature of the series is the short publication time and world-wide distribution. This permits a rapid and broad dissemination of research results.

Advisory Board

More information about this series at http://www.springer.com/series/11156

Andrey Filchenko · Zhanna Anikina
Editors

Linguistic and Cultural Studies: Traditions and Innovations

Proceedings of the XVIIth International Conference on Linguistic and Cultural Studies (LKTI 2017), October 11–13, 2017, Tomsk, Russia

 Springer

Editors
Andrey Filchenko
Nazarbayev University
Astana
Kazakhstan

Zhanna Anikina
Research Centre Kairos
Tomsk
Russia

ISSN 2194-5357   ISSN 2194-5365   (electronic)
Advances in Intelligent Systems and Computing
ISBN 978-3-319-67842-9   ISBN 978-3-319-67843-6   (eBook)
DOI 10.1007/978-3-319-67843-6

Library of Congress Control Number: 2017952860

Printed on acid-free paper

This Springer imprint is published by Springer Nature
The registered company is Springer International Publishing AG
The registered company address is: Gewerbestrasse 11, 6330 Cham, Switzerland

Preface

Dear Reader,

This book is the XVIIth edition of the International Conference on Traditions and Innovations in Linguistic and Cultural Studies (LKTI 2017), which will be held on October 11–13, 2017, in Tomsk, Russia. The conference aimed to focus on a variety of issues such as cultural studies, linguistics, interdisciplinary pedagogy, language teaching and learning. The following topics were discussed in detail:

- Educational environments
- Lifelong learning
- Engineering education
- Evaluation and outcomes assessment
- Tertiary education
- Language teaching and learning
- Linguistic studies

We believe that we are mature enough to experiment and do innovative things. So, this year we did our best to enhance equality through partnership at different levels:

- LKTI 2017 was organized by the Department of Foreign Languages at the Institute of Electrical Engineering, Tomsk Polytechnic University in conjunction with Research Centre Kairos. This allowed to provide collaboration of governmental and non-governmental institutions for research purposes;
- English is a dominant language for international publishing. This fact makes non-native scholars experience considerable difficulties. We sought the ways to occupy this gap being convinced that contributions from non-native scholars are crucial for the global world, on the one hand, generating new insights into research questions and, on the other hand, retaining local voices which is meaningful in order to launch regional research and provide diverse contribution to knowledge;

- Our authors are expert scholars and early-career researchers from various regions of the Russian Federation who demonstrated their achievements, outlined further prospects and shared valuable experience.
- This project will contribute to the PhD research exploring the ways to engage Russian academics into the worldwide academia, identifying the key challenges, and finding the solutions. The research was started at the University of Sheffield (Sheffield, UK) in January 2016 and continued at the University of Westminster (London, UK) in January 2017. We hope to make a substantial contribution to academic context in Russia.

We are grateful to *Advances in Intelligent Systems and Computing* for this precious opportunity to share our research results and will be happy if this book becomes the start of new collaborations.

We wish you enjoyable reading and new discoveries!

Zhanna Anikina

LKTI 2017

Program Committee

Chair

Yuriy Kobenko — Tomsk Polytechnic University, Tomsk, Russian Federation

Co-chair

Zhanna Anikina — Research Centre Kairos, Tomsk, Russian Federation

International Advisory Board

Terry Lamb — University of Westminster, London, UK
Elizaveta Kotorova — University of Zielona Góra, Zielona Góra, Poland
Andrey Filchenko — Nazarbayev University, Astana, Kazakhstan
Andrey Nefedov — University of Hamburg, Hamburg, Germany
Joe Sykes — Akita International University, Akita, Japan

Home Advisory Board

Nikolay Baryshnikov — Pyatigorsk State University, Pyatigorsk, Russian Federation
Marina Bovtenko — Novosibirsk State Technical University, Novosibirsk, Russian Federation

Nikolay Kachalov   Tomsk Polytechnic University, Tomsk,
                          Russian Federation
Alexander Shamov   Minin University, Nizhny Novgorod, Russian Federation

Organizing Committee (Hosting Institution)

Zoya Fedorinova   Tomsk Polytechnic University, Tomsk,
                          Russian Federation
Ksenia Girfanova   Tomsk Polytechnic University, Tomsk,
                          Russian Federation
Lidia Kazarina   Tomsk Polytechnic University, Tomsk,
                          Russian Federation
Peter Kostomarov   Tomsk Polytechnic University, Tomsk,
                          Russian Federation
Veronika Rostovtseva   Tomsk Polytechnic University, Tomsk,
                          Russian Federation
Liubov Sobinova   Tomsk Polytechnic University, Tomsk,
                          Russian Federation
Ekaterina Tarasova   Tomsk Polytechnic University, Tomsk,
                          Russian Federation
Denis Tokmashev   Tomsk Polytechnic University, Tomsk,
                          Russian Federation
Irina Sharapova   Tomsk Polytechnic University, Tomsk,
                          Russian Federation
Victoria Vorobeva   Tomsk Polytechnic University, Tomsk,
                          Russian Federation

Contents

Educational Environments

Analysis of a Lesson as an Educational Resource of Enhancing the Quality of Future Teacher Training

Lyubov A. Nikitina[✉][iD]

Altai State Pedagogical University, Barnaul 656031, Russian Federation
nikitina.fnk@rambler.ru

Abstract. The article considers the text composition as a method to shape future teachers' pedagogical reflexive component, i.e. their ability to analyze a lesson. On the basis of the survey the author reveals such problems in analyzing a lesson as lack of motivation, its formality, detachment in analysis grounds and its realization, absence of analysis results in improvement of teaching methods. The article introduces the technique which includes reconstruction, an analytical comment and analytical generalization to teach the lesson analysis via the text. By reconstructing a lesson or its episode, students recover methodological joint activity at the lesson thus expressing their perception including interest, surprise and attitude to the lesson itself. The analytical comment presents the second text which means that students, having preliminary stated the grounds, analyze the situations using the information from reconstruction and formulate questions. Analytical generalization includes a valid assessment of joint work organized by the teacher during the lesson, detection of problematic issues and their solution. Comprising a textual analysis of the lesson is considered an educational resource in enhancing the quality of future teacher training methods. It means that students reveal methodological activity at the lesson as the way to organize cooperation and master their professional skills.

Keywords: Pedagogical activity · Lesson analysis · Research · Text composition · Educational resource · Quality of methodological training

1 Introduction

Modern requirements of the society to the teacher who is willing not only to be a translator of certain knowledge at the lesson but is also capable to build cooperative educational activity taking into account children's personal development define the changes in methodological training which includes the skills to analyze the lesson conducted by the students, their teachers and fellow students.

Lesson analysis as a research tool allows future teachers, on the one hand, to demonstrate their knowledge of methodological techniques in practice as they learn to correlate what they study with the reality, and, on the other hand, formulate their ideas of pedagogical activity and master their own methodological preparation. The survey conducted has demonstrated that respondents have ranked first the progress of a future teacher during pedagogical practice meaning that he *"conducts methodologically*

A. Filchenko and Z. Anikina (eds.), *Linguistic and Cultural Studies:*
Traditions and Innovations, Advances in Intelligent Systems and Computing 677,
DOI 10.1007/978-3-319-67843-6_1

accurate lessons", *"combines theory and practice"*, *"uses effective methods"* (a higher educational establishment – 83%, a college – 100%) [13]. At the same time, it has been noted that a pedagogical reflexive component which means *"the ability to find the proper way out of any problematic situation during a lesson"* is one of the quality indexes of good training (a higher educational establishment – 43%, a college – 39%). It is possible due to the analysis of methodological organization of joint activity, avoiding the gap between methodological techniques and cooperation with children. Nevertheless, in practice students face a great deal of obstacles in lesson analysis which is determined by lack of practical experience in lesson conduct, absence of reflexive skills, formal attitude to analysis grounds and lack of personal judgments, inability to state the ways of improving methodological organization of a lesson, etc.

The purpose of the article is to consider the opportunities to involve students in analysis research of the lesson via the text which will become an educational resource of enhancing the quality of teaching methods.

2   Literature Review

The approach to lesson analysis by future teachers in foreign publications is rather interesting. Stanford University actively develops and employs the so-called "video analysis" or "micro teaching" [1, 2]. This method claims that "analysis is impossible during live observations and education is a cultural activity which is only effective when the process of education is slowed down and critically analyzed" [16].

With the development of technical aids and availability of information technologies at universities in late 1990s and early 2000s this method is still up-to-date [3, 9, 10].

For instance, during training Santagata et al. demonstrated video recorded lessons of other teachers to students and then analyzed them before future teachers start to conduct their own ones [16].

As a result, students tend to express more critical ideas and explain the choice of their pedagogical strategies [16].

Research papers of Trippa and Rich are also dedicated to studies of video analysis influence on tactic changes in teachers' behavior [18].

Along with video analysis foreign sources depict such a model of lesson analysis as joint team work on planning and introspection of the lesson by students [11]. In foreign theory of teacher training special attention is paid to «Lesson Study» which represents "a form of research activity at the lessons that is aimed at mastering knowledge in teaching practice" [5].

Russian pedagogy has thoroughly investigated and actively employs various ways of lesson analysis: event-related, content-related, conceptual and phenomenological, elementary, causal, logical, structural and functional, etc. [6–8]. Nevertheless, many scientists and experts agree that educators should possess the schemes of lesson analysis and apply them in accordance with various purposes and objectives [4, 14, 17]. While teaching students lesson analysis, the worked out schemes are used as methodological basis.

3   Basic Assumptions

A great variety of types of lesson analysis and schemes gives rise to a formal approach to its realization from the view point of both students and teachers. Our survey conducted in 2016 among the 4th year students getting education in the field "Pedagogical Education" (educational specialization "Primary Education" including 30 people) and primary school teachers (25 people), gives evidence of it.

The majority of respondents mentally subdivide lesson analysis into its components (stages) (87% of students and 47% of teachers).

However, for 35% of teachers to analyze a lesson means to single out positive and negative moments (*"what went right – wrong", "analysis of positive and negative moments", "considering all the parts of the lesson from the point of view of successful and unsuccessful strategies"*). While teachers note that lesson analysis helps them give *"an adequate assessment of their activity at the lesson"*, technique usage, *"reveal effectiveness"*, students have problems with a reflexive component which is determined by lack of practical experience.

Answering the question *"What steps do you think should obligatory be taken to prepare for lesson analysis?"* students (75%) and teachers (60%) choose a plan (scheme) of analysis, aspects and criteria. The respondents' reaction makes it possible to draw a conclusion that the search for a ready plan that can allow them to analyze a lesson is still the major act, while the objective, aspects and grounds of analysis are not evident.

Responding to the question *"What do you more frequently pay attention to when analyzing a lesson?"* the major part of students (91%) has respectively named those components that are generally mentioned in the analysis scheme, i.e. the lesson objectives, tasks, techniques, methods, expected results, metadisciplinary results, and lesson stages (structure). Teachers rank first the correspondence between "objective and result" (40%), ("whether the expected goal has been successfully achieved"), search for positive and negative moments (33%), participants' cooperation and a teacher's performance (15%), methodological conception of the lesson (10%).

Answering the question *"What makes it difficult for you to analyze a lesson?"* 65% of students have mentioned methods of teaching, 30% – lesson assessment from the point of view of realization of desired goals and results, 20% – looking for methodological errors, understanding and distinguishing universal training actions. Teachers have noted difficulties in presence of logic in lesson analysis (60%), the choice of analysis aspect (20%) and assessment of teachers' performance at the lesson (20%).

However, responding to the question *"What should you learn to analyze lessons?"* students mention the stages connected with the lesson planning that namely include such aspects as being able to formulate lesson goals and tasks and point out a number of educational skills to be formed (56%), *"to accurately plan"* (conduct) the lesson, *"to formulate ones' questions"*, and *"to have good cooperation with children"* (35%). It means that for students lesson analysis is connected with its planning. It is quite an explicable fact because their main task is to improve the skills of lesson planning, while analysis appears a formal (time remote) step which has not yet become essential in their professional activity. Teachers answering this question consider it necessary to be able to single out the grounds of analysis (67%), to conduct lessons using a systematic and practical approach (20%).

The survey results make it possible to draw the following conclusions:

- the specific scheme of lesson analysis is relevant for students and teachers because it explains what should be done;
- however, the scheduled logic of analysis does not always make it possible to express one's own view of the lesson and to assess cooperation between a teacher and pupils. The analysis itself becomes bulky (*"a person tends to pay attention to all the positive moments"*, *"to consider all the lesson requirements"*);
- the choice of analysis aspects causes difficulties for both teachers and students which actually means the starting point in analytical activity. In other words, one and the same lesson can be viewed from various angles that can help improve it.
- Long term practice of teaching students lesson analysis demonstrates the following:
- using ready schemes, students formally state the ground, without explaining their judgment;
- the analysis turns into retelling of the lesson progress *"the teacher greeted the children, asked to write down the date in the notebooks…"*, *"the children opened the textbooks and read a rule"*, *"the teacher asked the children "What have you learnt at the lesson today?"*;
- it is difficult for students to express their opinion on this or that analyzed technique in reference to children's activity because they have not recorded in their observation report what the children were doing, whether they actively participated in discussions or demonstrated their initiative. Students always focus on teachers' actions (their questions and tasks), which is quite understandable as they are not yet ready "to see" children;
- lesson analysis primarily presupposes its recording, in other words, making a report. Students employ different ways to do it: they write down the teacher's questions, sometimes they record the children's answers. Stated differently, students observe only the evident aspect of the lesson attended. They do not consider the other situations at the lesson: the teacher asks children a question that has been answered by a number of pupils, but only one response has been recorded; the teacher listens to children's answers but dwells on the one he has been expecting without taking into account the others. There may be many similar situations in class but a student does not consider them because he is concentrated on recording questions, tasks and directions of the teacher. Thus, children's actions and their participation in the lesson remain unobserved and, consequently, unmarked in the report.
- While analyzing a lesson, a student who uses the scheme (aspects) given by the teacher, more frequently enumerates teaching methods but is unable to assess them from the point of view of accuracy, relevance and necessity in children's activity. The analysis of the method technique presupposes not only its naming but evaluation of its contents and effectiveness that is reflected in children's actions;
- self-analysis of the conducted lesson turns to be even more problematic for students and remains formal because they arrange it in accordance with the goal formulated: *"if he has had enough time to complete the tasks or not"*, *"if he has managed to do everything he has planned or not"*, *"if all the children have been working or not"*;

in addition, they evaluate the lesson from the point of view of children's discipline at the lesson (*"the children did not maintain discipline, therefore not all the exercises had been completed"*).

4 Methodology

The ability to analyze a lesson is being formed not only during subject study of teaching methods throughout the training process at a higher educational establishment. Nevertheless, this skill is complicated firstly, by the fact that primarily students learn the unique features of teaching methods which do not always accompany direct observation of real school practice. From the start, students accumulate theoretical knowledge of specific nature of lessons, techniques and so on, and so forth. Secondly, they often use ready schemes (plans) to analyze a lesson which, of course, will be rather helpful but such an analysis quite frequently turns into a formal act that singles out positive or negative strategies but does not reveal the causes of these actions. In addition, such analysis "lacks" children and their participation in the activity organized by the educator.

Within the formation of an ability to analyze a lesson, from our point of view, students should be able to master such research strategies as reconstruction (reproduction), an analytical comment and analytical generalization [12]. It should be based on composing the text of phenomenological description as a way and means of understanding teaching methods and a model of educational innovation – "personal presence in education" [15].

Reproduction is a reconstruction of the lesson seen (conducted) in the form of a text which is subject to analysis. Text composition is a complicated task because it is necessary not only to render the teacher's actions in written form, which more frequently takes place in oral form (questions, tasks given by the teacher to children), but also children's reaction (not formal in reference to the task offered but an emotional response, initiative, drawbacks, questions, etc.). Due to this a student should emotionally perceive how the lesson is being conducted as such a personal act demonstrates presence of educational sense of the strategies carried out. The point is that students learn to compose the so-called "retelling" of the lesson fragment in university classes. Reconstruction differs from a report by describing the participants' activity (not only questions and answers).

An analytical comment also presupposes text composition where the lesson is characterized from a certain point of view: methodological organization of joint activities at various stages; usage of certain techniques according to the lesson goal and type in mind, etc. At the same time, the comment contains text "quotations" that is reconstructions to illustrate the lesson structure. While commenting on the fragment or the lesson itself from a certain point of view, the text acquires not only evaluative judgments (*"whether the work has been arranged wrong or right"*) but proven judgments from a reconstruction text (the evidence itself). The comment gives rise to a number of questions to one's activity: *why the technique has not worked, what factor has prevented to manage the children* (didactic material, question formulation, the type of work), *how to determine children's problems at the lesson*, etc. The questions set are a reason to consider teaching methods, this or that technique (its relevance, children's orientation, effectiveness).

Analytical generalization is a final step in analysis in which there is a general idea of the lesson and its evaluation according to the criteria stated earlier. Such triple "pronouncing" of the lesson gives students an opportunity, on the one hand, to acquire teacher's specific methodological vocabulary, and on the other hand, to learn "seeing" techniques in action and assess them from various angles, furthermore, it develops written speech and contributes to formation of self-analysis.

5 Data Analysis

Approbation of the offered technology for teaching lesson analysis as a research tool at higher educational institutions and during pedagogical practice has demonstrated the fact that students' idea of teaching activity begins to change. Thus, analyzing their own lesson out of 30 students who have been participating in the research for 3 years, 58% (41% in the 4th year, 30% in the 5th year) single out various grounds for analysis: while they are in their first practice, they look for successful situations in their performance mainly connected with "following the work book" and adhering to "all the tasks and techniques at the lesson". Further, 36% (in the 4th year) and 40% (in the 5th year) of students pay more attention to the problems in personal actions and try to find their causes (organization of children's activities). In addition, 23% (in the 4th year), 30% (in the 5th year) of students observe the discrepancy in the techniques of joint activities and look for the ways to reorganize them [13].

Data analysis has allowed detecting that lesson methodological organization becomes meaningful for students during teaching practice which gives evidence of changes in their methodological training. This alteration has become possible due to the students' ability to use lesson analysis as a research tool.

6 Conclusion

Mastering a research skill of lesson analysis is an educational resource that helps students realize the idea of pedagogical activity as they reveal its goal that is organizing joint educational cooperation. Text composition leads to:

- understanding its purpose by a student, writing the text;
- forming a subjective position to one's professional activity;
- reconstructing and reforming one's activity;
- using research as professional technique to organize and realize collective educational activity at the lesson.

References

1. Allen, D.W.: A new design for teacher education: the teacher intern program at Stanford University. J. Teach. Educ. **17**(3), 296–300 (1966)
2. Allen, D.W., Ryan, K.: Microteaching. Addison-Wesley, Reading (1969)
3. Carpenter, T.P., Fennema, E., Franke, M.L., Levi, L., Empson, S.B.: Children's Mathematics: Cognitively Guided Instruction. Heinemann, Portsmouth (1999)
4. Churakova, R.G.: Tekhnologiya i aspektnyj analiz sovremennogo uroka v nachal'noi shkole [Technologies and aspect analysis of the modern lesson in elementary school]. Akademkniga, Moscow (2011). (in Russian)
5. Dudley, P.: Lesson Study: Professional Learning for Our Time. Routledge, London (2014)
6. Kodzhaspirova, G.M.: Pedagogika [Pedagogy]. KNORUS, Moscow (2010). (in Russian)
7. Konarzhevskiy, Y.A.: Analiz uroka [Lesson Analysis]. Centre "Pedagogicheskyi poisk", Moscow (2009). (in Russian)
8. Kul'nevich, S.V., Lakotsetina, T.P.: Analiz sovremennogo uroka [Analysis of the modern lesson]. Uchitel, Rostov-on-Don (2006). (in Russian)
9. Lampert, M., Ball, D.: Teaching, Multimedia, and Mathematics. Columbia University Press, New York (1998)
10. Mumme, J., Seago, N.: Examining Teachers' Development in Representing and Conceptualizing Linear Relationships Within Teaching Practice. American Educational Research Association, Chicago (2003)
11. Myers, J.: Lesson study as a means for facilitating preservice teacher reflectivity. Int. J. Scholarsh. Teach. Learn. **6**(1), 15 (2012). https://doi.org/10.20429/ijsotl.2012.060115. Accessed 01 July 2017
12. Nikitina, L.A.: Ot umeniya analizirovat' urok k refleksii sobstvennoj deyatel'nosti [From lesson analysis to reflection of one's own activity]. In: Savelieva, L.V., Shegoleva, G.S., Gogun, E.A. (eds.) Proceedings of the International Conference on Linguistic and Cultural Education, pp. 111–117.BBM, St. Petersburg (2016). (in Russian)
13. Nikitina, L.A.: Stanovlenie issledovatel'skoj kompetecii v metodicheskoj podgotovke studentov pedagogicheskogo vuza v usloviyah innovatcionnogo razvitiya obrazovaniya [Formation of research competence in methodological preparation of pedagogical students within innovative education] (2014, unpublished thesis). (in Russian)
14. Pichugin, S.S.: Analiz sovremennogo uroka v nachalnoj obshcheobrazovatel'noj shkole [Analysis of the modern lesson in primary school]. IRO RB, Ufa (2015). (in Russian)
15. Prozumentova, G.N.: Strategiya i programma gumanitarnogo issledovaniya obrazovatel'nyh innovacij [Strategies and schemes of academic research of educational innovations]. Tomsk State University, Tomsk (2005). (in Russian)
16. Santagata, R., Zannoni, C., Stigler, J.W.: The role of lesson analysis in pre-service teacher education: an empirical investigation of teacher learning from a virtual video-based field experience. J. Math. Teacher Educ. **10**(2), 123–140 (2007)
17. Selevko, G.K.: Tekstovyi aspektnyi analiz uroka (in Russian) [Text and aspect analysis of the lesson] School technologies, vol. 2, pp. 105–119 (1997). (in Russian)
18. Trippa, T.R., Rich, P.J.: The influence of video analysis on the process of teacher change. Teach. Teach. Educ. **28**, 728–739 (2012)

Collaborative Learning in Engineering Education: Reaching New Quality and Outcomes

Zoya V. Fedorinova[1]([✉]) (iD), Svetlana I. Pozdeeva[2] (iD),
and Alexandra V. Solonenko[2] (iD)

[1] Tomsk Polytechnic University, Tomsk 634050, Russian Federation
favl@rambler.ru
[2] Tomsk State University, Tomsk 634050, Russian Federation
svetapozd@mail.ru, erutpac@yahoo.com

Abstract. The article presents the research on the means to enhance collaborative learning practices, to reveal its potential for engineering students' soft skills formation, as well as to trace the connection between purposeful organization of collaborative activity by a teacher and new level of learner's participation in collaborative activity. The recommendations on the choice of the appropriate interaction model for teachers organizing collaborative learning are provided and the principles of open educational teaching activity with the space for sense-making and student initiatives are generated. The research approach based on phenomenological description of the precedent (class organized with the use of case-study method) allows to reveal and analyze the educational potential of the method used and its effect on the quality of collaborative learning.

Keywords: Soft skills · Case-study method · Collaborative learning

1 Introduction

Today an engineer is to work in an extremely heterogeneous and fast changing world, the subjects of which are interdependent. The basic competency of contemporary engineering graduates is to carry out their activity in the area of the latest innovative technologies and developments, with account for global trends, varying social structures and modern people's needs [9]. The given ability is determined not only by mastering **professional competencies** in a broad sense of the term (so-called *hard skills*), but also **universal strategic competencies** (so-called *soft skills*) which can be revealed in professional activity in the area of interpersonal communication. The soft skills include:

1. ability to communicate: to negotiate, discuss, argue, prove, build communication for effective achievement of the set objectives, and solve conflicts;
2. critical and creative thinking;
3. skills of team work when doing an interdisciplinary task;
4. self-learning process management due to individual education-career trajectory (*autonomous learning*) and others.

© Springer International Publishing AG 2018
A. Filchenko and Z. Anikina (eds.), *Linguistic and Cultural Studies:
Traditions and Innovations*, Advances in Intelligent Systems and Computing 677,
DOI 10.1007/978-3-319-67843-6_2

Soft skills formation and development are pedagogical problems being solving by the active methods of learning organization (case-study method, problem-based learning, project learning, etc.). Referring to active learning methods, the first is related to changes in learning objectives, transition from ready knowledge translation to learning and processing of a completely new, team-generated knowledge. The second one targeted at the building of educational relations between a teacher and students and organization of non-didactic interaction between them (not only teacher teaches a student, but also a student can teach his teacher something) [6]. In our research, we would like to point out the main pedagogical principle of the already existing method - principle of cooperative learning organization for the teacher and the students - and to consider it over, i.e. enhancing its humanities, communicative and interpersonal components. Furthermore, we provide a set of recommendations herein for class organization, which contribute to performing a collaborative learning activity at a new level, facilitating a successful soft skills development for future engineers.

First ideas about changing the learning type for students to get more opportunities, have a dialogue with each other, learn from each other, look for solutions to various problems cooperatively, were expressed in the 1950–60s by teachers of secondary schools and medical institutes of the United Kingdom (Ch. James, L. Smith, M. Abercrombie).Those ideas were supported in working practice by American universities' faculty in the1970s, as a solution to a big gap in knowledge and academic skills among the students, an attempt to form cooperation and mutual assistance between weak and strong students for gradual "leveling" of an academic group (D. Bremer). Collaborative learning became a part of common educational practice in the 1980s: it has evolved into a methodological principle of learning management rather than a simple technology. This principle does not rely much on the student group work under a studied problem, but focuses on group participation in the process of "intellectual talks", collective decision-making, and the discovering of an opportunity to get control of their knowledge and later to be independent in their application [2]. This principle also changes the view about position and functions of the pedagogue. To organize the conditions updating collaborative learning, teachers should think over their role in the learning process and transform the position of the ready knowledge translator into the organizer, facilitating mini-communities in the classroom that aspire students' self-development and self-learning [1].

Collaborative learning is also applied in engineering education, despite a number of difficulties: difference between the classic engineering education and active educational unreadiness as well as low motivation of some students (especially the strong ones) hinder successful collaboration in the group [8]. Collaborative learning constantly demonstrates the qualitative changes in the results of contemporary engineering education: development of teamwork competency, communication skills, a new level of the subject knowledge (*knowledge obtained through participation as result of action*), acquisition of responsibility for own learning (both individual and shared by team members) [8, 12].

Collaborative learning in Russian pedagogy should be perceived within the Post-Soviet tendency towards democratization and transparency in education. Integration of foreign educational methods and principles with Soviet pedagogical science resulted in several approaches, based on democratization and humanization of educational process, active learning methods, developments of the Soviet period psychology (Leontyev's activity theory, Elkonin-Davydov theory of learning activity and developmental teaching). One of such approaches we draw upon is pedagogy of collaborative activity, authored by the Russian scholar G. Prozumentova.

Within the framework of collaborative activity pedagogy, the focus of research is directed on collaborative activity as a factor of influence on a person's education, as the subject of special efforts by the teacher who organizes, directs, facilitates, and summarizes students' collaborative activity. Thus, we take a great interest in *organization* of such *collaboration* (beginning, development, ending, teacher's role at every stage of collaboration), its *qualitative assessment* (nature and effectiveness of collaboration, the degree of students' engagement and its effect on a final result), tools of *experience reflection* on taking part in this kind of activity by the students themselves to develop their critical thinking, to form a conscious and responsible position towards one's own education.

2 Materials and Methods

Materials and methods used in our research are presented within the logic of the humanities research approach developed by Prozumentova [11]. The approach is distinguished by participation, involvement, and introduction of the persons into their development and education. The humanities approach distances itself from "external nature" of science as a body of knowledge and makes subjectivity a cornerstone viewed as participation, understanding, interpretation and description. Participation in the collaborative activity is related to acquisition of new experience, requiring understanding, correlation with a person's own values and needs, interpretation, analysis for further studying by the participants (students, teacher). These requirements determine selection of the research method: it is based on the phenomenological approach which is focused on direct study of experience with the emphasis on importance and sense of subjective view of the situation, view of consciousness as a given, active and filled with sense, as well as recognition of new knowledge generation from subjectification of life experience, reflection on personal experience [10, 13]. The act of personal action of self-description, the interpretation, and discovery by the person of the sense of their actions is important for us [7]. The next step is based on textual descriptions of phenomena (teachers describe their own experience in collaborative learning organization) where specification of repeated typical signs of phenomenon are observed, then the phenomena are systemized and systematically interpreted. These procedures bring the stated research method closer to the grounded theory by Strauss and Corbin, based on the unique research movement "vice versa"; from facts description to their explaining, from understanding to the notion, from presentation to justification, from the given to its conceptualization. By explaining the necessity of such research movement "vice versa" (traditionally the research moves from building the concept to analysis of particular cases), Strauss, Corbin note that such research allows investigating "stories of life and behavior...organization activity" and thus create the "grounded theories" [4].

3 Situation. Application of Case-Study Method for Collaborative Learning

The description of case-study method application at the stage of new topic introduction, module "Ecology" is presented. The following case was discussed in class: "Industrial waste is a global problem of our time", the case author is the 2nd year student.

3.1 Description of the Phenomenon

At the first stage i.e. to involve the learners into communication, they were suggested to participate in the process of cooperative formulation of the coming class topic. To engage the maximum number of participants, the association method was used: students had to choose the associations to the term *"energy sources"*. Different opinions were expressed and various types of energy sources were named (*coal, wind, water, uranium, etc.*), some adjectives (*natural, expensive, dangerous,* etc.), and set expressions (*power plants, generate energy,* etc.) were listed. Thus, we managed to involve *all* participants in the communication. Associations with the suggested expression were so different and unexpected that, for example, some adjectives suggested in the list of associations e.g. *"innovative"*, "provoked" a question from other participants about what such association is connected with, followed by the student's explanation that today innovations in different spheres are actively discussed and in future a new source of energy could be discovered and named innovative. It was not only the number of participants that increased but also directions of interaction, characterized by questions to each other referring to personal experience and interdisciplinary knowledge.

Then participants were suggested sharing information about new, just discovered energy sources that can be considered as innovations nowadays. Communication organization in such a way is characterized by increase of participants' number, number of questions to each other, multiple and various statements, involvement in the discussion relying on the personal experience, available knowledge, and here the communication is not anonymous but is presented in I-statements, on behalf of a particular person. In addition, one of the features of such communication is multiple and various topics mentioned within one assignment during a relatively small amount of time. The students discussed such topics as various types of energy sources, innovations in the energy sector, environmental problems, connected with operation of this or that energy type.

The second stage of the class became the stage of collaborative activity development. To involve the students in the collaborative activity, prior to beginning of work with the case each group was suggested discussing and making a list of rules everyone should stick to so that discussion didn't get into arguments among its participants. The discussion was planned and organized by students themselves, they discussed and decided who would write down the coming suggestions, offered the rules, selected the most interesting suggestions from their point of view and approved the final list of rules.

Having obtained the material for work, students themselves planned their work with case materials, showed their initiatives in goal-setting, determined the work sequence with the material, organized discussion, calculated the time, made decision and chose the speaker. Actualization of learning process sense-making for the participants was observed during collaborative activity and communication: there were the questions to each other about the point and sequence of actions in working with the suggested material, about nature of this or that opinion and its application in this particular case. Thus, we could observe the transition from "forced", "prescribed" speaking by the communication and collaborative activity conditions to "natural" and "free" communication.

The third and final stage of the class became the stage of reflection results and process of students' activity. To involve all the students in the feedback and reflection, each of them was suggested characterizing the class by one adjective, for instance, "interesting, boring, informative, etc." and explaining their opinion if they wish. Then each of them was offered to answer three questions in writing, such as: *"What new did you discover during the class?", "Where and how could you apply the experience from the lesson?", "What do you wish to improve in work in the class using such case-study method in future?"*.

3.2 Analytical Comment

The suggested model of a class using case study is based on the collaborative activity organization developed by G.N. Prozumentova. We point out three main stages:

The first stage: *Immersion in collaborative activity.*

Key objectives: forming motivation for collaborative activity; demonstration of initiatives of discussion participants. The following options for work are possible at this stage:

1. A case can be handed out to students before the class for independent study and preparation of answers. Knowledge of case by students and their interest in case and its problems is monitored at the beginning of the class.
2. The class can start from prepared questions, actualization and systemizing the students' knowledge on the topic to be discussed, interdisciplinary connections, and in such a way leading students to the case.

The second stage: *Organization (deployment) of collaborative activity on problem solving.*

Key objectives: organization of group and inter-group discussion, results preparation and presentation. The following options for work are possible at this stage:

1. Students are divided into groups for collective discussion, preparation of answers to questions, making decision during the given period of time.
2. If the case is for self-study at home, then the group compares individual answers, adds them and comes to a unified position, which is shown in presentation. Each group chooses a "speaker" who will present the group decision. After a group presents its decision, an intergroup discussion is organized. Teacher organizes and leads the discussion.

The third stage: *Analysis and reflection on collaborative activity experience.*

Key objective: demonstration of educational and learning results.

At this stage the effectiveness of class organization is analyzed, the problems of collaborative activity organization are shown, the objectives for further work are set.

The teacher actions are to finish the discussion, to analyze the case discussion process and all groups work, and sum up.

Situation modeling allowed not only pointing out the stages of activity organization at the class using case-study method, but also determining the conditions, under which the potential of this method is developed. We found out that change in communication (growth in the number of learners involved in communication, diversity and number of topics discussed), change in quality of collaborative activity (participation in organization and planning, setting and achievement of collaborative activity goals) occur when class participants are given the right to choose topics for discussion, method of activity, opportunity for expressing the initiative is provided, and update of personal experience takes place. Particular emphasis is placed on the role of the teacher, who not only organizes the interaction but also takes an active part in it, proposes new ideas as well as encourages students' initiatives. Moreover, teachers should be highly- qualified, because they need to use a special format at each stage of their foreign language course: from the introduction of the new material to skills development [3].

3.3   Analytical Conclusion

Thus, among educational effects stipulated by the case-study method application, were revealed the effects, serving as evidence of changes in communication and collaborative activity quality, such as:

- *Emergence of personal initiatives and desire to realize them in teamwork and communication.* This is reflected in the way the students plan, arrange their teamwork and negotiate, which is evident from the texts: "I suggest that we first discuss …", "First, let everyone read, get acquainted with the information, and then we'll listen to everyone' opinion…", etc. During a regular class, the students' speech activity is usually confined to an answer to the question asked by the teacher, and making an exercise, following the model. Thus, students are deprived of opportunity to take the initiative, creativity as well as to be self-reliant in fulfilling the assignment.
- *Personal and emotional engagement in teamwork and communication.* It means that students are not only "involved" in the communication, they "get carried away" and "live" it. Furthermore, in the process of interaction the moments of emotional tension, excitement of students are constantly observed, that is due to existence of problem situation, which needs to be solved, number of students in the discussion, level of their engagement in the problem-solving process, their cognitive activity and availability of different points of view. This can be seen from the following texts: Students: "I was concerned by the fact that …", "Do not get excited, let the others say…"; Teacher: "… the situation was heating up, Anton insisted on his point", etc.

The quality of engagement provided by the case-study method should be mentioned specifically [5]. Besides personal and emotional engagement, we observe subject engagement, which is reflected in students' desire to build communication in a foreign language during the class. In communication process all students were involved in the discussion, even so-called "low-achieving" students, who preferred to be silent during the regular class. Being "involved" and "carried away" during discussion, students focused their attention more on the content of statement, gradually going to its quality, when number of grammatical mistakes reduced and the words were selected better, according to the context, there appeared idioms, idiomatic expressions and metaphors in speech. As a result, a conscious and therefore qualitative mastering of subject content took place [5].

- *Search for meaning of their actions, statements as well as of other students' actions and statements.* During communication there were multiple moments of "semantic tension", related to arising questions – students referring to each other: "Why?", "What for?", "By what criteria...?", "On what grounds...?", etc. Thus, students try to find and understand the sense of their actions and other participants' actions and phrases.
- *Value-conscious attitude to each other and themselves.* Another important indicator of change in the quality of communication is the value-conscious attitude of students to each other which is apparent from students' desire to listen to and consider each person's opinion as well as put themselves in place of the other. This is reflected in the following texts: "Imagine yourself in his place, what would you do in this situation ...", "...do not interrupt, it is hard for him to speak...", to hear and understand the other: "If I understand you correctly ...", "...by this you mean that ...", etc.

4   Results and Discussion

Thus, our empirical research has shown that application of case-study method results in the following effects:

- communication intensity increases while a solution to a problem is worked out together with students' emotional engagement;
- students as participants of collaborative activity define collaboration and communication as meaningful component of task completion, they are ablet o reflect upon their participation in communication and interaction;
- students take various roles in communication and change them while communicating;
- students perceive their activity as meaningful one (sense-making process and students' reflection upon it);
- students see the task they work upon as relevant to their future professional activity.

These effects contribute to development of such soft skills as communication, critical thinking, teamwork, and self-learning organization.

Furthermore, the mentioned above effects emerge when a teacher takes the role of collaborative learning organizer and takes the following steps:

1. coming up with the task which leads into collaborative activity, create a situation which provokes interaction and collaboration between students;
2. facilitating discussion process;
3. defining the interaction rules together with students;

4. transferring to the board and sum up the results of student discussion without imposing teacher's 'right opinion';
5. gathering feedback from students on the process of collaborative activity.

With those steps, taken collaborative learning acquires the following features:

- it is saturated with students' meanings of the actions, their aims and objectives of the task to be completed according to their own priorities and interests;
- various students' initiatives emerge aimed at modifying and the process of content of collaborative learning;
- task completion becomes an activity for the whole team in which each student can take his own role.

Those features can be detected through the feedback from students focused upon their experience of participation in collaborative learning. The questions we ask students can be divided into three main groups: *What?* (What happened during the entire work? Who was involved in work? What needs were met?), *So what?* (What did they think or feel? What did they learn from the experience they obtained in the class?), and *What now?* (What do they know now that they didn't know before? What attitudes and feelings do they have about the experience that they did not have before? Are they aware of any other changes that occurred in knowledge, skills, attitudes, or feelings as a direct result of this experience? If so, explain. How do they actually learn what is most important to them? What do they think they will remember or retain in other ways after the experience? What will they probably share with or demonstrate to others in the future? What changes would they suggest for future group experience?).

5 Conclusion

It is noteworthy that collaborative learning contributes to the engineering education development providing students with various opportunities to gain valuable experience of problem solving in collaboration and communication with peers, reflecting upon such experience and relates it to the aims and priorities of future professional activity.

References

1. Bruffee, K.A.: Collaborative learning some practical models. Coll. Engl. **34**(5), 634–643 (1973)
2. Bruffee, K.A.: Collaborative learning and the conversation of mankind. Coll. Engl. **46**(7), 635–652 (1984)
3. Buran, A.: How to use blogs in creating special opportunities for language learning. Mediterr. J. Soc. Sci. **6**(1), 532–536 (2015)
4. Corbin, J., Strauss, A.: Basics of Qualitative Research: Techniques and Procedures for Developing Grounded Theory. Sage Publications, Thousand Oaks (2008)
5. Fedorinova, Z.V., Vorobyeva, V.V.: Educational potential of case-study technology. Procedia – Soc. Behav. Sci. **206**, 247–253 (2015)

6. Felder, R.M., Woods, D.R., Stice, J.E., Rugarcia, A.: The future of engineering education II. Teaching methods that work. Chem. Eng. Educ. **34**(1), 26–39 (2000)
7. Geertz, C.: Thick description: toward an interpretive theory of culture. In: Martin, M., McIntyre, L. (eds.) Readings in the Philosophy of Social Science, pp. 213–231. Mit Press, Cambridge (1994)
8. Gol, O., Nafalski, A.: Collaborative learning in engineering education. Globa. J. Eng. Educ. **11**(2), 173–180 (2007)
9. Leicht-Scholten, C., Steuer, L., Bouffier, A.: Facing future challenges: building engineers for tomorrow. In: New Perspectives in Science Education Conference Proceedings, pp. 32–37. Libreriauniversitaria.it Edizioni, Padova (2016)
10. Lincoln, Y.S., Guba, E.G.: Naturalistic Inquiry. SAGE, Thousand Oaks (1985)
11. Prozumentova, G.N.: Perekhod k Otkrytomu obrazovatel'nomu prostranstvu. Fenomenologiya obrazovatel'nykh innovatsy [Transition to the Open Educational Space. Phenomenology of Educational Innovations]. Tomsk State University Publishers, Tomsk (2005). (in Russian)
12. Requena-Carrión, J., Alonso-Atienza, F., Guerrero-Curieses, A., Rodríguez-González, A.B.: A student-centered collaborative learning environment for developing communication skills in engineering education. In: Proceedings of Education Engineering (EDUCON), pp. 783–786. IEEE Press, New York (2010)
13. Van den Berg, R.: Teachers' meanings regarding educational practice. Rev. Educ. Res. **72**(4), 577–625 (2002)

Lifelong Learning

Supplementary Professional Education as a Socially Relevant Component of Lifelong Learning

Olga Yu. Makarova(✉) , Mariia I. Andreeva ,
Olga A. Baratova , and Anzhela V. Zelenkova

Kazan State Medical University, Kazan 420012, Russian Federation
mrs.makarova@yandex.ru

Abstract. The article specifies an issue of supplementary education and supplementary professional education, as its' constituent, viewed as alternative to traditional education. The authors define lifelong learning based on andragogy studies. The article presents the origin of the term 'supplementary education'. The article describes lifelong education from economic standpoint, i.e. its' contribution to overcome economic crisis worldwide. Moreover, the authors specify lifelong education in Europe and the USA, namely, study courses for employees implemented in the USA; cooperation between educational institutions and industries in Great Britain; development of universities' professional aspect in France, etc. In addition, an application of lifelong education in Russia is described. The programme "Translator in the field of professional communication" implemented at the department of foreign languages at the Kazan State Medical University proves supplementary education to be efficient and relevant.

Keywords: Supplementary education · Lifelong learning · Adult education · Overcoming economic crisis · Application of education

1 Introduction

Lifelong learning becomes more important educational domain within modern society. Nowadays, admittedly, basic education fails to provide a person with lifelong knowledge, skills, expertise, personal qualities and values which enable him/her to complete various social roles and develop as a person. Thus, an educational system functions and evolves with regard to lifelong learning perspective in every developed country in the world.

The significance and aim of the paper is to introduce a notion of lifelong education and supplementary professional education, as its' socially relevant component, specified within andragogy, into broad audience. Andragogy is defined as a branch of pedagogy that studies theoretical and practical issues of education of adults considering regional and professional aspects.

Lifelong learning is a holistic set of means, ways, and forms to acquire, extend and broaden general education, professional competence, and culture, to train personal civil and moral awareness.

© Springer International Publishing AG 2018
A. Filchenko and Z. Anikina (eds.), *Linguistic and Cultural Studies:
Traditions and Innovations*, Advances in Intelligent Systems and Computing 677,
DOI 10.1007/978-3-319-67843-6_3

The lifelong learning perspective was introduced by the greatest theorist P. Lendgrand at UNESCO forum.

According to him, every educational system prioritizes person, who needs to be given conditions to develop his/her abilities in full throughout whole life; traditional boundaries of study years, work and professional accomplishment are eliminated.

A fundamental work 'Learning to be' was issued by a special UNESCO committee headed by French statesman E. Faure in 1972.

Lifelong learning was stated to be the basic principle of educational reforms.

Recently a number of Russian and foreign researchers have been focusing on the lifelong learning perspective.

According to Zmeev, lifelong learning perspective distributes personal educational periods and work throughout life more rationally, subdivides education on phases of primary, or basic, and consecutive learning, provides person with necessary knowledge, skills, expertise, qualities, and values when needed [9].

It is to be emphasized that terminology which denotes lifelong learning differs in various countries, and reflects various aspects of the notion, namely, continuing education, lifelong education, adult education, postgraduate education, recurrent education, further education, etc.

2 Supplementary Professional Education Worldwide: Background and Dynamics

A modern alternative of traditional education includes models of lifelong learning and their social component, i.e. supplementary professional education, which is more flexible in content and more democratic in shape.

The supplementary professional education is an inherent, the most flexible and developing stage of lifelong learning. A number of researchers argue that 'continuous education is directly linked to supplementary education phenomenon and educational services' [3] and view lifelong learning as a supplementary professional education [7].

The development of a supplementary professional education has gained special value in Russia as it may be considered as the greatest backup to increase efficiency of social labour. Historic background proves wide scale application of educational system, primarily by higher school, to overcome critical economic situation. Americans, for example, believe education to provide 40% of economic growth and, thus it is considered as a basic component of investments to develop production. Efficient introduction of advanced technologies presupposes an increase in employee's education level. The specialists' training period lasts for 5 years in higher educational institutions, whereas, in fact, measures which improve higher school are efficient for 15 years. Modern conditions demand fast educational feedback. The issue can be solved by continuous development of supplementary professional education (hereinafter SPE) system. Worldwide SPE is regarded as the most flexible and fruitful constituent of specialists' continuous education.

It should be highlighted that nowadays period of specialists' training in higher education institutions equals the period of data deterioration. Data is considered to deteriorate in half by 3 or 4 years [1]. As a result, higher school cannot provide complex training of a specialist by itself. It becomes efficient when combined with SPE system. Prioritized SPE system in staffing country's economics is recognized worldwide. Back to the USA background, we may note that the SPE system covered 20 millions of people in 2000. Each of them took advanced study courses for 500 days per 10 work years (about 10–15% of work time). In Russia index is as follows: 30 days per 10 work years [8].

Nowadays, educational system in most countries prioritizes sciences, and trains a lot of practitioners, but not research workers. Pragmatic employers stress practical skills and combine them with fundamental knowledge only to meet the demands of general development. The higher educational institutions and industrial companies collaborate, and provide integral educational process: university performs general academic training and research methodology, whereas company shares special skills and knowledge. Nowadays, the USA spends 200 billions of dollars on advanced study courses for employees. An average American learner's image changes drastically. The number of senior learners increases. According to American researchers, the number of SPE students aged 18–24 increased by 34% between 1980 and 2008. The number of senior learners, who took the courses, increased by 170% within the same period.

Not satisfied with the educational outcome, Great Britain prioritized bonds between education and practice during the reform in 1988. As a result, according to the Confederation of British Industries, about 63% of schools had already kept regular contacts with businesses and 46% corporations maintained contacts with schools [8].

Reforms of higher education in France focus on professional competence of universities, known as training centers of fundamental research, which barely meet the demands of specific pragmatic professionalization.

An educational reform in China was started in 1985. It reconstructs higher education to meet the demands of economic development and focuses on enhanced establishment of institutions with short term training. In general, according to the obtained results, new higher educational institutions with full term training will not be established in the meantime [6].

3 Results and Discussion

Currently, every educational initiative rests on the idea of continuous education in Russia. The idea is implemented in two ways. Firstly, within postgraduate education, which implies advanced study courses, postgraduate studies, doctoral studies, and scientific internship. Thus a specialist upgrades professionalism within basic education. Secondly, an institute of supplementary professional education may be considered as a way to implement continuous education. It aims: to retrain the specialist, i.e. he/she changes the profession or obtains qualification additional to present professional education (both secondary and higher), and to provide various educational professional services.

Currently, an outlook is rather different in Russia. The state and society needs to develop brand new models of social institutions' activity within present economic environment. Thus, one of key tasks is to reform educational system, namely, to set potential for supplementary professional education.

These circumstances predetermine the need to train more qualified population to be re-professionalized.

The growing significance of SPE is a fact which goes without saying. Large scale research and technical projects introduced last century, and the greatest technological, social, economic, and cultural progress of modern days is possible due to the introduction of SPE system worldwide. It takes essential part in social forecasts of the XXI century. Primary audience of the system of supplementary professional education involves people who somehow combine learning, cognition and practice. As a rule, these are people of definite level of educational and professional training, motivated by education.

In other words, the SPE system was the first to face market demands, to train a competent specialist.

It is to be said, that literature research fails to define 'supplementary education' as a whole. The 'Education and pedagogic dictionary' defines children and adults' education separately as '... a branch of education, which meets the educational demands of people involved in professional activities' [5].

In general, education of adults encompasses all types of education which include an adult. An education of adults is a system of state and social activities to spread knowledge among adult population (specific educational institutions, forms and methods of study [5]). An adults' education is specified as a part of educational system, which aims to contribute to personal development during the person's independent life, i.e. after certain specific (professional) training [2].

The article does not aim to analyze in details a number of approaches to define this category of professional education. We will dwell on andragogy – a branch of pedagogy, dealing with theoretical and practical issues of education of adults considering their specifics. Based on its principles, E.V. Tikhonova states that, educational system, focused on adults needs to meet the following requirements:

- Unequal educational structures (systems, parts), varied educational process, its contents, tools, results;
- Individual education considering social and professional qualities of a learner;
- Dynamics of educational process;
- Democracy in relations among participants of education. Primarily, among teacher and learners, and their artistic collaboration;
- Adapted educational process, professional and personal mutual influence on subjects and objects;
- Maximal activity of learners, their encouraged professional reflection.

A notion of 'supplementary education' was introduced in pedagogy and referred to adult learners, as general and professional supplementary education, in works by Tallinn pedagogues H.Y. Liimets and M.Ya. Pedayas in 1976 [4]. The researchers suggested establishing one of the first faculties of adults' supplementary education in country.

They defined 'supplementary education' as an independent branch of andragogy. The supplementary education aims to gain new knowledge, form new skills and expertise, as a rule, based on general education, acquired in tertiary or higher educational institution. In other words, primarily, supplementary education aims to update and extend acquired specialists' knowledge. It is viewed as an essential part of modern changing and informative society.

H.Y. Liimets and his colleagues introduced a new system of supplementary adults' education, considering that it provides neither fundamentally new qualification nor advanced training, as opposed to a science degree or a higher category in the specialty. The authors emphasized that new professional requirements should adjust the existing level of qualifications. In their opinion, supplementary education is necessary during drastic social transformations, which cause knowledge update based on previous educational qualifications. Thus, supplementary education compensates for the insufficiency of what has long been known, and not adequately responded by the system [4].

A large number of works which focus on adults' professional advanced study courses and pedagogy of higher education has been performed in last two decades. Different approaches to the nature of supplementary education within adults' education were introduced. The approach developed by O.V. Kuptsov should be emphasized.

According to this approach, the system of supplementary education comprises two substructures: general supplementary education and professional supplementary education. General supplementary education, according to him, includes the following fields:

(a) Early childhood education;
(b) Informal education of children, aimed to develop individual creativity;
(c) Informal adult education, which compensates for deficiencies in basic education.

The researcher states that the supplementary professional education includes:

(a) Advanced training;
(b) Retraining;
(c) Students' simultaneous training for the second profession.

Supplementary professional education is represented by the fields of knowledge which are necessary for specialists as they face constant change of the tasks and working conditions.

The main task of supplementary professional education is lifelong professional development and retraining of specialists, development of the ability both to adapt to socioeconomic changes and to affect them.

A number of authors consider the institute of a supplementary professional education to be one of the educational institutions appropriate for adults. The necessity of thorough analysis and design of a supplementary professional education, as a subsystem, aim to elaborate the tasks of a new social ideology and philosophy, based on andragogy studies, to determine the initial positions, aims and principles of the forthcoming transformation, to select ways and strategies for a new educational system's formation.

Considering supplementary professional education, it should be noted that its purpose is to expand the core competencies by the new ones and to master new specialities. The purpose of such education is to help a working specialist to become more competitive, mobile, successful; "learn not to be afraid of ignorance and learn to work in complex, unstructured areas of concern, learn to define problems and ask questions" (Slastenin V.A, Mudrik A.V., and others). The system of supplementary professional education should provide a person with the competence to perform changing professional roles, personal and social development, creative self-realization and social adaptation to the market conditions.

The program of supplementary professional education "Translator in the field of professional communication", implemented at the Department of Foreign Languages of Kazan State Medical University since 2007 corresponds to the majority of these tasks. This educational project enables students, postgraduates and young doctors to master a new speciality. It forms learning environment that can increase their confidence in the future, help to interpret life opportunities competently, to promote individual consolidation based on common human values, and to determine their role in the society.

According to the survey, conducted among the students of supplementary professional education department, the respondents answered to the question "What motivates you to continue education?" in the following way (see Table 1 below):

Table 1. Continuous education stimuli.

Question: 'what motivates you to continue education?'	%
Answers suggested by respondents:	
1. Obtaining new knowledge	83%
2. Possibility of career growth	76%
3. Prospects for self-realization	54%
4. Increased demand for new quality	41%
5. Increase of a personal status	38%
6. A way to upgrade economic life standards	32%
7. A way to master communicative skills	28%
8. A way to make useful contacts	19%

4 Conclusion

Basically, the supplementary education program, implemented at the Department of Foreign Languages of KSMU aims to: closely collaborate with the field of professional education; ensure accessibility of a supplementary educational system; focus on continuous education, which results in a change of the attitude towards supplementary education; individualize and differentiate the system of supplementary education, based on the application of modern teaching handbooks, methods of educational material and computer technologies.

References

1. Khomutov, O.I.: Dopolnitel'noe professional'noe obrazovanie – tsivilizovanny put' resheniya problem zanyatosti naseleniya Rossii [Supplementary professional education - a civilized way to solve the employment problem of population in Russia] (in Russian). http://elib.altstu.ru/elib/books/Files/1998-01/11/pap_11.html, Accessed 13 May 2017

2. Makarova, O.Yu.: Nepreryvnoe obrazovanie v kontekste dopolnitel'nogo professional'nogo obrazovaniya [Lifelong learning in context of supplementary professional education]. Problems of professional training and specialist's activities. KSMU, Moscow, Kazan (2009). (in Russian)
3. Navazova, T.G.: Metodologiya nepreryvnogo professional'nogo obrazovaniya [Methodology of lifelong professional education]. Person and education, vol. 3, pp. 17–22 (2005). (in Russian)
4. Pedayas, M-I.Ya.: Aktual'nye ptoblemy dopolnitel'nogo obrazovaniya [Actual problems of supplementary education]. GPI, Tallinn (1983). (in Russian)
5. Polonsky, V.M.: Slovar' po obrazovaniyu i pedagogike [Dictionary of education and pedagogy]. Higher education school, Moscow (2004). (in Russian)
6. Postanovlenie TSK KPK otnositel'no reform narodnogo obrazovaniya (28 maya 1985) [Resolution of the CPC Central Committee on the reform of the public education system (27 May 1985)]. TSK, Beijing (1987). (in Russian)
7. Revyakina, V.I.: Rol' dopolnitel'nogo obrazovaniya v professional'nom razvitii pedagogov [The Role of supplementary education in the professional development of a teacher]. Information and Education: boundaries of communications. RIO GAGU, Gorno-Altaisk (2009). (in Russian)
8. Tikhonova, E.V.: Biblioteki v sisteme dopolnitel'nogo professional'nogo obrazovaniya [Libraries in the system of supplementary professional education: dissertation]. Piter, St. Petersburg (2006). (in Russian)
9. Zmeev, S.I.: Androgogika: osnovy teorii i tehnologii obucheniya vzroslyh [Androgogy: the fundamentals of theory and technology of adult education]. Per SE, Moscow (2003). (in Russian)

Participation of Primary School Teachers in Educational Innovations as the Groundwork for Their Professional Development: Organization and Management

Svetlana I. Pozdeeva[1,2](✉) (iD)

[1] Tomsk State Pedagogical University, Tomsk 634061, Russian Federation
svetapozd@mail.ru
[2] Tomsk Polytechnic University, Tomsk 634050, Russian Federation

Abstract. The paper is dedicated to the problem of teachers' professional development in the context of implementation of the Federal Educational Standard for Primary and Secondary Education. The objective of this research is to analyze the teacher-pupil collaboration work in the program "Development of open joint activity in the primary school". The purpose of the paper is to figure out management actions that form professional competences in this activity. The article is constructed in the context of humanities approach and involves the research, description of teachers' professional situations, analytical comments and final overview. Comparative analysis of two management approaches: administrative and humanities in the modern school conditions is carried out. The author makes an attempt to connect the professional development of teachers with their participation in preparing and realizing innovative programs. Finally, the author poses the professional development of teachers not only as a reproduction of the same functions but a real part in creating a new educational practice.

Keywords: Innovative educational program · Teachers' professional development · Management actions · Educational innovations

1 Introduction

Nowadays the Federal Educational Standard acts for Primary [3] and Secondary Education [4] and determines successful learning of a child. Thus, the objective of this research is analysis of the teacher-pupil collaboration work [8]. The purpose of the paper is to figure out management actions that form professional competences in this activity. As a result, such professional competences of primary school teachers as implementation of educational initiatives, the organization of different interactive forms (teacher-pupil, pupil-pupil), carrying out the trial and research actions (academic, research, project and reflective) are defined. The management experience at School of Joint Activity [11], functioning in Tomsk, is presented in the research (the school is run by Dr. Galina N. Prozumentova).

© Springer International Publishing AG 2018
A. Filchenko and Z. Anikina (eds.), *Linguistic and Cultural Studies:*
Traditions and Innovations, Advances in Intelligent Systems and Computing 677,
DOI 10.1007/978-3-319-67843-6_4

The paper is constructed in the context of the humanities approach and involves its techniques [12], in particular the comparative analysis of two management approaches: administrative and humanities. Based on this, the teachers' professionalism in the context of the administrative approach means the quantitative increase of outer competences, professional and personal qualities [5]. This opinion is in conflict with the author's interpretation and defined as a professional and personal development caused by the teacher-pupil participation in educational innovations. In this case, the author suggests naming it "open".

The author makes the connection between teachers' professional development and their participation in preparing and implementation of innovative programs [9]. Moreover, the results of humanities management experience are proved by author's vivid examples given in the article. Thereby, the main conditions of such development are the following: the positional principle of teachers' participation in educational innovations, the professional trials and their consequences, the andragogic support of teachers in their collaboration work [15]. Finally, the author clarifies the teachers' professional development not only as reproduction of the same functions but a real part in creating a new educational practice.

2 Research Basis

2.1 Professionalism of Teachers in the Context of the Administrative Approach

The teachers' professional development is constantly one of the leading management functions at school. This function is actively expressed to the young teacher attitude (tutorship, consulting and attending lessons of experienced colleagues). If a teacher has been working at school for more than 5 years, in this case management actions towards him are decreased with regard to the timely attestation and professional training out of the school. It is expected that crisis of professional adaptation is coped with by a teacher, and they form professional skills. The objective of school management among experienced teachers (work experience 10 years and more) is limited by their outer motivation of taking part in varieties of school professional events. Thus, working experience identified the level of collected professionalism [1]. Thereby, reproduction of ordinary professional functions' implementation (to teach, to go to staff meetings and fill in student record books) is considered to be teachers' professionalism. Following the famous proverb "The repetition is the mother of learning", we may admit that "reproduction of traditional functions is the core of professionalism."

As it has been found out, a steady position of any administrative school manager is a deep understanding of teachers' professionalism as the quantitative increase of outer competences, professional and personal qualities. Different evaluative scales of teacher's professional and personal skills are widely developed for this reason. It turns out that professional development is not aimed at changing of professional skills but a total increase: the more, the better. Unfortunately, professional development doesn't correlate with professional participation in educational innovations [13]. The teacher is not allowed to join the program of professional development: he is only its performer. The administration decides what a professional teacher should be, comes up with different activities for teachers to perform.

These management programs of professional development don't single out an innovative activity as a resource: teachers are not encouraged to change something in educational practice but attend the events developed by administration. The professional level of educators is reached by achieving the norms from directorship without any changes: their position, view on children and attitude to innovations. Under those circumstances, management at school should not be administrative (decision-making) but more humanitarian that helps educators become real participants instead of innovation performers. The next paragraph clarifies the management organization of the humanities approach, based on the example of the author personal experience from the joint work with primary school teachers.

2.2 Professionalism of Teachers in the Context of the Humanities Approach

Participation in the innovations is provided by taking part in preparing and implementing the innovative educational programs. At the same time, we differentiate concepts "innovation in education" and "educational innovations" [13]. In the first case, a teacher – a performer is left outside of innovations, in the second case – he is an active participant and developer. The active participation in innovative educational programs helps a teacher to become a real participant of changing the school educational practice. The management within the humanities approach at school means development of mechanisms to support teachers' initiatives and actions, implementing the innovative educational programs. Further, we find it necessary to consider the established management mechanisms in the practice of the innovative educational program "Development of open joint activity in the primary school".

2.3 Involving All Teachers in the Innovative Educational Program

Involving all teachers in the innovative program regardless of work experience, education and the attitude to the open joint activity is done on the basis of "positional principle of participation" [5, 9]. We are sure that an innovative educational program can be developed not only by local groups of teachers but a whole team of educators. There are established positions in the following practice (designer, participant, user, and trainee) that present ways of teachers' involvement into the innovative activity. It means a different scale, using the open joint activity, different kinds of pedagogical actions to shape it, different educational technologies, and different innovative products. "Positional principle of participation" allows every teacher to put themselves in educational innovation, not to implement innovations imposed from the outside without understanding their idea and ways but their own interpretation, methods and syllabus. *Teachers-designers* show *an* active attitude to educational innovations and idea of open joint activity at primary school. Due to these teachers, the innovative educational programs are developed. Teachers-designers not simply implement the productive educational technologies available in the culture but create their own technologies towards shaping of the open joint activity in the primary school. These teachers can design all kinds of innovative products: research, experimental and methodological projects. *Teachers-participants* usually don't create any technologies and experiment with educational technologies already known in the pedagogical culture. Moreover, they can use technologies from teachers-designers' experience, choosing the best methods and techniques. The scale of educational innovations and kinds of innovative products in this group of educators are limited. For example, methodological projects are created by them on the basis of teachers-designers' experimental projects and include some of the open joint activity techniques. *Teachers-users* demonstrate a neutral calm attitude to innovative ideas and dive into the program, enter it by copying actions of teachers-designers and teachers-participants. Typically, they are young teachers (trainees) or teachers who have previously worked at another school.

2.4   Creation of Innovative Products

Research, experimental and methodological projects allow teachers to reflect upon their own professional experience and its innovative component, to identify training and educational content in the classroom, to implement educational technologies and single out their educational resources in the shaping of open joint activity [6]. According to that,the following management actions are accompanied by the projects creation and expansion. Firstly, the projects *strengthening*. For example, a teacher made a methodological project with testing of some methods for open joint activity organization and then "grew" it up to an experimental project, having implemented several pilot actions and observing their results. As a rule, this way is applied by teachers-participants and teachers-users. Secondly, project *reconstruction*. A teacher-designer made an innovative product and then remodeled it into an experimental or methodological project for other teachers who want to give a similar lesson, having referred the appropriate innovative idea. This management mechanism works more effectively if a teacher-designer gives a lesson in another teacher's classroom: *teacher-participant* or *teacher-user* (in other words, "lives the lesson"). It should be stressed that the first option is needed for those who grow professionally, and the second way is very important for promoting innovative products and preventing their localization at school. Therefore, "positional principle of participation" means that teachers create different innovative products and professional development is the movement from methodological teaching materials to research and experimental projects.

2.5   Professional Experience at Different Stages of Innovative Activity

The academic year at our school consists of three stages: diving into the open joint activity, its deployment and analysis of results. At the diving stage (the beginning of an academic year) teachers test new methods, techniques and decide what they will do in the course of the year, what module of innovative program they will work in. The most effective form at this stage is the lesson-laboratory where the teacher shows all colleagues a new method or a technique. During preparation for this lesson the teacher consults with the program's manager or a teacher with a higher status in the innovative educational program who acts as an instructor-teacher. Moreover, the lesson-laboratory is discussed together with all colleagues in an active dialogue [10] and joint conclusion is made about educational resources of the technique, shown by the teacher. At the deployment stage, every teacher spends several open lessons, including teachers from other schools, and demonstrates his own vision of an educational technology and its place in the innovative program.

The main method at this stage is a mutual learning and support of teachers. In this case, the best lessons are formed in the written texts of research, experimental and methodological teachers' projects. Thus, the bank of innovative teachers' projects gradually accumulates and becomes a social asset belonging to all and everyone. At the stage of reflection and summing up (the end of an academic year) teachers analyze their learning and educational results and draw up the conclusions about their innovative activity in the program. Therefore, at this stage the diagnostic procedures are done to identify results and effects of an innovative activity at school. As it can be clearly seen, the organization of professional samples at all stages of school life is an important management mechanism of professional teachers' development.

3   Methodology

The paper is constructed in the context of the humanities approach and involves the research, description of teachers' professional situations, analytical comments and final overview [12]. This study employs the method of reconstruction of innovative experience as a basic method of research. This method assumes a description of real situations in educational practice and their analysis, description of precedents and particular cases in changing practice [13]. The main procedure for the research method is creation of self-description texts.

An example of the situation "Management of teachers' professional tests and their consequences".

The academic year at our school begins with the diving stage, when the emotional and communicative field of joint activity is restored. Children restore methods and forms of interaction, actively initiate, prepare and conduct lessons, plan events for the whole year. This stage can be called the maximum free and open for children and teachers who are not chained by the training regime. In terms of methodological work organization, the diving stage is not simply the stage of old learning form restoration but one for the new forms testing. If a new form is approved by teachers, we plan the program of actions and reflect it in the teacher's technical task.

Let's have a look at this way of the innovative activity on an example of a mutual learning lesson in the form of *Jigsaw* (the technology of critical thinking through reading and writing development). Why did this technology attract us and what open joint activity resources did we see in it? Firstly, the technology is based on the group cooperation and assumes a full interference of the adult. Secondly, a very interesting communication takes place in groups: at the beginning children together study a local part of a new informative text (every group studies its part) and then in teams of other children tell each other their parts in turns and discuss new information. Moreover, here the reading and verbal skills are developed very intensively.

The organization of professional samples in the form of *Jigsaw* was carried out as follows: the manager of educational innovative program presented this method; this technology was experienced by all the teachers in a real joint work with a text. In this case, it was important not only to learn a new process chain and predict children' difficulties but also see the shaping of open joint activity resources. At the next step, a methodological day with lessons-laboratories was organized: all together agreed to test this method in the fourth grades as the major one at the primary school. The manager of the innovative program provided educational and emotional support to every teacher, giving a choice of the subject, a theme, a text and so on. However, the scheme of trial lessons' self-analysis was made by all staff: *I would like to try... - I have done it... - I haven't done it...- Now I understand that....*"The result of reflection is selection and design of new objective tools ... for building activity" [14, p.151].

The next step is the analysis of lessons-laboratories when we took into account the abstract of the lesson, processing protocol and the protocol of discussion. The analysis showed the following: firstly, the educational effects of this technology were found (due to children's reflexive utterances), for example, the effect is not only training but also educational: "I understood today that in the group we have to be able to work not only with those people you know well but with any other person", "Today I was for the first time in group with Anton and I knew him better". Secondly, teachers not only identified the potential of the *Jigsaw* form for open joint activity and group's communication but also its powerful learning potential, associated with the organization of reading activity and text skills formation. However, not all of the teachers started to use this form; some showed a cautious attitude towards it.

To attract others to the new form, the following management actions were used: professional mutual learning and mutual support of the teachers; joint analysis and assessment of positive results of the *Jigsaw* form (development of the text, speech, communicative skills and group interaction skills), selection of educational effects (meeting various interlocutors, different in forms and contents of communication in the group, and inclusion of all kinds of speech activities). Thus, teachers-novices used the methodological projects of their predecessors as a teaching material. Due to the joint professional work the technology of the new form grew: using of algorithms for each step in the group work, tabular method of recording the new information by pupils, and lightweight preventive forms for younger students). Another result of teachers' professional tests is development of specific technological methods for organization of open joint activity. This is particularly effective for the *teacher-participant* or *teacher-user* as long as it is difficult for them to use complete and detailed methodological forms. These educators use work in pairs and teach children to read a new text together, find the keywords in it, ask each other questions about the text and retell the content to each other. These techniques help them to move the *Jigsaw* form for future work.

4   Results and Discussions

The established management mechanisms in practice of innovative educational program allow teachers to be the subject ("a host") of innovative activity and implement it in real situation. Therefore, management helps the teacher to organize his/her work on their own innovative trajectory. This is exactly what provides teacher's professional development as a change of his position, competencies and views. For example, the teacher proceeds from "permissive and allowing" actions (gives the opportunity to act, provides the freedom of choice, offers but doesn't impose, and allows to speak freely) to "observing and organizing" actions (maintains and develops initiatives, advises but doesn't tell, helps to choose a solution, and creates situations of the open activity as a member). In this case, a child is in the position of a significant and influential participant of this activity. We deeply believe, that this experience is fundamentally different from traditional one which we call the humanities approach. The teacher now is not a performer of outer innovative will but a subject of educational innovation who does what he/she is interested in and what is new for him/her.

Lines in Table 1 show the results of skill assessment during interaction in groups. We have identified 8 groups of interaction skills and assessed them by the expert observation. The data for all graduating forms at primary school is done (the average number is 90 pupils per year). Columns show the average results in the parallel grades in percentage. It can be clearly seen that there is a positive growth of skills at primary school during 3 years. It also demonstrates the relative level of certain skills. For example, the ability to hear others and the emotional perception of the rest participants are formed at a higher level more efficiently than breaking up the flood of nonproductive ideas and positional distribution in a group. If the trend continues, we may expect an overall increase in the rate of interaction skills formation not only in the primary school but upon pupil transfer to the next stage of training [2]. Generally speaking, this proves the effectiveness of teachers' participation in the innovative program.

Table 1. The expert assessment map of group interaction skills.

	Criteria	2012	2012	2012
1	Positional distribution in a group (organizer, designer, and time keeper)	55%	62%	65%
2	Expressing ideas in front of all group participants, and their justification	60%	65%	70%
3	Joint decision-making	58%	62%	67%
4	Breaking up the flood of nonproductive ideas	50%	55%	60%
5	The ability to hear others	70%	75%	80%
6	The emotional perception of the rest participantsin a group	82%	85%	87%
7	Presentation of complete and consistent results of the joint work	58%	60%	65%
8	The ability to ask and answer the questions during presenting a report	60%	65%	70%
	Average	61.6%	66.1%	70.5%

5 Conclusion

The management of teachers' professional development will be more effective if it is oriented not only to innovations, provided by the school administration but to creation of conditions for teachers in the joint innovative activity that changes educational practice. Firstly, the opportunity of implementing teachers' professional tests and their comprehension and making innovative products. Secondly, teachers realize innovations as their own professional sense and not as a task from the outside. Thirdly, a new type of professionalism becomes the result of this management: a teacher does not develop in playing the same professional functions but creates a new educational reality here and now.

References

1. Absatova, M., Ussenova, A., Ordahanova, A.: Social success of future primary school teachers in the process of professional training. Mediterr. J. Soc. Sci. **6**, 476–481 (2015)
2. Berkel, H., Scherpbier, A., Hillen, H., van der Vleuten, C.: Lessons from Problem-Based Learning. Oxford University Press, New York (2010)
3. Federal'nyigosudarstvennyiobrazovatel'nyistandartnachal'nogoobshchegoobrazovaniya [Federal state educational standard of primary education] (in Russian). http://window.edu.ru/resource/624/70624/files/373.pdf, Accessed 10 June 2017
4. Federal'nyigosudarstvennyiobrazovatel'nyistandartosnovnogoobshchegoobrazovaniya [Federal state educational standard of general education] (in Russian). http://window.edu.ru/resource/768/72768/files/FGOS_OO.pdf, Accessed 10 June 2017
5. Prozumentova, G.N., Pozdeeva, S.I.: Obrazovatel'noe soderzhanie sovmestnoj deyatel'nosti vzroslyh i detei v shkole: upravlenie i stanovlenie [Educational content of the joint activity of adults and children in the school: the management and formation]. Tomsk center for scientific and technical information, Tomsk (2015). (in Russian)
6. Plotnikova, N.N., Pozdeeva, S.I.: Forms of organizing collective educational activity in primary school. Procedia – Soc. Behav. Sci. **214**, 51–58 (2015)
7. Pozdeeva, S.I.: The collaborative teacher pupil activity as a condition of children communicative competence development. Procedia – Soc. Behav. Sci. **206**, 333–336 (2015)
8. Pozdeeva, S.I.: Innovatsionnoye razvitie sovremennoi nachalnoi shkoly: postroenie otkrytogo sovmestnogo deistvia pedagoga i rebenka [The innovative development of modern primary school: building an open joint teacher and child activity]. TSPU Publisher, Tomsk (2013). (in Russian)
9. Pozdeeva, S.I.: Osobennosti gumanitarnogo upravleniya innovatsiyami dlya stanovleniya professionalizma pedagoga nachalno shkoly [The humanities features of innovation for the teacher competence formation at primary school]. TSPU Bulletin **9**, 194–196 (2013). (in Russian)
10. Pozner, R.: Ratsionalnyiy diskurs i poeticheskaya kommunikatsiya: metodyi lingvisticheskogo, literaturnogo i filosofskogo analiza [Rational discourse and poetic communication: methods of linguistic, literary and philosophical analysis]. TSU Publishers, Tomsk (2015). (in Russian)

11. Prozumentova, G.N.: Shkola sovmestnoy deyatelnosti. Eksperiment: razvitie tseli vospi-taniya i issledovatelskoy deyatelnosti pedagogov shkoly [School work together, Experiment: The purpose of the development of education and research, school teachers]. TSU Publishers, Tomsk (1994). (in Russian)
12. Prozumentova, G.N.: Perekhod k otkrytomu obrazovatel'nomu prostranstvu, Fenomenolo-giya obrazovatel'nykh innovatsiy [Transition to the Open educational space, Phenomenol-ogy of educational innovations]. TSU Publishers, Tomsk (2005). (in Russian)
13. Prozumentova, G.N.: Obrazovatel'nye innovatsii: fenomen « lichnogo prisutstviya » I potentsial upravleniya (opyt gumanitarnogo issledovaniya) [Educational Innovation: the phenomenon of "personal presence" and the management capacity (experience of humanities studies)]. TSU Publishers, Tomsk (2016). (in Russian)
14. Shedrovitsky, G.P.: Sistema pedagogicheskikh issledovaniy: metodologicheskiy analiz. Pedagogika i logika [The System of pedagogical research: methodological analysis. Pedagogy and logic]. Kastal, Moscow (1993). (in Russian)
15. Vershlovsky, S.G.: Vzroslost' kak kategoriya andragogiki [Adulthood as a category of andragogy]. Voprosy obrazovaniya 2, 285–297 (2013). (in Russian)

Peer Mentoring as a Professional Test for Trainee Teachers in the Sphere of Deviant Behavior Prevention of Minors

Larisa G. Smyshlyaeva[1,2] , Lyudmila S. Demina[1] ,
Igor L. Shelekhov[1] , Dmitrij B. Nasonov[1] ,
Oksana I. Kravchenko[1] , and Svetlana S. Kalinina[2(✉)]

[1] Tomsk State Pedagogical University, Tomsk 634050, Russian Federation
ipiptspu@tspu.edu.ru
[2] Tomsk Polytechnic University, Tomsk 634050, Russian Federation
kalinina_ss@tpu.ru

Abstract. The article contains materials on peer mentoring as a professional test for trainee teachers. The specificity of a professional and pedagogical test action in the sphere of deviant behavior prevention of minors is identified. Modern peculiarities of peer mentoring as a practice of professional pedagogical activities are highlighted. The role of pedagogical practice in construction acme-oriented system of interaction between trainee teacher and children and teenagers including those from dysfunctional families is described. The possibilities of a professional test are presented as means of implementation the acme potential of educational process in Pedagogical Higher Educational Institution. Empirical research data is given needed for designing a model of organization peer mentoring practices based on the mainstreaming of idea about professional tests. The resource of peer mentoring is presented as a form of professional test implementing for trainee teachers. An author's vision for the creation of a space model is set to implement professional tests of the trainee teachers through peer mentoring in the sphere of deviant behavior prevention of minors. It has been proved that the most important tasks which actualize acme resource of professional test in teachers' training for the deviant behavior prevention of minors are: get acquainted students with peculiarities of demonstration deviant forms of behavior; assisting them in professional self-determination (orientation of the activities - prevention and correction of various deviations); development of motivation for professional social and pedagogical activities in accordance to the priorities of self-implementation. It has been shown that for these tasks to be solved successfully it is necessary to ensure: diversification of professional tests forms (its diversity, variations); individualization of professional tests; construction of smooth mechanisms for consultation and mentoring in the execution of professional tests; emphasizing objectives and content of pedagogical practices in the context of professional tests' implementation.

Keywords: Peer mentoring · Professional test · Teacher · Acme · Minors · Children · Teenagers · Prevention · Deviant behavior

© Springer International Publishing AG 2018
A. Filchenko and Z. Anikina (eds.), *Linguistic and Cultural Studies: Traditions and Innovations*, Advances in Intelligent Systems and Computing 677, DOI 10.1007/978-3-319-67843-6_5

1 Introduction

At the present stage, the practice of training teacher in Russia is undergoing a major transformation which mainly related to the need of finding a productive solution to the problems of "being introduced to the profession", problems of "occupational training" and problems of "holding down" in profession [6]. The solution of these problems requires a new vision of the aims, objectives, content and technological provision of professional teacher's training at all stages.

The topic discussed in this article identifies the problem of transformation higher education system of professional teacher's training, incorporating into educational process the experience of peer mentoring as a professional test for trainee teachers. The training of a modern teacher mentor requires serious comprehension not only because of need to implement changes related to the systemic transformation of professional teacher's training in higher education but also because of youngness of social pedagogy as a professional activity type.

The issue of peer mentoring is adjacent to the subject field of social pedagogy. In Russian Federation the profession of social teacher was approved at the legislative level in 1991. It is currently being introduced to the All-Russian Classifier of Specialties in Education. With regard to specific nature of social and pedagogical activities it should be emphasized that pedagogical personnel in this area always face the need of specialization, determination of leading (differentiated) activity directions, taking into account the peculiarities of work place and contingent of wards (their age, problems, socialization tasks). In a methodical letter "About social and pedagogical work with children" (1994) it is emphasized that the profile (specialization) of social and pedagogical activities is determined by regional and ethnic characteristics, needs of a particular society – city, district, village, and personal and professional opportunities of the specialists [9].

For solving individualization tasks and successful professional self-determination of trainee teacher oriented on social and pedagogical mentoring, in the process of his professional socialization, according to our point of view, one should clearly and fully identify his possibilities for professional development (acme-opportunities), depending on the chosen direction of activity [1, 4]. It is known that a successful professional is usually someone who has been able to accurately correlate his or her own personality with the trajectory of his or her own movement in a professional space. In order to properly defining oneself in this trajectory, it is necessary to try a wide range of different future professional activity types during the educational situation, be actually immersed in them, engage in understanding different contexts of their own emerging professional experience [2, 3, 11–13, etc.]. Therefore, existence of a well-functioning system of professional tests in the process of training a modern teacher during learning stage is an important factor of his "holding down" in the profession and successful continuous professional development in accordance with chosen trajectory of self-realization.

2   To the Essence of Professional Test

Professional test (PT) as a definition and as a phenomenon by overwhelming majority of experts in the area of continuous professionalization of a human during his life refers to both pre-professional and professional educational system, including the practice of additional professional training [6–8, 11, 13, etc.]. However, the gist of PT in all contexts of its consideration is clearly understood.

A professional test is a specially organized test (imitation situation) which models the elements of a particular professional activity type, as close as possible to the professional reality. From psychological point of view PT is a resource for individual search, professional self-determination and development which assumes:

- providing basic information on specific types of occupational activities;
- modeling main elements of different professional activity types;
- availability of organizational and methodical tools to measure the level of human preparedness for the successful performance of a particular professional activity type;
- creating conditions for full-scale "experience" of professional tests [3, 7, 8, 11, etc.].

In order to build a productive system of professional tests when preparing teacher to work, first of all, it is necessary to designate types of professional pedagogical activity to which future specialist will get acquainted both theoretically and practically. In the context of professional teacher's training to work in the sphere of deviant behavior prevention of minors it is due to the essential features and forms of deviant behavior of minors. In contemporary psychological and social and pedagogical researches deviant behavior is interpreted as aberrant from accepted, socially acceptable behavior in a certain society, from legal and moral norms (crime, alcoholism, drug addiction, prostitution, etc.). It can result to isolation, medication, correction or punishment of the offender [8]. In accordance with the forms of deviant behavior, a complex of professional tests is being built and is being tested by trainee teachers. Work as a peer mentor for a minor becomes a professional test for students from Pedagogical University who are involved in student construction team's work. During this process trainee teachers are given an opportunity to test different forms, methods and means of preventing deviant behavior of minors. Each test allows stimulating the process of self-determination for students and construction based on their choice of individual trajectory for further professional development. The forms and combinations of PT's implementation may be as follows:

1. Practical assignments that require designing (organization, conducting) of a completed event.
2. A series cycle of business imitative games.
3. Research or creative assignments.
4. A set of activities, for example, to develop and implement a programme for mentoring minors [3].

3 Materials and Methods

As an example, let us give you the experience of Tomsk State Pedagogical University (TSPU) in organizing the practice during the implementation of basic Bachelor's educational programs of social and pedagogical profile. In TSPU, bachelors are being trained in psychological and pedagogical education, profile: Psychology and social pedagogy, since 2011. The curriculum for the four-year training course includes six types of practices, which are united into two groups: training practices (2) and production practices (4). Immersion in practice begins with the first year of study. Overall the Bachelor's training practice is 20 weeks. The practices are conducted in external educational and social organizations or at departments and laboratories of the university, which possess the required personnel and scientific and technological potential [5].

In order to determine whether the practice provides the opportunity for "experience" of professional test, an empirical study was conducted (an interview method of 27 respondents who are the students enrolled in basic educational programs of a social and pedagogical profiled undergraduate, day-time department). Based on the results of analysis of the data for 2012 and 2013, 66.7% of the respondents did not evaluate the practice as an activity which provides "the effect of a professional test" in the context of self-knowledge tasks, career self-determination and self-development; 11.1% of the respondents noted that the practice fully provides this effect; 7.4% were difficult to answer, 14.8% felt that the practice partially realize functions of professional tests.

These data were considered as the basis for construction a model of organizing the practices based on the mainstreaming of the idea about professional tests, which in its turn served as a conceptual basis for creation of a new system of teacher's training for independent professional activity, including the field of deviant behavior prevention of minors.

As an important part of this system the construction of a mechanism for the interrelation of practice with other types of student's activities in the process of studying at a university that provides "the effect of a professional test" was defined. Implementation of this interconnection will provide the creation of workspace for realization of professional tests in the process of studying at a university for trainee teacher. The given workspace includes the totality of the following components in their synergies with the potential of pedagogical practices:

- training sessions themselves, purposefully focused on the imitation modeling of professional tests;
- student's involvement in social projects and volunteer activities of social and pedagogical orientation;
- work in student teams as peer mentors for minors, as upbringer of temporary children's groups in children's out-of-town camps and as camp counselor at school and courtyard playgrounds;
- participation in the work of student self-governing bodies in the faculties and in the university as a whole.

At the same time, it is important to provide counseling and mentoring support for students as a necessary condition for the effectiveness of their reflexive comprehension of pedagogical experience gained during the "experience" of professional tests. This can be achieved by purposeful inclusion in the curriculum the training course "Basic professional pedagogical tests in extracurricular activities: practical work".

We believe that the design and implementation of professional tests' system in the process of social teachers' training requires mandatory accounting of the pre-university individual experience of the students, reflecting their involvement in any types of pedagogical activity (quasi-professional activity).

In 2014, the experience of students on their involvement in professional tests of pedagogical orientation in the pre-university period (interrogation through structured interview) was studied. The survey covered 68 persons who are the students of TSPU. Based on the results of survey data's analysis, 5 students (7.4%) noted the *lack of experience* of professional pedagogical tests. Continuously and systematically (no less than six months, in the mode of permanent involvement) *students were engaged in activities of a professional pedagogical orientation*:

- camp counselor - 24 persons (35,3%):
- volunteer of pedagogical orientation - 25 persons (36.8%);
- practical activity in creative groups (editorial board, school newspaper, theater workshop, etc.) - 16 persons (23.5%);
- organization of events - 27 persons (39.7%);
- were activists of children's public organizations - 7 persons (10.3%);
- were the leaders of school self-government - 18 persons (26.5%);
- performed the functions of "nanny" in communication with younger brothers, sisters, nephews - 10 persons (14.7%).

Not long (one time or episodically) engaged in pedagogical activities 30 persons (44.1%).

In our opinion, the results of such studies should be reflected in design of the content of individual component of mentoring models [7, 10].

4 Conclusion

In conclusion, we outline the most important tasks, in our opinion, which update the resource of PT in the process of pedagogical personnel training for deviant behavior prevention of minors: to introduce students the peculiarities of deviant behavior; to help them in professional self-determination (selection of tools for mentoring technology and activity directions); to develop their motivation to professional social and pedagogical mentoring activities in accordance with the priorities of implementation. In order to successfully meet these challenges in higher educational system of personnel training for social pedagogy it is necessary to provide: diversification of PT's forms (its diversity, variations); PT's individualization; construction of smooth mechanisms for consulting and mentoring support during implementation of PT; emphasizing objectives and content of pedagogical practices in the context of PT's implementation (creation of conditions for providing the "effects of professional tests").

Thus, peer mentoring is a form of a professional test realization of trainee teachers. One of the required condition of peer mentoring effectiveness as a professional test of trainee teachers is the manageability of this acme practice. Working with minors and deviant teenagers allows trainee teachers to try the experience of involving into a wide range of pedagogical situations and gain experience in solving professionally and personally significant problems. The organization of peer mentoring for minors with deviant behavior on the basis of student units in TSPU is a trial and training site for the development of professional competences of trainee teachers, providing an opportunity, including, to place acme tasks and to solve them systemically.

Acknowledgement. The article has been prepared on the materials of scientific work in the project of the state mission No 27.4344.2017/NM "Improvement the mechanisms of peer mentoring with minors, including those from dysfunctional families" (2017) and the state mission of Russian Ministry of Education and Science No. 27.7237.2017/BCH.

References

1. Bodalev, A.A.: Vershina v razvitii vzroslogo cheloveka: kharakteristiki I usloviya dostizheniya (in Russian) [The Peak in Adult Development: Characteristics and Conditions of Achievement]. Flinta, Moscow (1998)
2. Buehler, C.: Parents and peers in relation to early adolescent problem behavior. J. Marriage Fam. **68**(1), 109–124 (2006)
3. Chistyakova, S.N., Rodichev, N.F.: Ot ucheby k professional'noy kar'yere (in Russian) [From Study to Professional Career]. Academiya, Moscow (2012)
4. Derkach, A.A.: Psikhologiya razvitiya professionala (in Russian) [Psychology of Professional Development]. RAGS, Moscow (2000)
5. Federal'nyy gosudarstvennyy obrazovatel'nyy standart vysshego professional'nogo obrazovaniya po napravleniiy podgotovki 050400 Psikhologo-pedagogicheskoe obrazovanie (kvalifikacii (stepen'): bakalavr)/utverzhden Prikazom Ministerstva obrazovaniya I nauki Rossiiskoi Federacii ot 22 marta 2010 g. № 200. Zaregistrirovan v Minyuste RF 28 apel'ya 2010 g. № 17037 (in Russian) [Federal State Educational Standard of higher professional education in the field of study 050400 psychological and pedagogical Education (qualification (degree): (Bachelor's degree)/approved by Order of the Ministry of Education and Science of Russian Federation on 22 March 2010, No. 200. Registered in the Ministry of Justice of RF on 28 April 2010, No. 17037]. http://fgosvo.ru/uploadfiles/fgos/5/20111115121912.pdf
6. Kontseptsiya podderzhki razvitiya pedagogicheskogo obrazovaniya, 14 January 2014. (in Russian) [The support concept for the development of teacher training education]. http://минобрнауки.рф/документы/3871/файл/2676/Концепция поддержки развития педагогического образования 11 12 13.doc
7. Pozdeeva, S.I.: Nastavnichestvo kak deyatel'nostnoye soprovozhdeniye molodogo spetsialista: modeli I tipy nastavnichestva (in Russian) [Mentoring as an activity support of a young specialist: models and types of mentoring]. Sci. Pedagog. View **2**(16), 87–91 (2017)
8. Pozdeeva, S.I.: Soderzhaniye I formy organizatsii obrazovatel'noy deyatel'nosti pri podgotovke pedagoga v vysshey shkole (in Russian) [The Content and form of Educational Activities in the Training of a Teacher at University]. Tomsk State Pedagogical University Publishers, Tomsk (2014)
9. Shakurova, M.V.: Metodika I tekhnologiya raboty sotsial'nogo pedagoga (in Russian) [The Methodology and Technology of a Social Pedagogue]. Academiya, Moscow (2004)

10. Shelekhov, I.L.: Sistemnyy podkhod kak metodologicheskiy bazis lichnostno-oriyentirovannykh psikhologicheskikh issledovaniy (in Russian) [Systemic approach as a methodological basis for personal-oriented psychological research]. Sci. Pedagog. View **2** (16), 9–20 (2017)
11. Ronald, L.S., Whitbeck, L.B., Conger, R.D., Conger, K.J.: Parenting factors, social skills, and value commitments as precursors to school failure, involvements with deviant peers, and delinquent behavior. J. Youth Adolesc. **20**(6), 645–664 (1991)
12. Smyshlyaeva, L.G.: Professional'naya proba kak pedagogicheskaya tekhnologiya (in Russian) [Professional test as pedagogical technology]. High. Educ. Russ. **4**, 65–69 (2015)
13. Verbitsky, A.A.: Kontekstnoye obucheniye v kompetentnostnom podkhode (in Russian) [Contextual learning in the competence approach]. High. Educ. Russ. **11**, 38–52 (2006)

Top Qualities of Great Teachers: National and Universal

Elena Yu. Ilaltdinova⊙, Svetlana V. Frolova⊙,
and Irina V. Lebedeva^(⊠) ⊙

Minin University, Nizhny Novgorod 603005, Russian Federation
ilaltdinova_eu@mininuniver.ru, lgeniy87@mail.ru,
lebedeva06.08@yandex.ru

Abstract. There are the results of comparative analysis of the specific features of great teacher qualities descriptions presented in nine regions worldwide. The main pecularities of the requirements to a teacher in different countries are systematized by the authors in the article. The rating of the qualities of a successful teacher allowed to single out a number of qualities which can be named as universal. Empathy belongs to this group of universal qualities.

Keywords: Ethics · Code · Teacher · Personality quality

1 Introduction

1.1 Building a Research Problem

Personality of the teacher is the embodiment of the national image where the past and the future are united. Teachers are guardians and communicators of national heritage including culture, ideals, traditions; they are mediators between generations. In the developed global educational systems, the image of the teacher reveals national specificity being the reflection of the national culture, aspirations, and hopes for the future.

Every region of the world is characterized by its authentic educational system, the key link in which is the image of the educator that embraces national specificity. Vivid global educational systems can be found in such regions of the world as East Asia (Japan, Singapore, Hong Kong, China, South Korea), South Asia (Philippines, India), Scandinavia (Finland, Sweden, Norway), Anglo-Saxon region (Great Britain, Ireland, Scotland), Eastern Europe (Russia and CIS countries), Middle East (Turkey), North America (USA, Canada), Latin America (Brazil), and Australia (Australia).

1.2 Basic Assumptions

The comparative analysis of professional and personal qualities of the teacher in different regions of the world reveals that the national image of the teacher is affected by the geography, national and cultural features, customs and traditions of a region. Yet, along with national specificity, it is obvious that great teachers display such qualities that are characteristic of every "ideal image" outside the national or geographic context.

© Springer International Publishing AG 2018
A. Filchenko and Z. Anikina (eds.), *Linguistic and Cultural Studies:
Traditions and Innovations*, Advances in Intelligent Systems and Computing 677,
DOI 10.1007/978-3-319-67843-6_6

Thus, in the world diversity of national images of the teacher, there is a set of personal and professional attributes that reflects, regardless of national and religious origin, the universal values prioritized for educational professional performance.

Rating of the personal and professional qualities of the teacher has allowed us to present the hierarchy of the most common, thus most universal, characteristics of educators in the world. The most demanded quality of the teacher in the world is the ability to feel empathy. The second position is occupied by such qualities as trust and the ability for cooperation. The third place in the hierarchy of the teacher's personal qualities is indicated by the ability for management and control, profound knowledge of the taught subject, creative approach to professional activity. The forth position is given to communicability, exemplary conduct, social activeness, individual approach to students, respect and patience.

2 Materials and Methods

According to the stated problem, the methodological base was a set of theoretical methods included comparative content analysis of authentic national, regional, local, and institutional regulations of ethical norms of teacher behavior specific to the different countries and regions. National professional standards comprise an important part of our research materials as well as informal ratings of valuable teacher qualities. The first stage of research was to make comparison to find specific features of great teachers in different countries. The second one was to make generalization and build a picture of universal portrait of a great teacher as an approach to description of learning outcomes of teacher education programmes.

3 Research Data

The image of the East-Asian educator incorporates such core characteristics as communicability, empathy, mobility, tolerance. In Japan, as in many other countries, special attention is paid to the professional training of educators, substantiation and provision of the conditions for the development of such professional skills that will ensure successful implementation of the educational process [6].

Thus, the Japanese teacher is, first of all, an active member of the society with a well-shaped social and civic position; a successful mature broad-minded person whose actions are characterized as being consistent and reliable. Undoubtedly, among his or her most significant traits are the ability for successful communication in society accompanied by the ability to adequately evaluate the character of interaction between people as well as foresee and plan its further development. Taken the above into account, we can state that one of the primary qualities of teachers in Japan is the ability to feel empathy towards other people, which is revealed in being able to go beyond the context of their own perspective on the surrounding reality, to consider a variety of viewpoints and opinions, to show respect toward the set of values of other people. Japanese teachers exert readiness to establish an effective interaction and positive relationships with administrators of educational institutions, colleagues, parents, making effort to promote this interaction as a part of their pedagogical aspiration towards a better school and a better society in general.

Japanese educators are capable of self-organisation and self-development, they exhibit high ethical standards and the ability of self-control. They are viewed as carriers of a special mission and role models in society being the possessors of the exemplary self-consciousness and elevated moral ideals.

The Japanese teacher needs profound content knowledge and in-depth understanding of the taught discipline; he or she must be capable of applying various types of academic activities with a focus on each individual in the classroom. Of indisputable importance in the educator's competence in Japan is the ability to manage a group of students as well as effectively participate in and contribute to school management.

Being active members of the Japanese society, teachers must be able to apply their adaptive skills to properly adjust and respond to changes in society realizing the importance of their social mission. Their professional activity must be performed with love toward their students, understanding and accepting of their growth and development, respect of their human rights [8].

In South Asia, the issue of professional training and professional skills development of educators is of special importance. In the Philippines, graduates of pedagogical educational institutions take the oath collectively and publicly; this oath reflects their strong beliefs and aspirations to follow the way of their professional and personal development in their pedagogical practice. It is worth noting that the Philippines is one of the few South Asian countries that has developed and implemented the "Code of Ethics for Professional Teachers" [5]. This code for educators regulates and describes professional and personal characteristics of teachers in the key spheres of interrelation including the teacher and the state, the teacher and the society, the teacher and the school, the teacher and the colleagues, the teacher and the parents, the teacher and the students. Needless to say, that the Philippine teacher is also given a special social mission – mission to translate the exemplary behavior and be a positive model in society.

To perform this crucial and responsible task, the Philippine teachers must be professionals who possess dignity and reputation, demonstrate active social position and intellectual leadership in society. The Philippine teacher must be able to exert self-control and control over a group of learners, meanwhile, building harmonic and cooperative relationships and facilitating each student's needs. The notion of educational professional activity in the Philippines is closely associated with the notion of professional commitment and dedication incorporating the ability to cooperate, support, and self-sacrifice for common good.

The specific traits of the Philippine teachers not typical of the educators' image in other global educational systems of the world are that they should worship God and exercise freedom of conscience and religion. Meanwhile, a teacher is free to have any religious influence on students but he or she must not affect the formation of a student's political outlook and attitude. All in all, the Philippine teachers are obedient to God, politically indifferent social leaders capable of finding ways to establish harmonious relationships in society and promote national values and norms.

The South Asian portrait of the teacher is harmoniously complemented by the colorful India introducing its own specificity. The image of the teacher and his professional and personal qualities are the focus of attention of the Indian researchers, as well as of the National Policy on Education that defines the strategy for the development of the educational system. The image of the teacher, according to Kingdon, is a determining factor that contributes to the effectiveness of student learning and success [7]. On August 20, 2013, the "Three-Pronged Strategy for Improving the Quality of School Teachers" was adopted in India. It aims to improve the quality of education and includes several areas for the pedagogical education modernization:

– strengthening of pedagogical educational institutions and their development;
– modernization of the pedagogical educational program in accordance with the national educational program of 2009;
– establishment of the minimum qualification requirements for educators and their continuous professional growth [12].

As evidenced in some research, the most important professional qualities of the Indian teacher include the ability and aspiration for continuous professional self-development. Scholars believe, that one of the manifestations of this aspiration is expressed in the desire and readiness of a teacher to obtain master's degree as a qualification requirement for the professional educator. Other significant professional and personal attributes are commitment and passion for teaching, professional motivation, ability to maintain a steady and effective interaction with the learners, and profound knowledge of the taught subject [2]. To succeed in their professional practice, the Indian teachers employ such qualities as organizational skills, communication skills, kindness and wisdom.

The guardian of the Western world values and cultural heritage, including pedagogical practices, is the educational system of Western Europe. The vivid representative of Western educational system is Finland known for holding the top position in the ranking of the best educational systems in the world (as for 2012). The dominant aspects of the Scandinavian teacher portrait are based on the qualities that contribute to the formation of a positive school atmosphere and effective learning environment.

Professional educators in Finland seek ways to establish equal conditions for every participant of the educational process in school or in classroom environment, striving to provide accessible quality education to any student regardless of their development peculiarities or any other individual features. This uneasy task becomes feasible due to the ability of the Finnish teachers to use individual approach in their classroom.

To help students acquire the subject content effectively, the Finnish teacher should focus in the educational process on the possibility of applying theoretical knowledge to practice in the surrounding world. The secret behind the success of Finland's educational system also depends on the teacher's commitment to develop cognitive autonomy of the learners, which cultivates students' own feeling of responsibility for learning progress, ability for critical thinking and independent "retrieval" of information and its analytical processing. The principle of "learning how to learn" employed by the Finnish teachers ultimately results in high educational outcomes with a minimal workload of the learner. In the assessment of the academic achievements of students, objectivity and impartiality are the most important teacher's qualities.

As important personal attributes of the teacher contributing to the formation of a comfortable and safe educational environment we can emphasize the ability to build a trusting relationship and to treat students with respect, to put no pressure and eliminate coercion toward the learners (principle of voluntarism). These principles are the basis of interaction between the student and the teacher. Most researchers that focus on the phenomenon of the Finnish education effectiveness recognize that the leadership in the quality ranking of the world's educational systems is also achieved due to the personality of the teacher that incorporates these personal and professional qualities.

The classic educational system of Great Britain identifies the following important professional and personal qualities of educators:

– trust to students;
– self-belief (the ability to handle a loss of confidence or crisis in professional practice despite any failures);
– patience (a patient attitude toward students implies the ability to wait patiently or even "fight" for student's learning success and progress);
– genuine compassion for students (the ability to feel empathy, compassion, create friendly relationship, not to be indifferent to students, to be eager to help and sympathize);
– understanding (the ability to perceive life from the other perspective);
– striving for professional growth;
– feeling proud of students' success;
– joyfulness.

In Ireland, the teacher personality is also seen as an important factor for the improvement of education quality [3]. The most significant teacher's characteristics are as follows:

– respect - teachers should encourage each learner in their spiritual and intellectual development; they should be able to recognize the accomplishments of each student and create the conditions conducive to their growth. In their educational practice, Irish teachers demonstrate respect to the spiritual and moral values of a student; they possess the values of freedom, democracy, they strive for social justice and care about the environment.
– care - professional motivation of the Irish teacher is primarily determined by learner's educational interests being his or her main concern. In practice, it manifests itself through teacher's positive influence, professional judgement and empathy;
– trust - relationships between teachers and students, their parents, colleagues must be based on trust. The leading personal qualities in this respect are justice, openness, and integrity;
– tolerance - moral personal qualities of the Irish teachers are embodied in the integrity of their professional style and practice.

The Eastern European region is represented by the Russian educational system. In Russian pedagogy, the image of the teacher, his or her personal and professional attributes, aspirations, professional skills and competences occupy one of the leading positions in scientific investigations.

The scholars of the Minin University (Minin Nizhny Novgorod State Pedagogical University, Russia) developed the Code of Ethics for Educators that outlines the most important personal and professional characteristics of the Russian teacher portrait [4]. As envisioned in this code, the highest professional value of the teacher is the personality and individuality of the learner. The Russian teacher represents the Russian pedagogical community, thus, he or she should exercise high standards of conduct and be the role model for the society. The important personal and professional attributes of the Russian teacher are integrity, openness, impartiality, feeling of responsibility, respect to students, colleagues, and the pedagogical community as a whole.

A special principle that distinguishes Russian teachers in the global palette of educational systems is gratitude and respect to their teachers: "I am an educator as long as I am studying myself."

The ability and aspiration for continuing professional development of the Russian educator is regarded as the key idea in the concept of the Professional Standard for Teachers: «In a fast changing open world, the main professional quality that teachers must constantly demonstrate to their students is the ability to learn" [11]. This aspiration is based on the obligatory personal and professional characteristics of a teacher including readiness for change, mobility, the ability for non-standard professional actions, responsibility, and independence in decision-making. The condition for the formation and revelation of these qualities is based on the expanded space for pedagogical creativity where teachers are free to take initiatives.

The Northern American region and its education system sets out the personality of the teacher and its characteristics as a key factor that contributes to learning success. The Association of American Educators (AAE) has developed the AAE Code of Ethics for Educators, which focuses on the manifestation of the personal and professional characteristics through four basic principles:

− ethical conduct toward students;
− ethical conduct toward practices and performance;
− ethical conduct toward professional colleagues;
− ethical conduct toward parents and community [1].

The professional educator in the USA accepts personal responsibility for teaching students' personal qualities that will help them evaluate the consequences of and accept the responsibility for their actions and choices. The code envisions parents as the primary moral educators of their children. Yet, it is stated that all educators are obliged to help foster the civic virtues including integrity, diligence, responsibility, cooperation, loyalty and respect for the law, for human life, for others, and for self [3].

The professional US educators, accepting they are trusted by community, measure success not only by each student's progress toward the realization of his or her personal potential, but also as a citizen of the country.

The portrait of the US educator is revealed in the conduct toward students based on the sustainable rules of ethical conduct:

− the professional educator deals considerately and justly with each learner and strives to resolve professional problems, including the problem of discipline, in accordance with the law and policy of the educational institution;
− the professional educator does not intentionally expose the student to disparagement;

– the professional educator does not reveal any confidential information about the students provided it is not required by law;
– the professional educator makes constructive efforts to protect students from the conditions that threaten their learning, health, or safety;
– the professional educator endeavors not to distort facts and present them without bias, or personal prejudice.

The American educator assumes responsibility for the quality of his or her performance and constantly strives to demonstrate competence, meanwhile trying to maintain the dignity and social prestige of the profession.

The AAE Code contains several rules with regard to manifestation of the ethical conduct in professional practice:

– the professional educator continues professional development.
– the professional educator does not distort the official policy of the school; he or she clearly distinguishes these views from his or her own personal opinions.
– the professional educator does not use institutional or professional privileges for personal advantage.
– the American educator accepts that education quality is the common goal of the community, education boards, and educators, and to attain these goals, a cooperative effort of each of these groups is required.
– the professional educator makes coordinated efforts to communicate to parents all information that should be revealed in the student's interest.
– the professional educator endeavors to understand and respect the values and traditions of various cultures presented in the community or in his/her classroom.
– the professional educator demonstrates an active and positive position in social relations [1].

The top personal and professional attributes of the Northern-American educator include personal style and individuality, ability for effective classroom management, ability to establish open and cooperative relationships with parents, belief in better abilities of the student, masterly command of the subject, engagement (passion) for and commitment to teaching, justice, empathy, creativeness.

In Australia, the most crucial personal characteristics of a teacher include such quality as communicability that is seen as the ability to create open and effective dialogue with all the participants of the educational process. A teacher must also demonstrate:

– the ability to design educational programs;
– the ability to create a promising educational environment;
– the ability to cooperate with the student;
– communication skills;
– the ability to control;
– the ability for professional partnership [8].

The portrait of the Lain American educator has been extensively studied by the Brazilian professor Vincente Martins. According to him, one of the key missions of an educator in Latin America is to teach and train students to succeed in life, to promote and contribute to happiness and peace in the world. Vincente Martins pinpoints the following important personal and professional qualities of an educator:

- the ability to cultivate human dignity in students;
- creativeness;
- the ability to establish effective cooperative relationships with students and colleagues;
- empathy and compassion to students;
- the ability for management and control;
- cooperativeness;
- respect toward students [9].

Personal and professional qualities of educators occupy one of the central place in social and pedagogical studies in the Middle East countries. An educator of the Middle East region (Turkey) is envisioned, first of all, as a talented, communicative, creative, open-minded person with a broad outlook. The Turkish teacher treats students with love and respect, he is considerate and patient. Therefore, one of the crucial attributes of the Turkish educator is the ability to feel empathy [10].

4   Conclusion

In the world national diversity, teachers must be able, first of all, to exert empathy to their students, which means to be able to feel proud of students' personal achievements, to rejoice at their success, to show sympathy with them on their failures. Overall, it means "to accept" a student. Empathy is revelation and manifestation of the most elevated pedagogical feeling – pedagogical love incorporating the whole diversity of the essence of the educator's profession. "Reasonable" pedagogical love assumes equality in the relations between the teacher and the student, proclamation of mutual trust and cooperation. However, professional qualities of teachers are of no less importance than their personal attributes. It is a well-known fact, that knowledgeable and competent educators are capable of infecting students with love toward the subject they teach, hence they can awaken and foster students' cognitive activity, their natural curiosity and autonomy skills. In addition, great teachers must possess a strong will power, they must be able to manage a group of students in and outside the classroom. In the world diversity of national specificities, the teacher is a role model and a moral ideal that students should look at and follow.

The conducted comparative analysis reveals the necessity for the development of pre-service and in-service teacher training programs based on the rational balance of the objective-value component of educational programs aimed at combining the activities directed at subject and methodological training of educators along with the development of their personal qualities. The Russian experience and tradition of training teachers in line with the social requirements for the personality of the teacher and understanding of their high social mission have always been centered on the ethical constituent of educational profession. Under the influence of the Western pragmatic traditions throughout recent decades the priority has been shifted toward the efficiency of the teacher. Modern situation allows integration of the world achievements (for example, developments in the context of the multiple intelligence theory) and the Russian traditions in understanding of the teacher's mission. This approach is considered to be the basis for the revision and reinterpretation of the objective and technological components of teacher training models that exist in Russia.

References

1. AAE Code of Ethics for Educators. http://www.aaeteachers.org/index.php/about-us/aae-code-of-ethics. Accessed 24 May 2017
2. Azam, M., Kingdon, G.G.: Assessing teacher quality in India. IoE, University of London & IZA, London (2014)
3. Code of Ethics for Educators. https://www.aaeteachers.org/index.php/about-us/aae-code-of-ethics. Accessed 24 May 2017
4. Code of professional conduct for teachers. http://www.teachingcouncil.ie/en/Publications/Fitness-to-Teach/Code-of-Professional-Conduct-for-Teachers.pdf. Accessed 24 May 2017
5. Code of ethics for professional teachers of Philippines. http://teachercodes.iiep.unesco.org/teachercodes/codes/Asia/Philippines.pdf. Accessed 24 May 2017
6. Dzhurinsky, A.N.: Chemu I kak uchat v Yaponii (in Russian) [What and How They Teach in Japan]. Publishing house Narodnoye obrazovaniye, Moscow (2001)
7. Kingdon, G.G.: Teacher Characteristics and Student Performance in India: A Pupil Fixed Effects Approach. Global Poverty Research Group, Oxford (2006)
8. Kozyreva, V.A., Rodionova, N.F.: Rasvitie kvalifikazionnih trebovanij k professionalnoj dejatelnosti v obrasovanii (in Russian) [Development of Qualification Requirements for Professional Practice in Education]. Herzen University Publishing, St. Peterburg (2008)
9. Martins, V.: Educador por Vocação, Educandos em boa Direção. http://sitededicas.ne10.uol.com.br/artigo20.htm. Accessed 01 June 2017
10. Ozsevik, Z.: The Use of Communicative Language Teaching (CLT): Turkish EFL Teachers' Perceived Difficulties in Implementing CLT in Turkey. College of the University of Illinois, Urbana-Champaign (2010)
11. Professionalnij Standart pedagoga (in Russian) [Professional Standard for Teachers]. http://минобрнауки.рф/. Accessed 24 May 2017
12. Three-Pronged Strategy for Improving the Quality of School Teachers (from 20 August, 2013. Press Information Bureau, Government of India). http://pib.nic.in/newsite/erelease.aspx?relid=98428. Accessed 28 May 2017

Engineering Education

Methodical Procedures of Scientific-Technical Scope Development as a Condition of High-Quality Students' Training

Galina I. Egorova[1] , Andrey N. Egorov[1] ,
and Nikolay A. Kachalov[2(✉)]

[1] Tobolsk Industrial Institute, Tyumen Industrial University, Tobolsk 626150,
Russian Federation
egorovagi@list.ru
[2] Tomsk Polytechnic University, Tomsk 634050, Russian Federation
kachalov@tpu.ru

Abstract. According to the new requirements to the contents and level of
professional training of university graduates, integration into the world educational environment, it is necessary to pay attention to the personal development
of students in the process of education. Peculiarities of the system aimed at the
development of bachelors' scientific-technical scope as a flexible subsystem of
professional training in a technical University are theoretically and methodologically proved. The purpose of the research is to develop scientific-technical
scope as a condition of increasing the quality of professional training and
professional competences. The results of the research: (1) the definition to the
term "scientific-technical scope" is given and the conditions of its development
are described; (2) functional-didactic meaning is defined from the point of view
of psychology and pedagogy; (3) the system of methodical procedures of
science-technical scope development, required in the process of professional
training is developed and tested. Methods of the research are theoretical,
empirical, experimental, methods of mathematical analysis.

Keywords: Poly-paradigm understanding · Scientific-technical scope ·
Education

1 Introduction

The authors have determined the term "scope", its polysemy is shown as well as its
relation with cultural world view, scientific world view, and humanitarian part of the
education. The peculiarities of scientific-technical scope of bachelors are shown in the
system of their diversity. Key components of the term "scientific-technical scope" (the
scope of relevant professional knowledge and skills) is very important in regards to the
Federal State Educational Standard (FSES 3+), and is considered as a key factor of
students' professional competences formation [7].

The new generation formation, integration of the Russian high school into the
world educational system is possible in case of the robust training system of "elite
bachelors" [4]. It is the quickest and the most effective way of cultural and spiritual

© Springer International Publishing AG 2018
A. Filchenko and Z. Anikina (eds.), *Linguistic and Cultural Studies:*
Traditions and Innovations, Advances in Intelligent Systems and Computing 677,
DOI 10.1007/978-3-319-67843-6_7

society renewal. Russia has taken a priority place among European countries, but many traditions, innovations in elite education have been lost and this fact affected the system of bachelors' training.

2 Materials and Methods

We have analyzed positions, definitions and functional characteristics of the term "scientific-technical scope" which are used in modern socio-cultural system: criteria and result of educational activity; criteria and quality indicator of students' readiness, criteria of students' personality development. We have determined and studied functional-didactic meaning of the concept "scientific-technical scope" which shows its diversity in educational system (result, means, indicator, and personality development grounds). Every sphere of studied notions has a certain model, regularity. However we suppose that not a competition of ideas is important, but the methodology aimed at polyparadigmal, comprehensive understanding of the concept "scientific-technical scope", differentiated methodological approaches, showing its functional importance.

From the pedagogical point of view, we understand "scientific-technical scope" as the result and indicator of personality development growth at different age stages (from development of mental activity, new knowledge acquisition, understanding of the studied material, up-bringing of intellectual elite, morality development), that is especially important for the modern world. From the point of view of psychology, we should take into account mechanisms of interrelation of science-technical scope with personal qualities, peculiarities of high psychological functions, and age of students. From the point of view of sociology, we distinguish the peculiarities of "science-technical scope" of a bachelor as the basis of the society development in XXI century. We should note key peculiarities of scientific-technical scope related to the process of morality development, spirituality of a person, that is currently the basis for solution of global challenges. From the point of view of cultural studies, we define the peculiarities of scientific-technical scope as a part of culture on the whole. It is important to note here that the scientific-cultural scope has become a central idea of professional training. Taking this fact into account we should teach bachelors to develop their scope of knowledge via acquiring new knowledge and skills. Having analyzed the above points of view we suggest the following definition of the notion "science-technical scope": it is a resource-innovative indicator which provides professional training of a future specialists. It is both a process and result able to enhancing professional culture of future bachelors. Let's study the methods of scientific-technical scope development of students as a condition of high quality training. The first technique is studying local cultural material via independent work with study aids, scientific literature of local character, archives, materials of Tobolsk state historical-architectural reservation-museum, "SIBUR Tobolsk" museum. The purpose of local cultural material use is intellectual development, interest in chemistry, revealing chemical aspects in the objects of the world, developing patriotic feelings to the home region.

The aim of studying chemical disciplines helps to acquire the system of local cultural knowledge: main characteristics of Tyumen region's history, Tobolsk as an ancient Siberian capital; regional peculiarities of regional fuel-energy complex;

chemistry role in regional issues; regional raw materials and their rational use; chemical factors in stable development of region; intellectual sphere and peculiarities of its formation in the region. Students' skills developed upon acquiring and using local cultural material: systemization and generalization of the material according to different aspects; solving various problems with the knowledge of local cultural material; using local cultural material in internship. A special role is given to the value orientation [1]: forming of regional thinking and developing their own point of view on the main issues of the region; understanding the value of regional chemical material; understanding possibilities of chemistry and technology concerning rational use of natural resources as a perspective direction of region's development; understanding intellectual value of every person in solving main problems of the region. Using local cultural material in the system of professional training has a strategic importance.

We would like to study the second methodological technique of developing scientific-technical scope, with the leading role played by oil and chemical industry and chemization of regional economy. Studying petrochemical processing enterprises bachelors analyze regulations and process schemes of "SIBUR Tobolsk" LLC. The main aim of the second methodological technique is comprehensive knowledge on industry development, Regional chemical school, which was in the scientific basis of modern production originated from D.I. Mendeleev's ideas [2].

The third methodological technique of developing scientific-technical scope we connect with the integrated system "high school - enterprise - region" [6] which guarantees high-quality professional training of bachelors in chemical-technological disciplines by studying key enterprises of Tobolsk industry, such as: nitrogen, oxygen production; propane production, butane production, i-butane, pentane production, i-pentane; liquefied lower paraffin hydrocarbons production.

Revealing scientific principles of production: production process continuity; extension of the reagents surface; back current; using optimal temperatures; pressure, catalyzers; mechanization and automation of the processes; main installations and equipment. Students see practical issues of their future professional activity, discuss their peculiarities in dialogues, in teams with production specialists: operators, technologists. Issues of ecology and industrial safety are very interesting for students, as well as investment issues, enhancing quality of products and raw materials, equipment (reactors, catalyzers of different kinds, process schemes, monitoring and measuring instruments, flow schemes).

In butadiene production students determine the key blocks of contact gases fractionation of (dehydrated butane separation), aroma concentrate obtaining. Comparing theoretical and practical knowledge, students understand key issues influencing the selection of optimal conditions of the process. What is the role of catalyzers and types of reactors? What is the temperature regime, raw materials quality and production? What are the peculiarities of the process scheme and what are the common factors of n-butan dehydration process? What are the conditions of i-butane dehydration process? Solving these problems involves creating students' own system of work. Dialectical connection "student- teacher-engineer" provides interaction in solving key issues.

The fourth methodological technique takes into account psycho-pedagogical concepts person-centered and practice-oriented study [3]. Chemical material is studied combining inductive and deductive approaches that show ideas of fact, phenomenon, and process in more details. Contents is studied in modules divided in several stages. First, the content of key concepts is defined. We pay attention to the main ideas of the course, their role in forming understanding of technology, production, scientific image of the world taking knowledge from other disciplines. The students learn to work with schemes, tables, plans, notes. Classes are conducted in different forms: combined lectures, and practical classes with different forms of assessment (tests, questioning, chemical dictation, working with process schemes, cards) which provide interrelation of the course components.

The fifth methodological technique of scientific-technical scope development is connected with research practical training during classes and extra-curricular activities [9]. We have developed the research practical training curriculum (2 h a week). Additional practical classes appear due to the modular arrangement of the material. Practical classes in general and inorganic chemistry include local cultural material on the following themes: "Irtysh water in crystallohydrates", "Studying thermal decomposition of natural salt crystallohydrates", "Determining thermo-chemical characteristics of the process of dissolving salts in natural water", "Determining water carbonate hardness and ways of its elimination", "Comparing chemical composition of rivers Irtysh, Konda, Ob", "Preparing chemical solutions for leather treatment", "Preparing salt solutions of different density for fish salting at home", "Producing dyes from plants (berries, moss) and their extraction", "Determining chemical composition of moss as a nutrient of northern animals", "Carbonates and hydro carbonates providing hardness of natural water". "Non-metals burning in oxygen", "Experiments with dry ice".

The experiments in organic chemistry are conducted in the following themes: "Pyrolysis of Siberian wood", "Cranberry extraction", "Distillation of benzoic acid", "Determining biochemical composition of animals blood (deer, elk, fur game)", "Natural clathrates of perma-frost", "Thermal decomposition of roots and determining their composition", "water extraction of ammonia from animals urea", "Pine and fur-tree oil". "Soap production". "Dry distillation of wood". "Receiving turpentine from galipot".

The research experiment on analytical and physical chemistry involves the following themes: "Receiving carlook from fish substance", "Determining chemical composition of jelly from fish husk", "Adsorption of died substances by charcoal".

The sixth methodological technique of scientific-technical scope development, is connected with different forms of lessons [5]. We used game-forms of the classes. Information game "Genealogic tree of chemical elements and their role in people's history", live journal "Mendeleev's table in discoverers' faces", etc.

The seventh methodological technique, oriented at the development of scientific-technical scope is connected with quizzes on local cultural material [8]. The authors have developed several quizzes. Below you can see a part of such a quiz "Do you know the geography, history, chemistry of Tyumen province".

1. What is the year of Tumen region foundation and what is its square? (Tumen region is the largest in Russia, founded in 14.08.1944).
2. What countries can be placed on its territory? (We can place such countries as Spain, France, Italy, Great Britain, Austria, Germany – their total area is – 1 mln. 435 thousands km^2).

3. What regions and districts are connected with Tyumen region by means of oil and gas pipelines? What are the most important of them? Oil pipelines: Shaim – Tyumen (1965), Ust – Balyk – Omsk (1967) 1100 km, Samotlor – Kuibyshev (1976) 2000 km. Oil pipelines: Punga – Serov – Nizhniy Tagil (1966); Samotlor – Kuzbass; North – Center; Urengoi – Chelyabinsk; Urengoi – Novo-Pskov; Urengoi – Uzhgorod 4500 km; Yamburg – Yelets.

4. What are the fossil fuels of this region except for oil and gas? (Wood species: pine, lime tree, birch, fir-tree, silver-fir, cedar); mineral rock (clay, sand, mineral water, peat, diatomite, brown coal, sapropel clay).

5. How wood industry waste is used, what do we make from wood? What industrial waste do you know? (railway sleepers, chipboards, sap, galipot, turpentine).

6. What is the role of oil and gas in fuel, energy supply of the country and what are the rational ways of using them? (Oil and gas – are valuable hydrocarbon products. Natural, associated gases have certain advantages comparing with other raw materials: cheapness, high temperature of burning, richness of chemical composition, low amount of additives).

7. Name scientists who made a great contribution into the development of petrochemical industry in Russia, what do you know about their ideas? (D.I. Mendeleev, A.V. Topchiev, I.M. Gubkin and others.).

8. What are and will be the products of oil-gas industry of Tyumen region? (lubricants, commercial monomers, liquefied gases, nitrogen, oxygen, bleachers, rubber products, polymer materials).

9. When and where were the first oil and gas deposits discovered in Tyumen region? (The first oil – Shaim village (1960), the first gas deposits – Berezovo village (1953).

10. Who named oil a"black gold"? Who gave it? Is there "White oil"? (D.I. Mendeleev, White oil contains light hydrocarbons (0,65–0,87 r/sm^3).

11. What is the longest northern pipe-line of a "blue fire"? What cities does it connect? (Igrim – Serov – Nizhnii Tagil (505 km), Yamburg – Yelets).

12. What is the biggest gas pipeline connecting the North of Tyumen region with Europe, which goes through the territory of Russia. Name it. (Nadym – Ukhta – Austria – Hungary – Italy).

13. Name the first plants and factories in the region. (Arms plant founded by Peter the Great Decree (1700), gun powder plant (1703), paper factory of merchandisers Medvedevs (1744), later V.Ya. Kornilev (1751), linen, silk (1790), town pharmacy (1762), winery and brewery plant of merchandisers A.A. Syromyatnikov (1850), leather-processing, candle-making plant of merchandiser Ershov (1830).

14. What are the study years of D.I. Mendeleev in gymnasium? Who influenced him the most? Study years (1841–1849 yy.). (Mikhail Ivanovich Dobrokhotov – Geography and History teacher; Petr Pavlovich Ershov – teacher of literature, inspector, poet; Denis Petrovich Shelutkhov – Arts teacher, calligraphy; Ivan Karlovich Rummel – Maths and Physics teacher).

15. What was the role of the Mendeleevs family in the development of town Tobolsk, who were D.I. Mendleev parents? (Father – Court Councilor Ivan Pavlovich Mendeleev, teacher (1807–1818 yy.), Director of Gymnasium (1827–1834 yy.). Mother – Maria Dmitrievna Kornilyeva, the Head of the glass-making factory).

3 Conclusion

The system of methodical techniques for scientific-technical scope development is being incorporated into professional training of high – school pupils of the city and districts (University Saturdays, makertones, creative workshops, research laboratory of natural objects studying). All these enhance the level of knowledge among the university applicants. To determine the efficiency of scientific-technical scope development we conduct an annual monitoring according to the following criteria: educational criteria (quantitative and qualitative criteria of the educational level), personal criteria (students' motivation to study the subject) (P.I. Tretiakov Methodics), and social criteria (social success of the graduates) (employment results, interviewing).

References

1. Agranovich, B.L., Chuchalin, A.I., Soloviev, M.A.: Innovatsionnoe ingenernoe obrazovanie (in Russian) [Innovative engineering education: engineering education], vol. 1, pp. 11–14 (2004)
2. Aitov, N.A.: Vyscheye tekhnicheskoe obrazovanie v usloviyakh NTR (in Russian) [Higher Technical Education in the Conditions of STR]. Higher school, Moscow (1983)
3. Egorova, G.I.: Intellektualnaya kultura v sisteme professionalnoi podgotovki spetsialista (in Russian) [Intellectual Culture in the System of Professional Training of a Specialist]. IOV RAO, St-Petersburg (2004)
4. Egorova, G.I.: The efficiency of formation of professional identity of bachelors of technology in technical education. In: Proceedings of the 3rd International Congress on Social Sciences and Humanities, pp. 88–95. East West, Association for Advanced Studies and Higher Education, Vienna (2014)
5. Yenikolopova, E.V.: Dinamicheskaya organizatsiya intellektualnoi deyatelnosti (in Russian) [Dynamic Organization of Intellectual Activity]. MSU, Moscow (1992)
6. Efimov, Y.Y.: Intellektualizatsiya vuza – bazovyi istochnik rosta intellekrualnogo potentsiala regiona (in Russian) [Intellectualization of University – Basic Source of Intellectual Potential Region Growth]. SGAE, Samara (2003)
7. Kachalov, N.A., Kornienko, A.A., Kvesko, R.B., Nikitina, Y.A., Kvesko, S.B., Bukharina, Z. A.: Integrated nature of professional competence. Soc. Behav. Sci. **206**, 459–463 (2015)
8. Onushkina, V.G.: Funktsionalnaya negramotnoct i professionalnaya nekompetentnost kak factory riska sovremennoi tsivilizatsiyi i rol nepreryvnogo obrazovaniya vzroslykh v ikh preodoleniyi (in Russian) [Functional illiteracy and professional incompetence as risk factors of the modern civilization and the role of life-long learning of grown-ups in their overcoming]. NRI OOV, Leningrad (1990)
9. Fokin, Y.G.: Prepodavaniye i vospitaniye v vysshei shkole: Metodologiya, tseli, soderganiye, tvorchestvo (in Russian) [Teaching and Training in a University: Methodology, Purposes, Contents, Creativity]. Academy, Moscow (2002)

Humanization and Humanitarization of Engineering Education

Evgenia T. Kitova$^{(\boxtimes)}$ (iD)

Novosibirsk State Technical University,
Novosibirsk 630073, Russian Federation
kitovaet@mail.ru

Abstract. The paper deals with the problem of humanization and humanitarization of technical education. The relevance of these processes for training highly qualified engineers is shown. Students of technical universities must acquire the humanitarian knowledge to realize the mission of science and technology in the life of humanity and for identification of their own role in realization of the mission. One of the ways to do it is interdisciplinary education; social and humanities in technical education should be treated as fundamental, including advanced language study, changing stereotypes of thinking, acceptance of humanitarian culture and development of communication competence. The new state educational standards in Russia for all technical programs require a new approach to designing curriculums. The proposed research studies the curriculums designed for engineering fields of study at technical universities. The results of research show that elective courses have been added to curriculums which can be chosen by students. They are aimed at developing general cultural humanitarian competences and personal qualities important for a future engineer. New curriculums include an increased amount of credit hours for foreign language study and many new humanities elective courses such as Culture of Scientific and Business Communication, Culture and Personality, Basics of Personal and Communicative Culture, Social Technologies, and Organizational Psychology. Methodological system is developed based on polyparadigmal approach. The results of comparative analysis of programs for Electrical Power Engineering and Electrical Engineering at technical universities of Siberian region are provided.

Keywords: Engineering education · Humanitarization · Polyparadigmal approach

1 Introduction

Values of engineering education today are defined by the modern world issues. High technologies development requires students to be creative and have designing abilities. The world ecological crisis causes changes in ecological education for future engineers, cultivation of professional morality as they should develop environmentally friendly technologies and industries. The revolution in information technologies development and information transformation of the society needs higher informational culture,

A. Filchenko and Z. Anikina (eds.), *Linguistic and Cultural Studies:*
Traditions and Innovations, Advances in Intelligent Systems and Computing 677,
DOI 10.1007/978-3-319-67843-6_8

critical thinking and introduction of IT in educational process. But changes in minds are not as quick as global problems growth.

Contemporary engineering education programs represent a significant change from earlier programs that had specific orientation, as they generally propose a wider concept of what engineering education should be concerned about. One of the concerns in most engineering curriculum documents is an emphasis on values. That is, the values that students should explore, be exposed to and understand, as a result of involvement in an engineering education course.

However, implementation of any policy depends on teachers understanding the policy and having strategies suitable for implementing it. Two assumptions underpin and provide part of the rationale for the study. The first is that teachers need to have an understanding that values are involved in engineering education. The second is that teachers need to understand and be able to use strategies aimed at the development of particular values among students.

Therefore, vocational training should provide engineering students with planetary thinking, new disciplines such as system modeling, prognostics, globalistics to give them ability to understand global issues, to acquire new values based on general humanity principles. Humanization and humanitarization are the tasks for higher educational institutions. There is a great need for technical institutes and universities to speed up the fulfillment of the tasks.

2 Materials and Methods

The term humanization of education refers to the process of creating conditions for self-realization, self-determination of the student's personality, creating a humane environment at technical universities, promoting the development of creative potential of the individual, the formation of noosphere thinking, value orientations and moral qualities and their future implementation in professional and social activities. We should not forget that knowledge itself is neither good nor evil - a man makes it good or evil.

Today, we can realize the Academic Vernadsky's words about scientific knowledge: "Will a man use this power for a good purpose and not for self-destruction? Is he clever enough to use this power given to him by science?" [5]. Students of technical universities should acquire human knowledge with the aim of realizing the mission of science and technology in the life of humanity and their own place in the realization of this mission.

Humanitarization of education, especially technical, involves expanding the list of humanities, deepening integration of technical and humanitarian knowledge.

To solve the problem, technical institutes should achieve the penetration of humanities knowledge into natural sciences and technical disciplines. The concepts of humanization and humanitarization may include:

- Humanities teaching and training technologies;
- Training on the border of humanities and technical areas (on the border of the animate and inanimate, material and spiritual, biology and technology, etc.);

- Interdisciplinary education;
- Social and humanities disciplines in higher school should be treated as fundamental;
- Advanced language study;
- Changing stereotypes of thinking, acceptance of humanitarian culture.

Unfortunately, today in Russia the humanities are sometimes opposed to technical fields.

Nowadays, an engineer without a good humanitarian training cannot become a good specialist. That is why the humanization of technical education is a priority for the Russian higher educational system.

Speaking about reforms in higher education in Russia, it should be mentioned that there have been some changes towards humanization and humanitarization. The state educational standards appeared introducing disciplines, which did not exist in previous curriculums.

The method of comparative analysis of educational programs developed for training engineers was used in the proposed research.

The research studied the curriculums designed for engineering fields of study at technical universities. The results of research show that elective disciplines have been offered and can be chosen by students, and which are aimed at formation of general cultural humanitarian competencies and personal qualities important for a future engineer. For example, the curriculum for "Electrical Power Engineering and Electrical Engineering" program at Novosibirsk State Technical University includes an increased amount of hours for foreign language study (10 credit hours compared to previous curriculum with 3 credit hours) and a number of new elective courses aimed at developing communicative competence. These disciplines include: the culture of scientific and business communication, culture and personality, the basics of personal and communicative culture, social technologies, and organizational psychology. These disciplines are aimed at development of personal qualities, required by engineers for realization of professional functions in modern conditions.

In global engineering there is a trend to the similarity with other service businesses and industries. Therefore partner collaboration management and communication are important, and value is produced on the basis of overall experience and atmosphere at all stages of the engineering service or product designing and manufacturing.

In Russia the key stakeholders involved in skill building policy include academia, governmental agencies, private training organizations, industries, and a young and dynamic new generation student community. The young generation is more confident, has excellent access to education and information, tends to be more independent, at the same time more tolerant to alien ideas, values and thought streams, and highly informed and conscious of social, political and environmental issues.

The important function of an engineer is not only a technological function such as designing and maintenance of equipment and technological process, but also social, socio-economic, administrative and cultural functions. Highly qualified engineers must possess professional and social, personal and interpersonal competences. The European Union documents require the following interpersonal competence: the ability to work in a team, have interpersonal skills, ability to work in an interdisciplinary team, the ability to accept the cultural differences, the ability to work in an international context, and commitment to ethical values.

Today's market and changing working conditions require that engineer had special "soft skills" in addition to "hard (technical) skills", they should be able to carry out their work, using all available resources, including personal. "Hard skills" can be easily identified and determined for a particular speciality. These are technical knowledge and skills in a specific technical area. "Soft skills" are a collection of personal and inter-personal qualities that may not necessarily be measured quantitatively, but nevertheless will determine the effectiveness of human labor. "Soft" skills are also referred to interpersonal skills, the ability to interact and solve professional problems with their colleagues. Examples of "soft" skills are the ability of critical thinking, leader skills, ability to make decisions, ability to work in a team and others.

It is a well-known fact that there is a great demand for people with relevant soft-skills to address this complex requirement. Because of various reasons it is increasingly difficult to allocate trained resources on the job without necessary soft skills training. Engineers, as professionals, tend to be less trained soft-skill than some other disciplines. The brilliant mathematical, analytical brain may not be much of a communicator. Whether you are in the United States or in the Russian Federation this is a common issue. In recent period many engineering programs have included com-pulsory subjects that develop soft skills, and also other supplementary subjects that reinforce such skills, including project works in teams, study groups and so on. In Russia a continuous work is being done on curriculum development and deployment of courses that addresses soft skills for future engineers. Some behavior profiling instruments have been developed that profile people and identify their dominant style and behavioral patterns. These assessments are getting more and more refined, based on years of cumulative data.

The research has clearly shown that the softer and practical attributes of engineering work have grown across generations. These attributes include such areas as self-management, being proactive, always keeping records, being clear and concise in documentation, tips for an engineering manager, and professional ethics.

At various levels of the career, the soft skills requirements keep changing. The areas and impact of roles keep increasing as careers progress, and therefore the specific interventions and methods to augment soft skills need to be carefully planned. A learning map, explained to employees, helps to realize and be aware of behavior expected at their level and at the next higher level. At lower stages in career, com-petency gaps can be reduced through training courses, but at senior levels of career, dominant strengths have to be focused on and developed. The approach of bridging competency gaps may not work out at senior levels.

Recent challenges of globalization are proving that the share of technical excellence towards overall effectiveness has reduced making way for newer skills like knowledge of interacting with trans-national cultures, business etiquette, expected and acceptable behavior in new geographies, graphic communication including use of annotations with pictures, conducting walk through Web sessions and others. Traditional soft skills continue to be relevant and these include adaptability, open-mindedness, problem solving, decision making, communication skills, self-learning and knowledge discov-ery, empathy and team work, motivation, attitude and a spirit of enquiry. The thinking workforce continuously tries to explore the "why", not just "what". Going beyond "why", the doors of innovation are opened up through imagination and asking the question "what if" and then following it through in a structured way.

The changing business scenario today has resulted in specific soft skills importance.

Professional organizations in many countries define soft skills needed by engineers. Association of Engineering Education in Russian Federation (AEE RF) among the "soft" skills highlights ability to communicate with colleagues which includes literacy in oral and written native language for professional communication, as well as ability to communicate in at least one of the most common foreign languages; knowledge of psychology and ethics of business communication; and management skills [3].

The Federation of European National Associations of Engineers (FEANE) highlights the ability to freely express opinions on technical matters on the basis of scientific analysis and synthesis; correct oral and written speech, the ability to make reports; fluency in one of the European languages, in addition to their native language; the ability to use statistics; and ability to mobilize human resources [1].

American Board of Engineering and Technology (ABET) defines socio-cultural competence as follows: the ability to communicate effectively, ethical behavior and communication [2].

There is not much material available on the skills demanded in the highly time-conscious global engineering sector; this article intends to bridge some of the gaps.

In addressing these universal tasks education plays a crucial role. It is well known that the discipline Foreign Language offers unlimited opportunities for development of "soft" skills or interpersonal skills as it is communicative in its nature.

Implementation of the modern federal state educational standards, based on competence and subjective scientific approaches requires the development of innovative methodical systems.

The application of the methodological system in teaching humanities is carried out in a complex multi-parameter, dynamic and social learning environment. The result depends on a combination of factors: age and individual abilities of students, facilities and equipment, curriculums, programs and didactic support, application of innovative methods, regional factors and others.

In our study the methodological system in teaching humanities refers to purposeful, content meaningful and organizational aspects of emotional and intellectual interaction between teacher and student, aimed at the conscious acquiring of humanitarian knowledge, as well as acquiring the ability to apply it in practice.

The methodological structure of the system includes goals, objectives, subjects and objects, the content of education, organization, administration, pedagogical diagnostics and continuous pedagogical monitoring.

The purpose of the methodological system is to create adaptive educational environment and comfort conditions for effective teaching of humanities to technical students. Methodological system solves the following main interrelated problems: formation of the required level of humanitarian competence for future engineers; promotion of self-educational activity in the process of studying the humanities; acquiring skills for effective use of information resources in the learning process; developing students' self-reflective abilities and tolerance in relations with group mates.

In the development of methodological system for teaching humanities to technical students, we used polyparadigmal approach as a set of several paradigms, assuming the dominant role of the leading paradigm that is not opposed by other paradigms, but add to it on the basis of synergy [4].

Polyparadigmal approach is an interdisciplinary approach, based on a common set of programs, implementing several scientific approaches, theories and principles. The developments in pedagogy, psychology, ergonomics, cybernetics, and computer science show that application of a variety of scientific approaches and theories is a specific methodological requirement of polyparadigmal approach.

3   Results

The results of comparative analysis of curriculums for the field of study 13.03.02 "Electrical Power Engineering and Electrical Engineering" offered by universities in the Siberian region are shown below. The data has been received from the official sites of the universities which are open accessed (Table 1).

Implementation of polyparadigmal approach provides a common methodological and organizational framework. It requires the involvement of experts from different fields, coordination of efforts to achieve a common goal of developing methodical system for teaching the humanities. Development of methodological system requires collective thinking activity, the leading role in which is taken by the teacher of particular discipline.

Table 1. Humanities at Siberian technical universities.

University	Basic humanitarian disciplines	Elective humanity courses
Novosibirsk State Technical University (www.nstu.ru/education/edu_plans)	Foreign language (10 credits)	Fundamentals of personal communication culture (3 credits) Psychology and technology of social interaction (3 credits)
Kemerovo State Technical University (www.kuzstu.ru/learning/curriculum/program/ list)	Foreign language (8 credits)	Culture science (2 credits) World and native culture (2 credits) Business correspondence (2 credits)
Irkutsk National Research Technical University (www.istu.edu/structure/57/4308/ 4449/?parent=73043)	Foreign language (9 credits)	Russian language and culture of verbal communication (2 credits) Culture science (2 credits) Sociology (2 credits)
Far East Federal University (www.dvfu.ru/about/information-about-educational-organization/education.php)	Foreign language (8 credits)	Russian language and culture of verbal communication (2 credits)

4 Discussion

Nowadays there is no method to solve all of issues. But we can draw a very important conclusion: the process of effective acquiring of communication skills depends on acquiring three fundamental competencies: "Communication skills", "Psychology" and, of course, "Technical knowledge". For example, if you try to speak to a young generation on the need for a "green" lifestyle, the approach you follow in communicating this message has to draw on "excellent communication skills", include the consequences of decisions they can relate to, and they are technically convincing, supported with data and evidence. This is a great area of work and only a shared effort can bring the desired result. The proposed poliparadigmal approach can add to achieving the common ultimate goal.

5 Conclusion

The world needs qualified and competent engineers, and in order to educate them, universities have to put the rigid requirements for optional courses available to them. Trainers should show them how to discuss and support their ideas or other theories. Though this may seem unimportant to an engineering education, it is precisely why they should be offered in the curriculum. Being smart in mechanics and computing isn't any good unless they can communicate with others, and a methodological and mechanistic world isn't any good unless they can improve it.

References

1. Chuchalin, A.: Modernizaciya bakalavriata v oblasti tehniki i tehnologii (in Russian) [Modernization of bachelor education in engineering and technology]. High. Educ. Russ. **10**, 20–29 (2011)
2. Chuchalin, A.: American engineer and Bologna model: comparative analysis, problems in competence education. Educ. Issues **1**, 84–93 (2007)
3. Pokholkov, Y.: Razvitie nezavisimoi obschestvennoi akkreditatsii inzhenernyh obrazovatelnyh program v Rossii (in Russian) [Development of independent public accreditation of engineering educational programs in Russia]. Eng. Educ. **12**, 42–49 (2013)
4. Skibitskii, E.: Poliparadigmalny podhod – teoreticheskaya baza dlya razrabotki poleznyh obrazovatelnyh instrumentov (in Russian) [Polyparadigmal approach – a theoretical basis for the development of useful educational instruments]. In: Skibitskiy, E. (ed.) Quality of Education in the Context of National and Global Issues: Materials of International Scientific Conference 2014, pp. 119–134. SibAFBD, Novosibirsk (2014)
5. Vernadskiy, V.: Nauchnaya mysl kak planetarnoye yavlenie (in Russian) [Scientific Thought as a Planetary Phenomenon]. Nauka, Moscow (1991)

Interdisciplinary Convergence
in the University Educational Environment

Tatyana I. Suslova(iD), Elena M. Pokrovskaya(iD),
Margarita Yu. Raitina[(⊠)](iD), and Alena E. Kulikova(iD)

Tomsk State University of Control Systems and Radioelectronics,
Tomsk 634051, Russian Federation
raitina@mail.ru

Abstract. Competence profile of a technical university graduate should be formed in the logic of the needs and interests of modern information society. The authors consider the main directions of interdisciplinary convergence of competences in the research and educational projects of the university. The methodological basis for this is the interdisciplinary, comprehensive and competency-based approaches, practical usefulness principles focusing on personality. The article considers specific features of introduction of interdisciplinary mechanism of competences formation in the scientific and educational projects of the Faculty of Humanities in Tomsk State University of Control Systems and Radioelectronics (TUSUR), such as formation of graduates' intercultural competence and the study of TUSUR students' motivation potential in the learning process. Practical significance of the work lies in the fact that relevance of a modern specialist depends on the degree of development of the cultural identity and ability to successfully interact with representatives of other cultures, as well as internal personal factors that interfere with learning and encouraging learning, are defined. Socio-educational projects, initiated and developed by the Faculty of Humanities of TUSUR as a coordinator for the interdisciplinary convergence of practices, are aimed at creating the atmosphere of tolerance and interethnic harmony in the region. The authors conclude that interdisciplinary educational effect goes beyond the cycle of the humanities disciplines, allowing us to talk about qualitative changes in training.

Keywords: Transdisciplinarity · Interdisciplinarity · Intercultural competence · Motivational potential

1 Introduction

Today competence profile of a technical university graduate (TUSUR) has to be formed in the logic of the needs and interests of modern information (communication) society. An important component is a productive realization of personal and social trajectories; this is largely due to implementation of the Federal State Educational Standards based on the competency approach, which is the trend of the study of applied aspects of education in relation to the needs of the economy and society.

The research problem, stated in the work, focuses on solving the key challenges of the university related to the increasing influence of TUSUR on the development of

© Springer International Publishing AG 2018
A. Filchenko and Z. Anikina (eds.), *Linguistic and Cultural Studies:*
Traditions and Innovations, Advances in Intelligent Systems and Computing 677,
DOI 10.1007/978-3-319-67843-6_9

economy and social sphere of the region on the basis of implementation, including interdisciplinary approach, and necessity of purposeful educational strategy. Achievement of goals and objectives involves the harmonious integration of classical and innovative trends and technologies in educational space of the university.

2 Theory and Methods

The paper presents the main directions of interdisciplinary convergence of competences in the scientific and educational projects of the university, where special attention was paid to such areas as:

- improving the implementation of methodological approaches to formation of competences on the basis of interdisciplinary synthesis in the conditions of innovative economy into the educational process;
- organizing and holding of scientific and educational activities of the integral interaction, representing the practice of interdisciplinarity;
- using the technology for interdisciplinary research in scientific and educational activities of the university departments in the framework of the applied research and scientific and educational projects.

Compliance with such requirements leads to understanding of the educational thesaurus of a young specialist. The processes of globalization in the modern educational space and a larger degree of involvement in Russian education determine the necessity of mutual integration of educational programs. In the context of globalization in connection with the modern environmental, socio-ethical and socio-biological problems, there is a new type of modern vision science that brings us closer to the ideal of a single universal scientific knowledge, synthesizing the Sciences and the Humanities methods of cognition. Therefore, in terms of information flow, increasing the emergence of transdisciplinary complexes of knowledge becomes topical for the search of new ideas and approaches to modern educational institutions.

Therefore, for implementation of effective professional activity it is more important to apply not the disparate knowledge and skills, but generalized, allowing to go beyond the subject field. However, the competence model of various TUSUR departments focuses mainly on in-depth specialization in the areas of training. This trend may limit the possibilities of cooperation in innovative projects and the sectors oriented to continuous interdisciplinary interaction, leading to leveling of the humanities component, which is included in the systemic methodology for students' professional qualities formation. In this case, introduction of the humanities plays the role of an integrating factor in formation of interdisciplinary educational environment.

The main purpose of education is to ensure high quality of educational process in educational programs and implementation of scientific research, aimed at improving the system of professional education activities and research [3, 4]. This is due to implementation of the following tasks:

- Development and improvement of methodological support of educational process, introduction of new interactive educational technologies and teaching methods, emerging at the intersection of interdisciplinary research;

– Development of communication skills in various cultural situations in accordance with the principles of tolerance, respect for other cultures and universal values. This challenge is reflected in the set of common cultural competencies of the Federal State Educational Standards on various aspects of training;
– Active inclusion of employers in educational process of the profile areas of training;
– Integration of theoretical knowledge and practical skills, development of scientific research; organization of research activity for students, integrating academic and professional orientation.

The methodological basis for this is the interdisciplinary, comprehensive and competency-based approaches and the practical usefulness principles, focusing on personality [1].

The main direction of scientific and educational activities of the departments is the interdisciplinary synthesis of approaches used in students training in the disciplines of basic and optional parts for most of the training areas of TUSUR [5]. The direction of university courses is of integrative and interdisciplinary nature, and requires the integrated research, formation of knowledge system and development of integrated courses, enhancing interdisciplinary connections.

Scientific and methodological activity of departments of the Faculty of Humanities in the applied aspect solves the problem of social and humanist environment modernization at the university.

The use of multidisciplinary synthesis technology is actively implemented in the framework of applied research, scientific and educational projects conducted by the Faculty of Humanities. This experience is extrapolated to the educational environment of the University in all faculties.

The methodological basis for interdisciplinarity implementation is a learner-centered approach, taking into account the graduates' competence model. In this case, interdisciplinary connections are, in fact, the applied aspect of the integration processes in the field of education for educational process design. Since formation of competences in all Federal State Educational Standards can be carried out not by one, but several disciplines of the basic educational program curriculum. Interdisciplinary synthesis of knowledge, abilities and skills of students is also necessary due to the fact that undergraduate programs in all areas of training become more practice-oriented with the general decline in the number of lecture course hours [2].

3   Results and Discussion

The systematic work of the Philosophy and Sociology Department and Foreign Languages Department on interdisciplinary implementation study is associated with the analysis of conditions of the graduates' intercultural competence formation. An educational module for the academic discipline, contributing to formation of intercultural competence, methodological recommendations and practical tasks were developed and piloted.

The concept of "intercultural competence" is interdisciplinary, it is the object of study in various sciences. The relevance of the study due to the fact that intercultural competence is a key component of modern student training at any university, and is incorporated in basic educational documents of the Federal State Educational Standards. This is confirmed by the importance of the intercultural aspect in professional activity of modern specialists, which is associated with different cultures interaction and effective intercultural professional communication in a multicultural world.

The proposed methodological approach of the research is based on interdisciplinary experience of philosophical, cultural, sociological, psychological and historical analysis of the dialogue of cultures. This basis makes the competence-based approach highly productive as it allows identifying the structural, content and functional components of intercultural interaction.

Processes in a globalizing world strongly suggest that cultural integrity is accompanied by emergence of new, previously unknown problems. Extension of intercultural contacts leads to crises, destructive phenomena, world transformation and innovative searches in every culture. Interaction of cultures creates the need to reassess cross-cultural contacts and cultural identity, based on the ideas of intercultural tolerance and adequate perception of cultural differences [6].

Analysis of the questionnaire results allowed us to clearly assess the performance of students on the notion of intercultural competence, which we formulated earlier, intercultural competence is considered the ability of a person, based on specific knowledge, skills, personal attitudes and strategies, through which it is possible to successfully implement professional communication with representatives of other cultures.

It should be noted that a large proportion of the respondents understand this concept as a set of the following basic elements: cultural relativism, self-actualization, reflection, prestige, education and psychological component. Thus, there is a loss of important elements, such as knowledge, based on their own experience, analysis, ethnocultural background, ability to engage in dialogue and motivation.

It is obvious that the problem of investigation of students' motivational sphere is relevant for any faculty, since one of major tasks of education is formation of the student competence profile and professional development.

An interdisciplinary study, associated with the study of TUSUR students' motivational potential, was conducted in the Scientific and Educational Center of the Faculty of Humanities under supervision of E.M. Pokrovskaya, the Director, and D.V. Ozerkin, Dean of Radio-Design Faculty, based on the need to define a balanced policy to motivate students to a better academic performance, and, therefore, a better preparedness for future professional activity based on a portfolio of competencies.

Novelty of the work is the calculation of the integral of motivational potential by the formula (1)

$$ИМП(N,y)=\frac{1}{N}\sum_{i=1}^{N}МП(i,y),\qquad(1)$$

where МП (i, y) – motivational potential of the i-th student at time y; ИМП(N, y) is the integrated motivational potential of the totality or sample of students at time y; N – the number of students in the sample.

Motivational potential is the need, interest and positive attitude to ongoing activities. During the study a motivational potential of each participant was calculated by the formula, similar to the formula of the Hackman – Oldham method, in relation to the learning process (2).

$$ИМП(N,y) = \frac{1}{N} \sum_{n=1}^{N} \frac{(PУ(n, y) + OУ3(n, y) + 3O(n, y))}{3} \times A(n, y) \times OC(n, y), \qquad (2)$$

where N is the number of students in the sample; PУ (n, y) – diversity training; OУ3 (n, y) – unambiguous assignments; 3O (n, y) – importance of education; A (n, y) – autonomy of educational process for students, and OC (n, y) – feedback presented on the above scale at time y.

The survey results were analyzed, and the integral of the students' motivational potential was calculated, both existing and undiscovered. Integrated motivational potential is an indicator of a students' team, which characterizes the overall level of the given set of students and gives the opportunity to generalize the characteristics of the sample.

Calculation of the integrated motivational potential will help to see what courses are on the decline, students' motivation and take correction measures, build an effective policy at the university in relation to the students' motivation.

The study revealed that majority of students, according to X theory of McGregor, are initially lazy and will try to avoid a vigorous activity, without a reward system they will usually be passive and try to avoid unnecessary responsibility.

Highlighting the internal, personal factors that interfere with learning, in addition to laziness, respondents identified such factors as inattention, inability to specify the main thing, and lack of interest in studies.

Among the personality traits that motivate learning such as perseverance, patience, responsibility, and conscientious attitude to learning, interest in their future profession were noted. It shows what part of the student's individual should be paid attention. To stimulate motivation in this case is much easier than in the case with dominating negative and hindering factors because students themselves seek the best execution of their duties.

Questions about external factors influencing the desire to learn brought such answers as financial aid, interesting forms of learning (games, workshops and debates), an interesting lecturer, comfortable, quiet audience, and good food in the dining room. According to the Herzberg theory, these factors relate to motivation, i.e. motivating to work needs, growth, development, and warning tardiness and truancy.

External factors that hinder the desire to learn, like a teacher's loud voice in the adjacent classroom, tutor's quiet voice, uncomfortable seats, good weather, inconvenient schedule, distant location of school from place of residence are the factors of anti-motivation, i.e. prevent the interest of students to learning.

Thus, it is possible to say that motivation of students, being average, is not a sufficient condition for a high student activity in the industry training, to increase the activity it is necessary to increase motivation. It is important to pay attention to internal and external factors influencing the growth or decline of motivation, and strive to change the students' motivation in a positive direction, upwards.

4 Conclusion

Finally, we note the effects of the scientific-methodological educational projects and events of the TUSUR Faculty of Humanities as a coordinator of the practice of interdisciplinary convergence:

- Number of research projects, application-oriented research carried out by the Faculty of Humanities was increased and allowed us to formulate a number of recommendations to extrapolate this experience into the educational environment of all TUSUR faculties;
- Inter-university education center for high schools interaction with national-cultural autonomies at the TUSUR Faculty of Humanities was created;
- Mechanisms of interaction of national-cultural autonomies with the universities in the city were formed;
- Socio-educational projects, aimed at creating the atmosphere of tolerance and interethnic harmony in the region, were initiated and developed;
- Activities of information and education awareness were organized, and they become precedents in the sociocultural space of the region;
- Interdisciplinary educational effect goes beyond the cycle of the humanities disciplines, allowing us to talk about qualitative changes in training.

The interdisciplinary nature of the university's research and scientific-methodological work in the modern educational paradigm aims at solving innovation problems of integration of education, research and practice-oriented components in the context of the competence approach.

Acknowledgements. The work is performed at support of the Ministry of Education and Science of the Russian Federation, Project No. 28.8279.2017/BP.

References

1. Budenkova, V., Saveleva, E.: Innovacionno-obrazovatel'naja programma «Formirovanie professional'nyh, lichnostnyh i obshhekul'turnyh kompetencij sredstvami kul'turologicheskih disciplin student XXI veka» kak opyt realizacii mezhdisciplinarnogo podhoda v uchebnom processe v klassicheskom universitete (in Russian) [Innovative educational program "Formation of professional, personal and General cultural competences by means of cultural disciplines, the student of the 21st century" as the experience of interdisciplinary approach implementing in the educational process in a traditional university]. Vestnik Tomsk State Univ. Philos. Sociol. Polit. Sci. 3(4), 123–128 (2008)
2. Hackman, J., Richard, O., Greg, R.: Motivation through the design of work: test of a theory. Organ. Behav. Hum. Perform. **16**, 250–279 (1976)
3. Meshcheryakov, R.V., Shelupanov, A.A.: Konceptual'nye voprosy informacionnoj bezopasnosti regiona i podgotovki kadrov (in Russian) [conceptual issues of information security in the region and training of personnel]. Trudy SPIIRAN **34**, 136–159 (2014)

4. Popova, N.: Mezhdisciplinarnaja paradigma kak osnova formirovanija integrativnyh kompe-tencij studentov mnogoprofil'nogo vuza (na primere discipliny Inostrannyj jazyk) (in Russian) [Interdisciplinary paradigm as the basis for formation of integrative multidisciplinary competences of students at the University (based on the example of Foreign language course)]. Unpublished theses abstract (2011)
5. Rezanova, Z.I., Romanov, A.S., Meshcheryakov, R.V.: Zadachi avtorskoj atribucii teksta v aspekte gendernoj prinadlezhnosti (k probleme mezhdisciplinarnogo vzaimodejstvija lingvis-tiki i informatiki) (in Russian) [Tasks of the author's attribution of the text in the aspect of gender identity (to the problem of interdisciplinary interaction of linguistics and informatics)]. Bull. Tomsk State Univ. **370**, 24–28 (2013)
6. Sadokhin, A.P.: Mezhkul'turnaja kompetentnost': sushhnost' i mehanizmy formirovanija (in Russian) [Intercultural competence: the essence and mechanisms of formation]. Unpublished theses (2009)

Problems of Education in the Context of Technoscience: Tradition and Innovation

Tatyana I. Suslova, Denis V. Ozerkin,
and Margarita Yu. Raitina[(✉)]

Tomsk State University of Control Systems and Radioelectronics,
Tomsk 634051, Russian Federation
raitina@mail.ru

Abstract. The article discusses the problem of the integrative nature of scientific, technical and institutional components interaction in engineering education. The authors consider the contradiction between the impersonal technologies of mass education, individual abilities and professionalism. In the process of graduates' preparation in the formation of the new technological system, techno-science, the role of social-humanitarian disciplines increases. Socio-humanitarian profile contributes to formation of critical and reflective thinking that is necessary for the convergence of knowledge that is in demand today. The authors distinguish between monodisciplinary and transdisciplinary organization of science and emphasize the problem of ethics in research, including, in particular, the limits of the permissible in a scientific experiment. The article shows the problem areas of ethical control in the field of NBIC-technologies and determines the need for theoretical and methodological reflection between new technologies and fundamental human values, and generating new versions of anthropogenesis. The semantic nodes between tradition and innovation, professional duty and responsibility, short-term and long-term goals, professional community and society as a whole is proposed as the basis for building the future of education. The key to this research is the model of engineering education for future seeks to consider the context of technoscience, ethics and responsibility.

Keywords: Technoscience · Ethics · Responsibility · Convergence · Transdisciplinarity

1 Introduction

Among the problems of higher education in recent years, there is a conflict between the need for integrative nature of the knowledge of the scientific and technical plan and institutional structure of engineering education. In other words, this contradiction is between a highly specialized training of engineers and convergent nature of now-dominant high-tech. Trends and requirements for competencies of graduates, aimed at formation of young specialists ready to work in a new technological system. This implies the possession of a much wider range of key competencies than development of highly specialized scientific, technical and engineering disciplines, willingness to learn

© Springer International Publishing AG 2018
A. Filchenko and Z. Anikina (eds.), *Linguistic and Cultural Studies:
Traditions and Innovations*, Advances in Intelligent Systems and Computing 677,
DOI 10.1007/978-3-319-67843-6_10

throughout life and to change their own professional settings. It is also necessary to strengthen the scientific component and the acquisition of research skills.

2   Methods and Results

There is, unfortunately, a big contradiction between the impersonal technologies of mass training and individual abilities, professionalism and intellectual breadth. In these conditions the role of social and humanities disciplines increases as the means for linking "nodes" between tradition and innovation, professional duty and responsibility, short-term and long-term goals, professional community and society as a whole. It is also necessary to improve the information technology influencing the fundamentals of mental and cognitive processes that are responsible for the ability to maintain personal integrity and identity. All this will help to develop the students' critical thinking and ability to improve personal qualities, responsible for self-organization, personal integrity and identity. It seems that these goals correspond to a number of disciplines, in particular, ethics and professional etiquette, fundamentals of human development, and responsibilities of the academic teaching of masters' and PhD students. As a result of adopting these disciplines, we form such competencies as willingness and ability to continuous professional self-development and self-improvement throughout life, the ability to work in team, tolerant perceiving of social, ethnic, religious and cultural differences. All these competencies are essential for modern graduates, as compared to the classical science; big science of the 20th–21st centuries has a completely different organizational structure. As a rule, large research teams, a variety of funding sources and research projects were created with the aim to attract investors, competition among the state-funded and private research teams. Science in the 20th and 21st centuries is based on different types of rationality. Science is one side of human activity [2]. However, note that the technical rationality of scientific knowledge is aimed at efficiency, and it distorts the man and his work, like a machine. In this regard, a famous scientist John Ziman offers a perspective on the ethics of science: he believes that a scientist must constantly maneuver between the moral imperatives of the scientist R. Merton and the realities dictated by the "scientific kitchen". This is the so-called locality principles: work at the firm, local work, authoritarian work, custom work and expert. In general the scientific-technical progress influences the society and the moral consciousness of people, science is presented as a means of coping with global challenges of our time: ecological crisis, economic crisis, international terrorism, dangers of cosmic and solar radiation, asteroids, etc.

Modern scientific knowledge is qualitatively different from the former type of production, which is organized by monodisciplinary science. In science the knowledge is produced in laboratories and other specialized research organizations. Knowledge translated to the society is becoming practical and applied. Moreover, parallel philosophical research shows that it is necessary to pay attention to the fact that knowledge is produced not only in science, but in the whole established social network that perceives scientific ideas, accumulating, distributing and applying them. Each of the social agents involved in this process, produces its own specific knowledge, which, on the one hand, allows running its functions more efficiently, and, on the other hand, to more successfully engage in cooperation with other agents. In our time, it becomes clear that to achieve the common good and tackle the most urgent anthropological and existential problems arising from the progress of all types of knowledge, a combination of a variety of previously disparate disciplinary knowledge is not only technical direction, but also scientific, philosophical, theological, legal, political, etc. [3, 4]. Moreover, this combination is possible only as a result of research beyond disciplinary boundaries in a transdisciplinary sphere of the lifeworld. The developers of this concept of new knowledge specify five of its features that will be useful to interpret the nature of this knowledge production. First, that knowledge is produced (more precisely— extended plays) in the context of practical application. Second, it is interdisciplinary and even transdisciplinary subject's intentions on integrity in the above sense. Thirdly, it is diverse and quality, and organizational variety of forms. Fourth, it does not just describe those aspects of reality, it inherently involves double reflection. Reflection on the modes of production of knowledge, and the value basis of the same ways. Reflection that produces knowledge of the subject itself is a form of his own active reproduction. Thus, with the production of knowledge, its producing subject is engaged simultaneously. Knowledge is produced by both individuals and their communities (firms, laboratories, etc.). Fifth, the classical knowledge enhances the standard scientific practice of assessing the quality of knowledge (in particular, the procedures for peer review), undertaken by one or another expert or experts. It (the community) assesses their ethical acceptability and social relevance (usefulness and efficiency).

Thus, modern science is implemented in three inextricably linked socially oriented educational systems: an interdisciplinary field of research, academic discipline and political institution. We are talking about the need to create ethical committees in the areas of contemporary knowledge that are people-oriented. Politics appears to be the sphere for harmonization of civil interests and goals for the common good. The famous Italian scientist Agazzi has formulated the main task of technoscience ethics: "Functioning of the technological system is essentially indifferent to the purpose; when certain facilities are made available, technology inevitably seeks their practical use; it is characterized by a tendency to pursue every opportunity" [2]. There is a danger of absolutization of technoscience, its rupture with morality. However, in the 21st century it became apparent that science merged with technology, the technology promises commercial wins at the expense of their applications. Science also includes promises, and this synthesis of science and technology has changed not only the classic division of labour between the explanation (science) and the promise (equipment technology), but has shifted to another aspect, the ethics of relationships in the scientific community. The dominant position occupied corporate ethics of the business community, where the working scientists carry out a specific order for development of a drug. That is, the science and the business began to promise, based on the tools using the technoscience: conceptual, mental, instrumental, and practical devices. Thus, the convergence of NBIC technologies (nano-bio-information technologies and cognitive science) is applied to the person by inhuman measures. It became the dominant technical understanding of the person. Repair, improvement, replacement, and it created a full human double made of non-aging material. The problem of death is reduced to a bodily existence. To overcome nature by turning the body whether into a machine, or a monster? There is not only one thing that exhausted our future. This also applies to the autonomy of the individual. In traditional and industrial societies, a different attitude to the problem of the autonomy of the individual [5]. Traditional society, the autonomy of the individual does not tend to implement; an identity can only belong to a corporation as an element of corporate bonds. In industrial society advocated the autonomy of the individual, allowing you to take in a variety of social communities and cultural traditions. Man is understood as an active, independent action and being. His extensive activities aimed at conversion and alteration of the outside world and nature. However, nature cannot be a bottomless reservoir for various kinds of man-made exercise, as human activity originally appeared as a component of the biosphere, but it is not as its dominant. The immediate impetus for the emergence of bioethics in the United States made a series of public revelations. These revelations explicitly demonstrated that the moral values that guide physicians, in some cases, are a direct threat to patients. All the moral issues of the day to refer to and especially not to solve, but let's be optimistic, looking at these issues and see the number of supporters of the preservation of the human species grow, enhanced by the ethics of responsibility of both scientists and business communities, and concerned government officials.

In 1997, the Council of Europe Convention "On protection of human rights and dignity of the human being with regard to biology and medicine application" was adopted. But it is clear that this Convention does not work, the agreements are violated. The concept of "man" and "human being" is not defined by the Convention and given to the legislators of the states. In this situation, the most difficult is the issue of patenting of genes transcribed medical centres, opening them and selling them to other pharmaceutical companies. Everyone agrees on one thing – to patent the natural properties of human beings is impossible. But it is this strive to make the business community developing new medical products based on the open human genome. The person wishing to establish an IR implant in his body should be informed about possible health risks and the possibility of unauthorized access to information stored in the implant. In the presence of a large number of pseudo-scientific clinics, there should be a careful monitoring of IR implants, which are offered on a commercial basis and are designed to "improve" man. Along with obtaining the informed consent from volunteers who agreed to participate in the study, you must ensure that they have not been caused any physical, mental or material damage. In addition, people should have the opportunity to withdraw from the research team at any time. It is also unacceptable to use the IR implants in the treatment of patients with serious neurological diseases, so that did not lead to discrimination or infringement of human rights. It is impossible to prevent their use and manipulation of mental abilities, changes of identity, memory, identity or perception of other people, dominating other people, for coercion directed against those who for any reason do not use IR implants.

The relationship between new technology and fundamental human values, problems of technoscience and development of global civilization increase social risks, degree of personal responsibility of scientists. According to some authors, it is possible to allocate at least ten fundamental relations, covering two main areas. The first is focused on the study of biological human nature; it is a project of new naturalism engaged in criticism of the introspective methods of human cognition and generating new versions of anthropogenesis. The second direction is connected with the phenomenon of transcendence, i.e. with the spiritual in man, and is determined by his moral aspirations. Both directions are causing heated debate in both the academic and social environment. So, the main opportunities, threats and risks to the future under conditions of uncontrolled development of molecular genetics and biotechnology are considered in the book by Nowotny and Testa "Naked genes: Reinventing humanity in the molecular age" [6]. They raise the problem of moral and legal responsibility of scientists-biologists of the future of human development. The famous Italian scholar Agamben draws attention to the fact that there is a process of erasing the distinction between public and private life and calls this process of the depoliticization of citizenship. A crucial role in this matter is the issue of security. Developed modern technologies of observation in the relevant state laboratories, institutes receive the state order. This ensures safety and reduces risks in business, politics, private and public life. Agamben says that "coverage for all citizens' identification by technologies developed for the criminals (later Jews), could not affect the political identity of the citizens" [1]. And further – "the most neutral and private information becomes a carrier of social identity, which thereby loses its public character" [1] – (the order of genes in the double helix, fingerprint for police passport card). Suspicion and checking became a normal attitude to us from the state. In other words, a state is a state control and surveillance over citizens. Thus, we see that security and democracy are incompatible, on the one hand. So, institutes, laboratories that develop and improve the technology of electronic surveillance and the business community, in fact, contribute to strengthening of the state control over citizens. Thus, education and science, fulfilling the state order, merged with technology and business.

3 Conclusion

To sum up: 1. Rationalizing and pushing the horizons of science cannot be unlimited, lost sense of proportion and values. 2. It is a tragic conflict of irreconcilable and incommensurable started science and professional ethics. It creates a border situation between man and animal, resulting from the crossbreeding of genetically different species, hybrids or chimeras. A wonderful goal to improve the man through overcoming his ultimate nature, transcending human capabilities, can lead to transformation into a machine or a cyborg.

Ethics is powerless in controlling genetic research which creates chimeras and cyborgs. The development of xenotransplantation leads to appearance of unknown virus, the animal nature that destroys the human genotype. We do not know where the animal ends and the man begins. 3. A border situation between man and machine is created. The device is not simply embedded in the human body, but it also affects heredity. 4. The therapeutic focus is one thing, but managing identity has eugenic transhumanist meaning. What are the limits of the invasion? It is use of converging technologies: co nano-, biomedical, information technology and cognitive science.

References

1. Agamben, G.: Jetika tehnologij bezopasnosti i nabljudenija [Ethics of security technologies and surveillance]. Man **2**, 57–65 (2015). (in Russian)
2. Agazzi, E.: Moral'noe izmerenie nauki i tehniki [The Moral Dimension of Science and Technology]. Philosophy Foundation, Moscow (1998). (in Russian)
3. Allhoff, F., et al.: Ethics of human enhancement: 25 questions and answers. U.S. National Science Foundation report (2009). http://ethics.calpoly.edu/NSF_report.pdf. Accessed 25 June 2017
4. Bess, M.D.: Icarus 2.0: a historian's perspective on human biological enhancement. Technol. Cult. **49**(1), 14–126 (2008)
5. Fukuyama, F.: Nashe postchelovecheskoe budushhee: Posledstvija biotehnologicheskoj revoljucii [Our Posthuman Future: Consequences of the Biotechnology Revolution]. AST, Ljuks, Moscow (2004). (in Russian)
6. Nowotny, H., Testa, G.: Naked Genes: Reinventing Humanity in the Molecular Age. MIT Press, London (2010)

Learning Outcomes Evaluation Based on Mixed Diagnostic Tests and Cognitive Graphic Tools

Anna E. Yankovskaya[1,2,3,4(✉)] ⓘ, Yury N. Dementev[2] ⓘ,
Danil Y. Lyapunov[2,4] ⓘ, and Artem V. Yamshanov[4] ⓘ

[1] Tomsk State University of Architecture and Building,
Tomsk 634003, Russian Federation
ayyankov@gmail.com
[2] Tomsk Polytechnic University, Tomsk 634050, Russian Federation
[3] Tomsk State University, Tomsk 634050, Russian Federation
[4] Tomsk State University of Control Systems and Radioelectronics,
Tomsk 634050, Russian Federation

Abstract. In this paper, we discuss the relevance of students' learning outcomes evaluation using computer-based testing. The learning process is based on mixed diagnostic tests. For the purpose of evaluation, we use the threshold, fuzzy logic and cognitive graphic tools. The construction of mixed diagnostic tests, representing a compromise between unconditional and conditional components, in order to develop students' knowledge evaluation is proposed for a number of disciplines. We suggest a technique for optimal mixed diagnostic tests construction based on the expert knowledge of the subjects for effective learning. The developed approach is used for a number of both the humanities and technical disciplines. One of useful outcomes of mixed diagnostic tests application is the learning trajectory design for each individual. We construct students' learning trajectory using the intelligent learning and testing system and suggest defining their inherent approach to the learning process within the problem area.

Keywords: Learning and testing intelligent system · Pattern recognition · Mixed diagnostic tests · Threshold logic · Fuzzy logic · n-simplex · Cognitive graphic tool · Learning trajectory

1 Introduction and Literature Review

In recent years, development of intelligent learning systems has been a relevant problem [5, 12, 14, 15]. New information technologies provide a number of innovative and very promising technologies for learning and testing. A significant group of these technologies is related to the online education or the blended education and training (blended learning), involving limited interaction between students and educators [4, 12]. Such technologies influence both the educational practice and the basic approaches of the distance education by enhancing collaboration, which is a very important component of both learning and testing processes. At the same time, the problem of

© Springer International Publishing AG 2018
A. Filchenko and Z. Anikina (eds.), *Linguistic and Cultural Studies:
Traditions and Innovations*, Advances in Intelligent Systems and Computing 677,
DOI 10.1007/978-3-319-67843-6_11

students' evaluation and efficacy increase of educational process becomes relevant due to absence of the teacher in such class. The concept of blended learning, combining the essential advantages of both face-to-face and distance learning approaches, was entirely introduced in the handbook [12]. Generally, the blended learning concept corresponds to an integrated learning environment, which combines online learning and traditional classroom teaching [12]. An essential challenge for higher education is development of approaches, methods and tools for effective learning and training of a large number of students, involving a wide diversity of tasks [3, 4, 12].

Success of the course realized through blended learning paradigm depends on its content, learning methods applied and software used to a sufficient extent [5]. Therefore, students need to learn how to use information in an effective way to maintain their knowledge level and to develop their intellectual and creative potential [8].

While designing a learning course, we use the basic components of blended learning process: face-to-face component (F2F) and the online one which can take 30 to 79% of the course time [10]. Providing a certain balance between F2F and online components by their mutual connection and dependence, both students and educators achieve the learning objectives in a more effective way.

Use of a variety of resources and methods in accordance with the structured syllabus could serve as an illustrative example of blended learning technologies. Despite the information technologies diversity, including the methods of artificial intelligence, the methods for obtained knowledge evaluation are not developed sufficiently. Development and design of such methods and evaluation tools are highly time-consuming and require considerable efforts [14]. Students with different knowledge level, skills and abilities have diverse preferences while learning, thus, achieving goals in their inherent manner. Using tests for students' abilities evaluation, such as initial level of knowledge, skills, and experience might serve as the basic step to efficient learning trajectory construction. An interesting approach to students' evaluation when they collaborate within a group and use clickers to answer multiple-choice questions of various levels of difficulty was proposed by Kulikovskikh et al. [6]. Thus, taking into account the students' knowledge level and the way they answer questions will make the learning process more effective. At each stage of the learning process, students are relying on the experience and skills obtained earlier to find the proper solution to the tasks given. For a better design of online courses, the teachers must take into account the fact that students should feel confidence at each stage of the learning process. This approach is entirely outlined in [7]. Feeling confidence is a source of motivation for students. Moreover, the knowledge students acquire and the skills they gain can also be wider and deeper than teachers can give [13].

Modern students need guidance in their learning process. Not all the teachers provide necessary feedback to their students to make them confident at each stage of the learning process. In this regard the ILTs are able to solve this problem.

The stages of learning process constitute a learning cycle [11], including: (1) background knowledge activation; (2) basic learning material presentation; (3) control of understanding and consolidation of the theoretical knowledge; (4) simulation and practical tasks performance; (5) application of acquired knowledge and skills.

The developed courses consist of learning objects (sections, modules) which have their own intermediate goals and form essential students' knowledge, skills and abilities within each module.

The growing interest to the design and the use of intelligent learning and testing systems (ILTS) [15] requires the design of special cognitive graphic tools as essential functional elements of ILTS. Graphic images on a computer screen facilitate sufficiently the new learning material understanding, alleviate user–computer interaction. That leads to a better perception and comprehension of the regularities in data, concerning the learning process. Visualization of information creates a "bridge" between computer data and human knowledge.

Mixed diagnostic tests are a versatile tool for self-education [15]. MDT are used for learning trajectory construction, effective user–computer interaction and assessment of the learning results.

We use an algorithm of data processing, based on simplex approach, to construct learning trajectory for each individual. Constructing special questionnaires which consist of interconnected and mutually dependent parts, the educator stimulates the learning process, predicting necessary questions and mistakes while answering the test questions.

The main advantage of MDT application is the opportunity to monitor the current level of students' achievements and correct their own learning trajectory. Learning trajectory is a broken line within the space of patterns. In our case, we divide the entire learning period into the intervals and assess the students' result at each interval. The evolution of the broken line gives the students information about their evaluation at a particular moment of the learning process, their achievements and the efficacy of their approach to education.

Cognitive graphic tools allow giving all necessary information about learning process both in statics and dynamics. Moreover, these tools motivate the students to create their own learning trajectory, guiding them to learn again and clarify the misunderstood topic if it is necessary. As a result, the students are able to achieve a higher level of professional development in the discipline under study.

Learning trajectory of each individual is formed via unconditional and conditional components of MDT, compromising between unconditional and conditional components, and is visualized by cognitive graphic tools. The compromise between the unconditional and conditional MDT components is achieved due to correct design of test questions leading towards the learning objectives and providing feedback if necessary. Conditional diagnostic tests contain internal feedback loops and provide majority of trajectories dependent on particular conditions.

Cognitive graphic tools, including 2-simplex, 3-simplex [15] and 2-simplex prism [18], proved to be effective assessment tools in a number of problem areas, including education. The tools combine data analysis algorithms and graphic representation of the results at the same time.

Cognitive graphic tools may give some advice to the student about the types of learning activities which should be improved, and which skills are effective enough. Moreover, the educator can give advice to students on how to make their approach better, which modules and topics of the discipline should be studied more deeply and what skills they might need to acquire for learning results improvement.

Cognitive graphic tools also represent the means for students' self-organization and learning behaviour, which are the key qualities for each individual.

Originality and relevance of the approach under study are confirmed by ILTS designed under supervision of Yankovskaya [16–18].

The purpose of the present paper is to analyze the learning outcomes of students' MDT-based testing within the framework of the course "Selected Chapters of Electronics", using cognitive graphic tools and, thus, revealing the MDT impact on the students' motivation to fully comprehend the discipline under study.

2 Mixed Diagnostic Tests Construction

To monitor students' knowledge, we have used various models: finite state machines and graphs (network, tree) [15–18].

Consider the basics of the MDT construction. We have a syllabus for a particular discipline, e.g. "Power Electronics". Each module of the discipline contains the teaching material and tests. An unconditional component of MDT contains the test questions (grouped characteristic features), which are introduced to students in a random sequence. The test results reveal the basic student's knowledge, skills and abilities. Then, if a student has succeeded in the unconditional component of MDT, he/she is allowed to pass a conditional component. A sequence of the grouped characteristic features submission of the MDT conditional component depends on the results of the previous step.

The conceptual diagram of the MDT construction is represented in Fig. 1 [15]. The table is given in the upper part of Fig. 1. Test questions correspond to rows of the table. The test questions are available to the respondent (student) only if the unconditional component of the test is passed successfully. A number of answers to each question correspond to columns of the table. In this case, n is a number of question, k is the maximum number of various answers to every question ($k \geq 2$). Some cells of the table can be empty. The element of the table a_{ij} is the weight of j-th answer to i-th question, $0 \leq a_{ij} \leq 1$.

The average weight of the answer to all questions is calculated as a result of passing the unconditional component of MDT.

If a respondent had received a predetermined value of the threshold assessment of learning material p ($0 \leq p < 1$) for the given module or didactic unit, they are allowed to pass the conditional component of MDT in case of training, otherwise the respondent is not allowed to pass the conditional component of MDT. This component is given in the diagram at the bottom of Fig. 1.

The respondent has to review the module material where the value of average weight of answers to all questions is lower than the predetermined value p_r ($p_r < p$) in case of training.

We consider the test to be successfully passed if the threshold p_r is achieved. The value of p_r is defined by an expert for each module of the discipline individually.

The framework of the paper allows representing only two levels of the conditional component of MDT in Fig. 1. In practice the number of levels is defined by the time

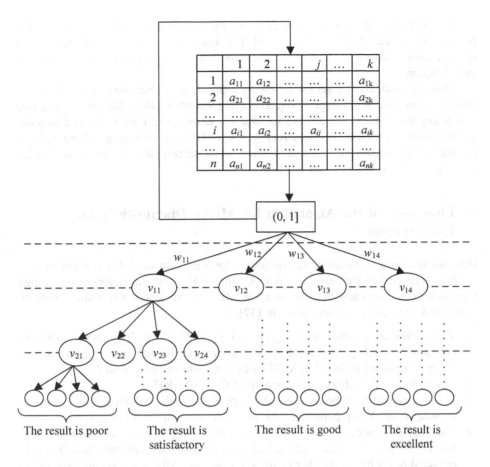

Fig. 1. Conceptual diagram of intelligent learning and testing system construction.

interval spent on the test or training. Indexing of questions of the MDT conditional component starts with number one.

Each edge of the graph matches the weight w_{ij}, which corresponds to the complexity of a test question where i is the level number, j is the number of the edge at the current level.

A test question corresponds to each node of the graph v_{ij}, where i is a level number, j is a node number at current level. After passing the conditional component of MDT, each student is able to observe visually an overall evaluation of their knowledge via the cognitive graphic tools.

Fuzzy logic is based on the fuzzy set theory, suggested and developed by Zadeh in 1965 [19]. Fuzzy logic has been applied in many fields from the control theory to artificial intelligence. Fuzzy logic is a form of many-valued logic. In contrast with the traditional logic theory where binary sets have two-valued logic: true or false, fuzzy logic variables may have real value that ranges in degree between 0 and 1. In this research we use four gradations of fuzzy logic variables: poor, satisfactory, good, and excellent.

One of the new directions in pattern recognition using fuzzy logic is represented in publications of Zadeh [19], Novak et al. [9], Batyrshyn et al. [2] and others. The models constructed using fuzzy expert evaluations and fuzzy systems are the bases for this direction.

Thus, respondents can get the following evaluations of their knowledge: it can be absent, below average, average, above average, good or excellent. Moreover, after data processing the respondents acquire information about their knowledge components, e.g. an ability of problem solving, circuitry design skills and knowledge of principles of circuits operation, and make decision which of these components should be emphasized in the future learning process.

3 Flowchart of the Algorithm for Mixed Diagnostic Tests Construction

Here we represent an informal description of the algorithm for MDT construction.

It is necessary to indicate the conditions for MDT operation (learning or testing). For each course or each didactic unit the MDT construction algorithm should be performed similarly to the one given in [17].

1. The teacher assigns the weight w_{ij} ($1 \leq i \leq n$, $2 \leq j \leq k$ for an unconditional component and $1 \leq i \leq m$, $2 \leq j \leq k$ for a conditional one) to each question (each component included in MDT) within an interval from 0 to 1.
2. Generate the unconditional component of the next MDT.
3. Find the weight S of the conditional component of MDT which equals to the sum of weights of all components w_{ij} of the MDT unconditional component.
4. Calculate the student's score S' using the components of the unconditional part of MDT. Find $p = S'/m$. Calculate p_r with each conditional component and divide the obtained result by the number of questions in each MDT component (conditional and unconditional).
5. If $p_r \geq 0$ then go to point 8.
6. In the training mode, if a student did not score the required weights for answers, then go to point 7. In case of testing, evaluate the result as poor and recommend that they should repeat the training attentively. Go to point 11.
7. Give advice to the student on the result of the MDT unconditional component. Go to point 2.
8. Generate the first component of the MDT conditional section.
9. Provide the next component of the MDT conditional section to the student.
10. Increase the score of weights of the student's answers by the weight of the received answer to the next component of the MDT conditional component in a question.
11. Verify whether all components of the conditional MDT have been provided to the student. If no, then generate the next component of the conditional MDT and go to point 9. If not all of conditional components are analyzed, go to point 8. Analyze the score of the student's answers according to the MDT result. Evaluate the student, i.e., make a decision as to whether they passed the MDT, based on the scored answer weights, or not.

Flowchart of the MDT construction is shown in Fig. 2.

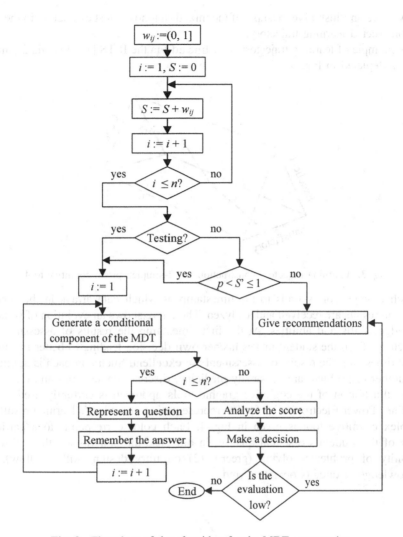

Fig. 2. Flowchart of the algorithm for the MDT construction.

While students do the MDT, they form their knowledge in the problematic area. After they pass the tests, the cognitive graphic tools tell them about the next steps in the learning process, providing an efficient approach to knowledge acquiring.

4   Learning Trajectory Based on the MDT Results

Here we give an illustrative example of the mixed diagnostic test evaluation in the form of a constructed learning trajectory.

An example of learning trajectory, constructed via the ILTS [16, 17], via 2-simplex prism, is depicted in Fig. 3.

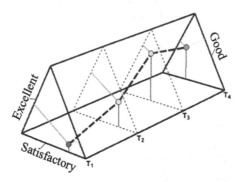

Fig. 3. Learning trajectory construction via 2-simplex prism cognitive tool.

Each triangle corresponds to the timestamp at which data arrays in the form of evaluation results are received and analyzed. The curve shows an evolution of students' knowledge, skills and abilities. At the first timestamp the results of assessment are satisfactory. Then the student makes his/her own decision leading to better results. At the last timestamp the results of assessment are excellent. Such approach is applicable to a number of problem areas, namely medicine, discrete mathematics and others.

The illustration of the cognitive graphic tools approach is currently used for the discipline "Power Electronics" [16]. The representation of student's learning results via 2-simplex cognitive tool is given in Fig. 4. Each color corresponds to a particular pattern of the student's essential quality. In our case the patterns are the following: (1) ability of problems solving (green), (2) circuitry design skills (yellow), and (3) knowledge of circuits operation (red).

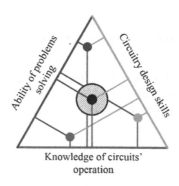

Fig. 4. Using 2-simplex cognitive tool to estimate the current level of knowledge. (Color figure online)

Green point at the top of the triangle, depicted in Fig. 4, corresponds to the student's abilities to solve problems and normal skills to design electric circuits. However, in this case the knowledge of electric circuit theory is poor. Therefore, according to this MDT results the student should learn electric circuit theory more thoroughly in future, giving necessary attention to problems solving and circuitry design. Similarly, the advice to the student will be formulated for yellow and the red points. The balance point is the black one within the shaded area. The area represents the admissible balance state of current student's knowledge level. In this case the balance between three patterns is achieved and the students have found their successful way of learning the discipline only when each pattern is evaluated as good or excellent.

5 Conclusion

Learning outcomes evaluation, based on MDT and cognitive graphic tools, is proposed for evaluation of students' knowledge, professional and personal skills and abilities. The results validation was proved on the basis of the courses "Informatics", "Discrete Mathematics" within the scope of VLSI design and "Power Electronics" [16]. ILTS with cognitive component is used to solve this problem. The proposed approach provides the design of an individual learning trajectory for each student. Such a trajectory is especially important since it allows personalizing the learning process, thus increasing its efficiency. ILTS is intended for students learning and testing, assessment of learning results and learning dynamics estimation for motivation purposes.

Problem background, its formulation and method of the MDT construction are given. The proposed algorithm of the MDT construction allows reducing the time to carry out learning and testing processes and to motivate students to study didactic units during semester.

Application of fuzzy logic, threshold function and cognitive graphic tools increases accuracy and quality of respondents' evaluation.

Cognitive graphic tools application to the problem under study allows making and justifying decisions about learning results at a fixed time or in time interval unlike any of the previously developed cognitive tools, based on n-simplex.

The approach used in ILTS increases effectiveness of blended learning by choosing the shortest ways to get the correct result and exclude the possibility of achieving it at a random path.

It is planned to apply the methodology to MDT development for other courses with different internal structures. Eventually, this approach might be used for students' learning, training and testing in both the humanities and technical disciplines, for example, mathematics, economics, computer science, etc. MDT can be constructed in such way to include explanations of all the answers to the test questions. That will enhance significantly the level of comprehension in solving complex practical problems.

The 2-simplex prism cognitive tool allows investigating objects dynamically within the time range of a user's interest. Cognitive graphic tools for the MDT results assessment allow performing mutually beneficial collaboration in learning process between students and educators.

Acknowledgements. This research is funded by a grant from the Russian Foundation for Basic Research (project No. 16-07-00859).

References

1. Ausburn, L.: Course design elements most valued by adult learners in blended online education environments: an american perspective. Educ. Media Int. **41**(4), 327–337 (2004)
2. Batyrshin, I., et al.: Theory and Practice of Fuzzy Hybrid Systems. Fizmatlit, Moscow (2006)
3. Bliuc, A., Goodyear, P., Ellis, R.: Research focus and methodological choices in studies into students' experiences of blended learning in higher education. Internet High. Educ. **10**, 231–244 (2007)
4. Bonk, C., Graham, C.: Handbook of Blended Learning: Global Perspectives, Local Designs. Pfeiffer Publishing, San Francisco (2006)
5. Brusilovsky, P., Knapp, J., Gamper, J.: Supporting teachers as content authors in intelligent educational systems. Int. J. Knowl. Learn. **2**(3/4), 191–215 (2006)
6. Kulikovskikh, I., Prokhorov, S., Suchkova, S.: Promoting collaborative learning through regulation of guessing in clickers. Comput. Hum. Behav. **75**, 81–91 (2017)
7. LeFever, L.: The Art of Explanation. Wiley, Oxford (2012)
8. Matukhin, D., Zhitkova, E.: Implementing blended learning technology in higher professional education. In: Kobenko, Y.V. (eds.) Proceedings of XVth International Conference on Linguistic and Cultural Studies, pp. 183–188. TPU, Tomsk (2015)
9. Novak, V., Perfilieva, I., Mockor, J.: Mathematical Principles of Fuzzy Logic. Kluwer Academic Publishers, Dodrecht (1999)
10. Obskov, A., Pozdeeva, S., Matukhin, D., Nizkodubov, G.: Educational aspects of interactive foreign language learning in high schools. Mediterr. J. Soc. Sci. **6**(4/3), 178–182 (2015)
11. Plekhanova, M., Prokhorets, E.: Modeling of electronic courses based on the reflective cycle. Int. J. Appl. Fundam. Res. **5**, 600–604 (2015)
12. Singer, F., Stoicescu, D.: Using blended learning as a tool to strengthen teaching competences. Proced. Comput. Sci. **3**, 1527–1531 (2011)
13. Sriarunrasmee, J., Techataweewan, W., Mebusaya, R.: Blended learning supporting self-directed learning and communication skills of Srinakharinwirot University's first year students. Procedia – Soc. Behav. Stud. **197**, 1564–1569 (2015). (Proceedings of 7th International Conference on Educational Sciences)
14. Uskov, V., Uskov, A.: Computers and advanced technology in education – perspectives for 2010–2015. In: Proceedings of 13th International Conference on Computers and Advanced Technology in Education, pp. 169–174. IASTED, Hawaii (2010)
15. Yankovskaya, A.: Logical Tests and Means of Cognitive Graphics. Lap Lambert Academic Publishing, Hamburg (2011)
16. Yankovskaya, A., Dementyev, Y., Lyapunov, D., Yamshanov, A.: Design of individual learning trajectory based on mixed diagnostic tests and cognitive graphic tools. In: Proceedings of International Conference "Modeling, Identification and Control" (MIC), pp. 59–65. IASTED, Innsbruck (2016)
17. Yankovskaya, A., Semenov, M.: Decision-making in intelligent training-testing systems based on mixed diagnostic tests. Sci. Tech. Inf. Process. **40**(6), 329–336 (2013)
18. Yankovskaya, A., Yamshanov, A., Krivdyuk, N.: 2-simplex prism – a cognitive tool for decision-making and its justifications in intelligent dynamic systems. Mach. Learn. Data Anal. **1**(14), 1930–1938 (2015)
19. Zadeh, L.: Fuzzy logic and approximate reasoning. Syntheses **30**, 407–428 (1975)

Tertiary Education

Siberian Arts and Crafts as Basis for Development of Cultural Traditions and Innovations of Bachelors

Galina I. Egorova[1] , Andrey N. Egorov[1] , Natalya I. Loseva[1] ,
Elena L. Belyak[1] , and Olga M. Demidova[2(✉)]

[1] Tobolsk Industrial Institute, Branch of Tyumen Industrial University,
Tobolsk 626150, Russian Federation
[2] Tomsk Polytechnic University, Tomsk 634050, Russian Federation
berezovskaya-olechka@yandex.ru

Abstract. The process of bachelor's training should meet such students' needs as learning the world, acquiring knowledge – on the one hand, and on the other hand adaptation to regional situation, which requires high level of personal development. The purpose of research is to increase the quality of bachelor's training by means of local cultural information. Methods of the research are theoretical, empirical, experimental, mathematical calculation (Fisher's criteria).

The results of the research: local cultural information about Siberian arts and crafts can be used to develop cultural traditions, professional competencies, and personal qualities of modern bachelors; chemical-technological bases of ancient crafts in Tobolsk province in XVII–XIX centuries, used in basic and elective parts of the curriculum are shown; efficiency local information on chemical bases of material and intellectual culture in development of bachelors' scientific and technical view is proved [1].

Key conclusions: the scientific concept on the possibility of bachelors' personality development in the process of studying humanitarian, scientific and special disciplines on the bases of local cultural material. The approach considers the example of Tobolsk province (Siberian arts and crafts) as the basis for development of cultural traditions and innovations of bachelors.

Keywords: Local cultural information · Professional competency · Bachelors' training · Siberian arts and crafts

1 Introduction

New educational standards (Federal Standards of Higher Education 3+) give a possibility to lecturers of universities to develop cultural traditions in the system of education. There are a lot of reasons for this, but the key ones are: increasing the quality of training and cultural level of a modern bachelor. That is why the simple answer to the question discussed at the seminars, meetings, conferences of different level – how to increase the quality of chemical training of schoolchildren and students in modern conditions? – is to study local cultural material from "ancient till modern times", to develop bachelor's skills in independent search for necessary information. Including

© Springer International Publishing AG 2018
A. Filchenko and Z. Anikina (eds.), *Linguistic and Cultural Studies:*
Traditions and Innovations, Advances in Intelligent Systems and Computing 677,
DOI 10.1007/978-3-319-67843-6_12

local material in to federal and regionally-oriented part of the curriculum (Siberian arts and crafts, chemical-technological peculiarities, their role in the development of the region) can serve as the bases for development of cultural traditions and innovations in bachelors' future professional activity. The peculiarities of our work are the following: the local cultural material is selected according to didactic principles of scientificity, availability, inter-discipline character, systematicity. Local cultural material was used for training students of different specialities during studying the subjects of federal and regionally-oriented part of the curriculum: general and inorganic chemistry, ecology, chemistry ant technology of organic substances, chemical technology of oil and organic substances of chemical technology, etc.

2 Materials and Methods

We use local cultural material in different ways but at the same time we pay special attention to chemical bases of ancient arts and crafts in Tobolsk province. In humanitarian disciplines we pay attention to the variety of natural resources and culture of Tobolsk province. We consider such issues as: Tobolsk province – is the province of Russian Empire (1796–1919); its role in exploration of the territory from the Urals to the Pacific coast; towns of Tobolsk province; Tobolsk as Siberian capital in XVII and XVIII centuries; pioneers of Tobolsk province (science, industry, culture, and technics), Tobolsk as a cultural center of Siberia. Studying this material we conduct various activities: scientific and creative workshops, maker-tones in studying the archives, inter-discipline seminars "My small homeland", "Cultural origins of the Historical Centre", open lectures "Personalized lectorium", "Tobolsk - a town of petrochemistry", seminars "Outstanding people of the town and region", binary lectures "Region's culture", lectures with intentional mistakes "Check your science-technical scope of knowledge". The contents for classes we obtain from our work with the archives. Firstly, the lecturer prepares the contents for classes. The first module is about outstanding people, who play the strategic role in the development of Tobolsk province and Russia. We cover the following issues: Tobolsk province is the home of A.A. Alyabiev, the author of the famous "Nightingale", outstanding artist V.G. Perov, N.V. Nikitin, the designer of Ostankino tower; B.P. Grabovski, the inventor of Soviet television, U.S. Osipov, the president of the Russian Academy of Science (RAS); life and work of P.P. Ershov, a fairy-tale poet; a scientist D.I. Mendeleev; Tobolsk province and Decembrists, writers: A.N. Radishev, F.M. Dostoevsky, V.G. Korolenko, N.G. Chernishevsky, M.I. Mikhailov, P.A. Grabovskiy. The role of Russian populists, Marxist-revolutionists, exiled into Siberia in different times. Life of the last Russian Emperor Nicholas II.

The second module is devoted to scientific and technical ideas contributing to the development of both Russia and the region. We pay special attention to scientific and technical achievements born in Tobolsk province. Tobolsk province and scientist D.I. Mendeleev. Scientific ideas of D.I. Mendeleev on the development of Siberian industry and culture. Role of Tobolsk province in the exploration of the North. Oil and Gas fields exploration in Western Siberia. Tobolsk province – a starting point in Russian North exploration. Transport lines, connecting Siberian High North with industrial regions of the country. Sledding, waterway, railway, first pipelines. Oil and gas pipelines, high-voltage electric lines, new chemical and oil producing enterprises. The contents of these modules are landmarks of Siberian towns even nowadays [2, 3].

The third module reveals the chemical and technological grounds of arts and crafts in Tobolsk province contributing to the development of the region and Russia. The contents of the module covers the crafts of Tobolsk province (in XVII–XX) centuries. The students showed great interest to the chemical and technological development of arts and crafts: leather craft, the art of bone-carving, glazing trade, gun powder, stationery industry, soap-making, wine distillery, metal-processing, tailoring, timber-manufacturing industry, candle-making industry, food industry, pottery and others. Crafts development and market demands. Regional goods in the markets of Central Russia and in the world exhibitions. Chemical-technological grounds and the first equipment [4]. The students were involved into the studying of modules it allowed us to form both their scientific and technical scope of view and interest to acquiring modern scientific information included into the curriculum. Determining cause-effect relations between the historical material and its contemporary significance the students understood the functional role of chemistry and chemical technology in modern civilization. Why were the goods demanded in Russia, Europe and America? Is it possible to restore the production technology of high-quality bricks, oil, clay, hemp fiber, soap? Why do classical ideas of chemists-scientists acquire new priorities? Numerous "why" give new lively idea to the principle of historicism "from history to modern life", "from a historical monument to the technology of its production", "from the fact to the personality of a scientist, architect, and craftsman". The development of brick-building industry in Tobolsk province can serve as example to it. Unique monuments of brick architecture – churches, houses, built in the past centuries are preserved. The reason is very simple – high-quality technology of brick manufacture. Working with the historical material the students revealed the stages of brick manufacture emerging and proved that the priority in brick-building belongs to bishops. In XVII century metropolite Korniliy built first two-storeyed brick metropolitan chambers. In XVIII century metropolite Pavel started building the St. Sophia Cathedral with bricks [5]. The students discover not only technology of high-quality building material and technology of production but do their personal discoveries. So, they found out that "S.U. Remezov-a famous map-maker took part" in organization of brick-manufacturing craft [6]. Studying the brick-manufacturing craft in 1699, S.U. Remezov established two building enterprises, and by studying chemistry and technology of brick-manufacturing process he was able to see its technological complexity. Studying these historical peculiarities the students trace their character in the Siberian brick in our time. We can see St. Sophia's Cathedral of the Assumption, Tobolsk Kremlin, Rentereia, and other monuments of the town in their original view. It makes them proud their small homeland, their patriotic consciousness [7].

Every student sets his/her own consequence "historical event – human factor – technological factor – final product – socio-cultural factor", i.e. every craft has certain nsequence in creating a final product according to ancient recipes, consequently the final product had wide socio-cultural significance. The final stage of work with local cultural material consisted of three stages. At the first stage (1–2 year) the students used the work pieces on the bases of archive materials. The activities were reproductive. At the second stage (2–3 year) the students completed semi-researching activities, and did some research on the bases of local cultural material. At the third stage (3–4 year) the students made the tables (chronological, synoptical, personal), fulfilled research projects "Chemical-technological grounds of Tobolsk paper production", "Chemistry and technology of ancient soap-making", "Chemistry and technology of Tobolsk glass", etc. Let's give an example of systemization of ancient crafts, which was prepared by senior students on the basis of archive materials (Table 1) [7, 8].

3 Research Data

Name of the enterprise, year, raw materials	Equipment, technological processes	Product and its realisation
Enterprise: paper-manufacturing factory of merchandizers Anton and Ivan Medvedev (1744), later V.Ya. Korniliev (1783) **Raw materials:** flax waste, hempen, fish oil, mixtures of wood and straw, hide mastic, Swedish aluminum sulphate, Berlin blue	Installations and apparatus: water presses, rolling spins, vapor engines, drier, rejection room Technological process: sorting the cloth, washing, pounding, sheets forming, drying, gluing, saturation	**Product:** 22 kinds of paper: for writing, wrapping and other **Markets:** Tobolsk, Tomsk, Irbit, Moscow, Tumen, St. Petersburg
Enterprises: leather-making factory of a merchandiser S. M. Tretyakov (1720) Leather-making factory of a merchandiser N. Semenov (1823) Leather-making factory of a citizen A. Povarnin (1826) **Raw materials:** tall bark, aspen ash, rye flour, fish oil, tar, lime, salt, sandal, aluminum sulphate	Installations and apparatus: tanning, ash pots, smashing press Technological process: soaking, degreasing, scrapping, dying, drying, stretching	**Product:** leathers: Russia leather, black, white, horse, veal, sheep, chamois **Markets:** Tobolsk, Tyumen, Omsk
Enterprises: soap and candle making plant of a merchandiser G.M. Ershov (1875) Soap-making plant of a merchandiser A.A. Mikhailov (1810) Soap and candle making plant of a merchandiser Nikolay Leo (1929)	Installations and apparatus: cast iron pots, infusions, wooden tubs, lockers, brick ovens, cast iron furnaces Technological process: Heating, mixing, greasing, cooling, forming	**Product:** household and cosmetic soap, white, plain, perfumed **Markets:** Tobolsk, Irbit, the Urals **Product:** cast candles, dipped candles, wax candles **Markets:** Tobolsk province

(continued)

(continued)

Name of the enterprise, year, raw materials	Equipment, technological processes	Product and its realisation
Soap making plant of a merchandiser A. Dranishnikov (1812) **Raw materials:** cattle fat, sheep fat, pig fat, fish oil, ash, lime, salt, soda salt, tin metal, wax, paper		
Enterprise: match manufacturing factory of a merchandiser G.I. Ershov (1875) **Raw materials:** pine tree, scrape, dyes, birch wood	Installations and apparatus: hand-driven, wooden, composing machines Technological process: work pieces production, inflammable mixture, saturation, drying	**Product:** phosphorus-scrape matches **Markets:** Tobolsk, Omsk
Enterprise: vine-making plant of a merchandiser of the 1st guild A.A. Syromyatnikov (1874) **Raw materials:** wheat, yeast, Siberian berries, sugar	Installation and apparatus: fermentation pots, brew cubes, spirit cooling cubes Technological process: brewing, detention, purifying, cooling	**Product:** bread wine, spirit, vodka, liqueur **Markets:** Tobolsk, Ob-Irtysh North, Tyumen
Enterprise: brewing and mead brewing plant of a merchandiser of the 1st guild A.A. Syromyatnikov (1875) **Raw materials:** barley, malt	Installation and apparatus: iron pots, brewing pots, mixers, crushing machines, dish-washing machines Technological process: malt making, mash, brewing, filtering, bottling	**Product:** Brew, mead **Markets:** Tobolsk, Tyumen, Ob-Irtysh North
Enterprise: clayware making plant P. Ponyushkin (1875) **Raw materials:** local clay, water, birch wood	Equipment and apparatus: clay mixers, furnaces, shafts Technological process: clay preparation, forming, drying, annealing	**Product:** clayware: red, black **Markets:** Tobolsk, Tyumen
Enterprises: silk – making factory of a merchandizer F.F. Kremlev (1793) The first cloth making manufacture of a civil servant M.T. Kutkin (1791) **Raw material:** cotton, flax, wool, silk, hemp	Installation and equipment: hand looms, wooden boards with patterns, furnaces, pots Technological process: raw material preparing, scotching process, rippling, cleaning, stretching, twisting, spinning, whitening	**Product:** cloth: silk, semi-silk, broad cloth, up to 6000 arshines a year, silk bands, napkins, head-cloths, paper ribbon **Markets:** Tobolsk, Omsk, Tyumen
Enterprise: the first Siberian free-lance printing plant of a merchandiser of the 1st guild V.Ya. Kornilev (1789)	Installation and equipment: Russian, foreign prints, printing machines	**Product:** post forms, art books, science books, magazine "Irtysh," which became

(continued)

(continued)

Name of the enterprise, year, raw materials	Equipment, technological processes	Product and its realisation
Raw materials: high-quality paper of different kinds		"Hippocrene" (1796 - issue 12 thousand books and journals)
Enterprise: glass-making plant of a merchandiser V.Ya. Kornileva (1749) **Raw materials:** ash, birch wood, aspen, chemical additives, clay, sand, potash	Installation and equipment: polishing machine, pottery room, furnaces, smithy, cast iron pots Technological process: clay mixing, sand mixing, melting, blowing out, dying, hardening, sorting	**Product:** green glass bottle of 1.23 l, choppin, glasses, pots, flasks, measuring glass, cups, window glass, inkstands, crystal **Markets:** Tobolsk, Moscow, Omsk
Enterprises: armoury (1700) of Peter the I decree; Nitric Plant S.U. Remezova (1703) **Raw materials:** steel, iron (from the Urals)	Installation and equipment: anvil, smithy Technological process: forging of an iron board, barrels' making, pipe welding, drilling	**Product:** gun barrel, fusees, swords, clubs, gun-powder **Markets:** Tobolsk province
Enterprise: creameries (8 plants in 1868) **Raw materials:** milk	Installation and equipment: wooden creameries, forms Technological process: creams making, cooling,	**Product:** butter **Markets:** Tobolsk, Tyumen, Omsk
Enterprise: bone-carving work-shops of Oveshkova-Melgunova (1883–1885) **Raw materials:** bobin, fossil ivory, whale tooth	Installations and apparatus: carving, polishing machines	**Product:** bone embroideries **Markets:** Moscow, Riga, Nizhnii-Novgorod, Kiev, S.-Petersburg, Europe
Enterprise: Fish cans plant (1883) **Raw materials:** river fish, spices, salt	Installation and equipment: furnaces, pots, wooden barrels, metal boxes Technological process: raw material preparing, pasterization, salting, marinating, packing	**Product:** fish cans from river fish **Markets:** Irbit, Omsk
Enterprise: salt-producing plant (1830–1854) Tobolsk province, Koryak deposits	Installation and equipment: hand salt breaking, hand carrying, carrying-out on horses Technological process: solution, thick solution preparing, assertion, filtering, packing	**Product:** gross salt, fine salt for fish, meat, vegetables **Markets:** North, Omsk

4 Conclusion

Experimental groups, which were trained according to this system, had the increase in average continuous assessment ranking and final assessment. The level of development was measured at every level of pedagogical experiment. If at the first stage the amount of learners at reproductive level was 65%, at the second level it was already 35%, at the third level the amount of learners of reproductive level was 9% (Fig. 1). Validity of the differences in the results, obtained in the experiment was determined by Fisher criteria for the significance level 0.05. The analysis showed the values of criteria Fcriteria > Fempirical (3.45 > 1.91; 3.43 > 1.41; 3.42 > 1.17).

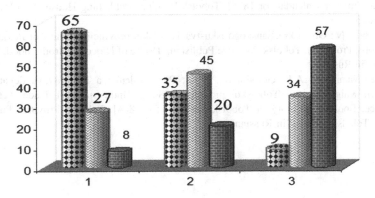

Fig. 1. Levels of students' development.

The first column shows the reproductive level; the second column shows the heuristical level; the third column shows the level of creativity.

Nowadays, there is enough information to study in detail different aspects of chemical-technological bases of technological aspects of ancient arts and crafts – local culture of Tobolsk province: level of engineering degree and technology development, peculiarities of materials treatment, cultural traditions. Possibilities of studying chemical grounds of ancient crafts as the source of local, chemical information can be enriched by archive materials: written, graphical, ethnographic. Studying chemical-technological bases of ancient arts and crafts gives information on the centers of production, range, technology of goods production, ways of distribution, as well as knowledge on material, culture of the region, its history, thus helping to develop patriotism and moral qualities of bachelors who will work in Tyumen region.

References

1. Egorova, G.I.: Regionalynaya istoriya [Regional History]. TumSNGU, Tumen (2009). (in Russian)
2. Vilkov, O.N.: Remesla i torgovlya Zapadnoi Sibiri v XVII veke [Crafts and Trading of Western Siberia in XVII Century]. Nauka, Moscow (1967). (in Russian)

3. Golodnikov, K.: Yarmarki, fabriki I zavody v Tobolskoi provincii Kalendar Tobolskoi provincii v 1889 [Fairs, Factories and Plants in Tobolsk Province Calendar of Tobolsk Province in 1889]. Tobolsk Province Publishing House, Tobolsk (1888). (in Russian)
4. Denisov, A.: Sibirskaya solevaya promyshlennost. Kniga pamyatiTobolskoi provincii 1861 I 1862 g. [Siberian Salt Industry. Book of Memory of Tobolsk Province for 1861 and 1862]. Tobolsk Province Publishing House, Tobolsk (1861). (in Russian)
5. Dunin-Gorkavich, A.A.: Potrebnosti Tobolskogo Severai ih udovletvorenie. Kniga pamiati Tobolskoi provincii 1908 g. [Tobolsk North Needs and Their Satisfaction Book of Memory of Tobolsk Province 1908]. Tobolsk Province Publishing House, Tobolsk (1908). (in Russian)
6. Kratkaya informacija po fabrikam I zavodam deistvujushim v Tobolskoi provincii v techenie 1891 g. bez ogranictnija ih proizvoditelnosti [Short Information on Factories and Plants, Acting in Tobolsk Province During 1891 Without Limiting the Amount of their Capacity. Tobolsk Province Calendar for 1895]. Tobolsk Province Publishing House, Tobolsk (1894). (in Russian)
7. Skalozubov, N.: Obzor krestjanskogo iskustva Tobolskoi provincii [Review of Peasant Craft of Tobolsk Province]. Tobolsk Province Publishing House of Diocesan Brotherhood, Tobolsk (1902). (in Russian)
8. Dmitriev-Mamonov, A.I., Golodnikov, K.M.: Promyshlennaya I torgovaya deyatelnost'v provincii Kniga pamyati Tobolskoi provincii 1884 g. [Industrial and Trade Activity of Province. Book of Memory of Tobolsk Province for 1884]. Tobolsk Province Publishing House, Tobolsk (1884). (in Russian)

Dual Training as a Condition of Professional Competences Development for Bachelors' in Engineering and Technology

Galina I. Egorova[1] , Andrey N. Egorov[1] , Natalya I. Loseva[1] ,
Elena L. Belyak[1] , and Olga M. Demidova[2(✉)]

[1] Tobolsk Industrial Institute, Branch of Tyumen Industrial University,
Tobolsk 626150, Russian Federation
[2] Tomsk Polytechnic University, Tomsk 634050, Russian Federation
berezovskaya-olechka@yandex.ru

Abstract. According to the Federal Standards of Education 3+ and new demands of employers there is an aim to enhance professional development of bachelors. In the scope of innovative knowledge-driven economy a single unit of training and bachelors' development based on the integration of the system "bachelor-lecturer - mentor" should be introduced. This dialectal interconnection may be fulfilled in the dual organization of bachelors' training where the practical orientation principle plays the key role both in personal development and enhancement of professional competences.

The purpose of the research is to increase the quality and level of professional competences. The results are: (1) selected individual educational trajectory for students allows developing general cultural and professional competences; (2) the importance of the concept is proved basing on scientific ideas, competence approach, as well as dual and project approaches; (3) it has been proved that the dual approach of bachelors' training helps to develop flexible and dynamic systems adaptive to the market demands; (4) the dual approach is fulfilled according to the integrational principle which allows studying the same processes many times (teaching and learning processes), but at the new level of development; (5) the system of dual training becomes effective when the university-based theoretical training is combined with practical training at the enterprise.

Methods of research: theoretical, empirical, experimental work, mathematical method (Fisher criterion).

The key conclusions are: the concept is created, the programmes are written and professional competences are included into the educational profile. The employers require the following professional competences: "ability to organize work and conduct the process of ethylene and propylene polymerisation", "ability to conduct technological process according to the regulatory documentation of an enterprise".

It is proven that the dual form of training allows to overcome the gap between theoretical and practical training.

Keywords: Dual training · Professional competences development · Techniques and technology · Bachelors' training

A. Filchenko and Z. Anikina (eds.), *Linguistic and Cultural Studies: Traditions and Innovations*, Advances in Intelligent Systems and Computing 677,
DOI 10.1007/978-3-319-67843-6_13

1 Introduction

1.1 Problem Statement

Nowadays availability of higher education in Russia is essential not only for students but also for development of intellectual state's resources [7].

The main curriculum for the major in 18.03.01 "Chemical technology" is designed for the students of oil and gas industry, and organic synthesis, that makes it important for the country on the whole [1].

We have developed and tried the concept of individual educational trajectory, which allows to train bachelors with profound professional knowledge in chemical technology and developed general cultural and professional competences according to the requirements of Federal Standards of Higher Education.

The main educational curriculum for 18.03.01 Chemical technology is a systemized set of documents describing the results of education, contents of training, work load, technologies of education, teaching and assessment. Its' aim is to develop graduates' competences in a certain sphere.

Here, two profiles are possible: chemical technology of organic substances and chemical technology of natural power sources and carbon-based substances. In the first profile bachelors are prepared for a wide range of enterprises of organic synthesis, including oil and gas production. The second profile is more specific and focuses on chemical technology of oil, gas and associated gas treatment [2, 3].

Taking into account several principles, scientific ideas and approaches, this concept was proved to be important both for an employer and university. Professional competence in chemical technology is understood as ability to act in compliance with the requirements: solve the problems and evaluate the results of the work methodically and independently. This approach allows to concentrate on actual achievements of students. Dual approach can create flexible, dynamic systems actively responding to the market demands. The peculiarity of dual approach is the integration, i.e. repeating the same processes of learning and teaching at new stage of development. In our research we understand a dual system as a means combining theoretical education in university and practical training at the enterprise [4, 6].

The main difference in comparison with the traditional academic system is the change in study schedule (e.g. one month is a theory training in university and one month is practical training at the enterprise under the guidance of a mentor). In our experiment we had special days in a week for training a dual group at the enterprise.

1.2 Literature Review

Studying literature on the issue we have found out that this system of training is widely used in professional training both in Russia and abroad. Taking into account the experience of foreign educational institutions (Department of Dual training, Atyrau University of Oil and Gas (AOGU), Forth Valley College), the ways of developing the system of dual training are determined for years 2017–2022 [5, 6].

Our experience with this system showed that it is important to develop such training concepts in regions where the economy is clustered, and unities of enterprises, research centers and educational institutions serve the key industrial branch of the region.

Oil-service innovative cluster is formed in Tyumen region, its aim is to unite enterprises and choose an asset management company. It is impossible to fulfill innovative projects without scientific research and competent specialists. That is why, economic clusters include research institutes and educational institutions. In our opinion, Tyumen industrial University should be included into Tyumen industrial cluster, as a key University of the region preparing specialists for oil and gas industry [6, 7].

1.3   Basic Assumptions

Dual training allows to solve the main problem of professional engineering education - the gap between theory and practice. Thus the following problems are solved:

- a trainee gains necessary experience. After graduation it will be easier for him/her to find a job;
- an enterprise will have a constant amount of qualified staff.

This program responds to the demands of leading regional oil-gas enterprises of Tyumen region: Limited Liability Company SIBUR Tobolsk, companies SIBUR, LUKOIL, and ROSNEFT, etc.

We have developed the syllabus, which includes a study plan, a matrix of competences, a passport of competences according to Federal Standards of Higher Education.

Every module has a basic compulsory part and variative part that allows to increase knowledge and acquire the skills, determined by major disciplines. Graduates competences determined by Federal Standards of Higher Education are formed with such modern technologies of education as project and interactive technologies. The key basis of project technologies is in study and research work of students in teams when fulfilling the course and diploma papers. Key issues of teaching interactive technologies, are developed by our staff at the department of Chemistry and Chemical Technology with the consideration of professional disciplines and developing teaching means.

The main ideas of the program comply with demands and conditions of high quality of training and include:

- Providing quality of graduates training with the help of employers;
- Monitoring and correcting the syllabi;
- Developing objective procedures of testing knowledge and competences of graduates;
- Providing high-quality teaching staff;
- Regular self-checking of the development and comparing it with the requirements of the university and employers.

As a result our graduates are highly demanded by state and private enterprises in oil and gas industry.

This syllabus is tailored to regional needs. To fulfill a further strategy, we attracted such potential employers as – SIBUR Holding, LUKOIL PJSC, ROSNEFT PJSC, SURGUTOILGAS PJSC and others. Within this experiment the set of documents was developed together with SIBUR Tobolsk LLC. This set of documents contains a model of future graduates. The table of competences, determined by an employer is given below (Table 1).

Table 1. Competences, required by employers from the graduates, who have completed the curriculum of a dual training program.

Universal Professional Competence -13 "ability to organize work and conduct the production process of polymerization of ethylene"		
Threshold level (as compulsory for students after graduating)	Knows influence of process parameters on synthesis, (structure, qualities) of polyethylene	Gives the definitions of a mechanism of radical ethylene polymerization
	Understands the influence of work parameters on polyethylene synthesis	Understands the reactor's work (pressure, temperature, amount of alarm initiators)
		Distinguishes main kinds of technological additives, metal deactivators, process additives and so on)
		Knows rheologic qualities of polyethylene (dependence on molecular mass and polymers polydispersion; on the degree of branching); their influence on human organism
		Analyzes the influence of isolators work parameters (temperature, pressure, time) and expulsion in silage according to the effectiveness of ethylene emission
	Can control the process of ethylene polymerization	Demonstrates control methods of ethylene polymerization process
		Can understand the degasation process, its influence on safety; ethylene extraction mechanism
Advanced level (comparing with the threshold level)	Knows process parameters influence on structure and properties of polyethylene	Understands the peculiarities of structures and qualities of polymer-composite materials, their use and influence on human organism; degasation process, its

(continued)

Table 1. (*continued*)

		influence on safety; mechanism of ethylene emission; influence of work parameters
		Analyzes the impact of ethylene separators operating parameters (temperature, pressure, time) and the purge in silos on the effectiveness of ethylene emission
	Understands the influence of reactor work parameters on polyethylene syntheses	Understands radical ethylene polymerization mechanism on all stages
		Understands the main kinds of filling agents of composite materials, their classification; structure and peculiarities of polymer-composite materials
	Can control the process of ethylene polymerization	Uses main kinds of technological additives (antioxidants, metal deactivators, processing additives and others) main kinds of composite materials and their classification

UPC -14 "ability to organize work and conduct the production process of polymerization of propylene"

Threshold level (as compulsory for students after graduating)	Knows influence of process parameters on synthesis, (structure, qualities) of polypropylene	Knows mechanism of propylene polymerization with complex metal and organic catalyzers
		Uses types of catalyzers, co-catalyzers, stereo regulating additives
		Knows how to prepare the catalytic complex and do phoro-treatment of a catalyzer by propylene in carbon dissolver or in liquid monomer
	Understands the influence of operational parameters on polypropylene synthesis	Uses the process of stereospecific polymerization of propylene in titanium-magnesium catalyzers in liquid monomers, carbon dissolver, gas phase
		Demonstrates the influence of reactor operating parameters (pressure, temperature, amount of so-catalyser, stereo-regulating additive, hydrogenium) on synthesis (stereocomposition, molecular-mass characteristics)

(*continued*)

Table 1. (*continued*)

		and extraction of polypropylene and co-polymers
	Can control the process of polypropylene polymerization	Uses methods of temperature regulation
		Uses methods of pressure regulation
		Uses methods of molecular mass regulation
Advanced level (comparing with the threshold level)	Knows process parameters influence on structure and properties of polypropylene	Understands the mechanism of stereospecific polymerization of propylene in complex metal organic catalyzers
		Distinguishes the types of catalyzers, co-catalyzers, stereo-regular additives; prepares a catalytic complex and phoro-treatment of a catalyzer by propylene in carbon dissolver or liquid monomer
	Understands the influence of reactor operating parameters on polypropylene syntheses	Determines the process of stereotypical polymerization of propylene with titanum – magnesium catalyzers in liquid monomer, carbon dissolver in a gaseous phase; influence of reactor operating parameters (pressure, temperature, amount of co-catalyzer, stereo-regulating additives, hydrogenium) on synthesis (stereo-composition, molecular-mass characteristics and extraction) of propylene and co-polymers
	Can control the process of propylene polymerization	Knows how to control the process of preparing a catalytic complex
		Knows the control methods of co-catalyzer amount and stereo-regulating additive
		Knows control methods of hydrogen amount

UPC-15 *"ability to conduct the technological process according to the regulations of the enterprise"*

Threshold level (as compulsory for students after graduating)	Knows interaction parameters of a technological mode with different parts of the process flow	Determines the norm parameters of a production process at the section under control
		Knows how to norm and dose raw materials and catalyzers

(*continued*)

Table 1. (*continued*)

		Knows main steps of the technological mode in the production process at the enterprise
		Knows the requirements to the raw materials quality; qualities of the used raw material
		Knows the requirements to the qualities of products according to the schedule of analytical control
		Knows possible fluctuations in the quality of released products
		Knows the causes and solutions of problems
	Knows the standard methods of products analysis	Knows test methods of released products
		Conducts the tests of products according to the analytical control schedule
Advanced level (comparing with the threshold level)	Knows interconnections of the production process parameters at different moments of it	Demonstrates the knowledge of further use of released products
		Conducts the test of products according to the standards
		Understands interconnection of parameters of the production processes at different moments of a technological mode; knows main components of a technological mode properties of raw materials, sorts and kinds of the products, standards, requirements to the released products
		Possible fluctuations in quality of the released products, causes and methods of correction
	Knows the standard analysis methods of the released products	Distinguishes the parameters of the production process at the enterprise
		Analyzes critically the process of raw materials dosing and initiators
		Knows modern physical-technical methods of analysis of the released products

2 Materials and Methods

During the experiment on the dual study form, control (academic bachelor) and experimental groups (bachelor of applied sciences) were distinguished. The comparative analysis of professional, cultural, competences was conducted on the basis of students' control works, tests, tests of the continuous assessment, assessment sheets provided by the employers. Qualitative mark was given on the basis of "Portfolio". Before the experiment the students of both groups were questionnaired, in order to determine their interest to the dual form of training. The opinions divided, the answers of the control group were indifferent "I don't need this", "Let me think", "I know everything" etc, while the answers of the experimental group were mostly positive (Table 2).

Table 2. Competences and indicators of requirements from the SIBUR LLC.

Competences	Students should know, be able to	Enterprise requirements to the students
PC-4 PC-8 PC-8 PC-18 UPC-13 UPC-14 UPC-15	**Know:** chemistry and process technology, production process of organic substances extraction, organic molecular species of materials required in fuel and energy fields; sources and qualities of carbons used in organically synthesized products, ways and technology of carbon extraction; rational methods of carbon separation **Can:** determine the purposes and aims of the production process, select the means, interpret and comment information; use the knowledge on different classes of organic substances in order to understand properties of materials; draw the schemes of production units; develop production processes; determine the composition of products, draw block-diagrams of hydrocarbon production **Manage:** technology of obtaining organic substances; ways and technological schemes in developing new production processes of oil refining and gas rectification; skills to fulfill all manufacturing steps within a project	**Pyrolytic action** **Knows and understands the processes:** pyrolysis, rectification, side-processes of pyrolysis, qualities of the applied raw materials, raw materials quality requirements, properties, composition of the target and by-side product, influence of operational mode parameters for the equipment in further process **Gas separation** **Knows and understands bases of processes:** rectification, hydration, absorption, adsorption, desorption, absorption and gas rectification, quality of raw material, requirement to the final product **Polymerization of ethylene and propylene Knows and understands the processes:** rectification, destruction, adsorption, influence of reactor operational parameters of a reactor (pressure, temperature, amount of initiators, type and quality of agents of the transfer chain) on the synthesis (structure, properties and quality) of polyethylene, main steps of the production process, homogenous polymers production, statistical and block-copolymers with ethylene

3 Conclusion

Indexes of professional competences formation of students in experimental groups are higher than in a control group. Increase (%) of professional competences for the experimental groups is 32%, for control groups - 1%, that proves the efficiency of dual form of training (Fig. 1).

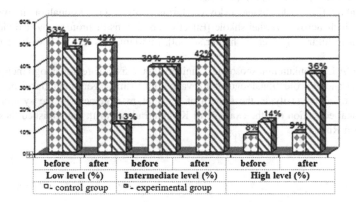

Fig. 1. Indexes of professional competences formation.

During the experiment the students expressed their opinion. On the basis the questionnaires and interviews it was evident that the students of the experimental groups adapted easier to the production conditions, do not think about the change of the professional career, are ready to acquire knowledge and skills, study with pleasure, are ready to study surrounding fields of the professional activity, ready to learn how to use new techniques and technologies, are in harmony with themselves and are ready to the professional realization and final employment.

The following can be attributed to the results of the students' research competency formation: (1) positive dynamics in the level of students' research competences development in the experimental group; (2) increase in the level of students' research activity at the final stage; (3) development of internal motivation for research activities in the university. These results indicate that the research activity becomes one of further educational and career steps for high school students. The presented main features allow to organize research activities and form students' research competency.

References

1. Alekberova, I.E.: Integratsionnii podhod v obrazovanii kak odin iz vazhnih aspectov lichnogo razvitija [Integrational approach in education as one of main aspects of personal development]. (in Russian). http://sociosphera.com. Accessed 28 May 2017
2. Bokov, L.A., Kataev, M.U., Posdeeva A.F.: Grupovoi proekt v universitete kak metod innovatsionnoi podgotovki specialistov [Group project in a university as a method of innovative specialists training]. (in Russian). https://science-education.ru. Accessed 20 May 2017

3. Egorov, A.N., Egorova, G.I.: Formirovanie innovatsionnoi competencii kak uslovie socializacii budushego inzhenera [Innovative competence formation as a condition of future engineer socialization]. Bull. High. Educ. Establ. Sociol. Econ. Polytics 1(44), 113–115 (2015). (in Russian)
4. Loseva, N.I.: Forma dvoinogo obucheniya v tehnicheskom Universitete [Dual Form Training in a Technical University]. TIU, Tumen (2017). (in Russian)
5. Malkova, I.U.: Proektirovanie v obrazovanii [Projecting in Education]. Methodological Materials. TSU, Tomsk (2006). (in Russian)
6. Martynenko, E.P.: Effektivhie usloviya formirovaniya professionalnoi individualizatsii budushih inzhenerov v visshei shkole [Effective conditions of professional individualization formation of future engineers in higher school]. Fundam. Res. 2(5), 1046–1051 (2015). (in Russian)
7. Udanova, A.L.: Konkurentnosposobnost' universiteta v sisteme regionalnogo mendzhmenta v sisteme visshego professionalnogo obrazovaniya na primere Kaliningradskoi oblasti [Compatibility of a university in the system of regional management in the system of higher professional education (on the example of Kaliningrad region)]. Unpublished thesis abstract (2007). (in Russian)

Language Teaching and Learning

On Motivation of Learning English as a Foreign Language: Research Experience in Russian University Context

Maria V. Arkhipova(✉) ⓘ, Ekaterina E. Belova ⓘ,
and Natalia V. Shutova ⓘ

Minin University, Nizhny Novgorod 603950, Russian Federation
arhipovnn@yandex.ru

Abstract. In this article we focus on the problem of motivation to learn a foreign language. Researchers show that regardless of the subject matter students with a high level of motivation tend to achieve better academic results, get a deeper understanding of a subject, feel more satisfied with their results, and strive for success. To explore the topic, we outline three components of motivation, namely, interests, emotions, leading motives, and study their levels. The empirical study is based on the Russian students majoring in Linguistics. The results prove that a low rate of any motivation component has a negative impact on academic achievement. The results of the experiment show that students' interest in learning English is episodic. The leading motive is the coercion motive, i.e. seeking to avoid criticism and academic problems. The majority of students have a low level of overall emotional well-being. We conclude with the issue of exploring the prospect for further research, based on the obtained data, as to find psychological tools for creating an emotionally comfortable atmosphere and means of fostering students' interest in English, cultivating their motivation, making them realize the value of learning; and, as a result, enhancing the quality of education.

Keywords: Education · Foreign language · Motivation · Leading motives

1 Introduction

1.1 Building a Research Problem

Enormous transformations are currently taking place in learning paradigms and education practices in Russia. The English language, being the international language of communication and learning, is included into a wide range of academic programs letting more doors open for career opportunities. Nowadays education in Russia is aimed at developing students' intellectual and personal resources with its emphasis on self-actualization and personal growth. The focal goal of education is to support development of students' learning, cognitive and creative motivation providing them with a more comprehensive understanding of the value of personal activity, realization and fulfillment of their talents and potentialities.

© Springer International Publishing AG 2018
A. Filchenko and Z. Anikina (eds.), *Linguistic and Cultural Studies:*
Traditions and Innovations, Advances in Intelligent Systems and Computing 677,
DOI 10.1007/978-3-319-67843-6_14

Motivation is sure to constitute the backbone of learning process. Motivation is a power that dynamises behavior of individuals for certain purposes, directs these behaviors and ensures those to maintain consistently [1, 7, 10, 14]. It is motivation that starts, energizes, sustains, and directs behavior and actions. The amount of effort people feel eager to put in their work depends on the degree to which they feel their motivational needs will be satisfied. On the other hand, individuals become demotivated if they feel that something prevents them from attaining good outcomes.

One of the main goals of education is to increase the level of students' academic performance by creating an effective learning environment. In this sense motivation is seen as a significant factor contributing to academic success [2, 16, 21, 22]. Pintrich and Schunkstate motivation as a continuous relationship between learning and achievement [26].

1.2 Literature Review

There are numerous theories on motivation and each theory emphasizes a different dimension of motivation. Motivation is one of the main terms used to explain driving forces of behavior and activities [4, 6, 8, 13, 23, 24, 32]. Researchers often contrast intrinsic motivation with extrinsic one, the first energizing and sustaining activities through the spontaneous satisfaction inherent in effective volitional action, the latter being governed by reinforcement contingencies [28]. Pintrich, Roeser, De Groot refer intrinsic motivation to participating in a learning activity for its own sake and extrinsic – to participating in an activity due to the enhancers(rewards or punishment components) like getting good grades, making others happy, success in competition and unwillingness to fail the class [26]. Traditionally, educators consider intrinsic motivation to be more desirable and to result in better learning outcomes than extrinsic motivation [28].

Keller divides motivation into two categories: trait motivation and state motivation [17]. Trait motivation is a general motivation and is stable across a particular time regardless of situational factors. In terms of student situation, student trait motivation is more enduring and refers to student's general motivation toward learning, not a particular course or subject content. State motivation, on the other hand, is not stable because it is significantly influenced by situational factors.

Utvær and Haugan distinguish Self-determination theory [31]. Different types of motivation are understood as different ways in which a person regulates his or her driving forces: from being externally motivated to becoming internally and eventually autonomously driven to perform certain behaviours. Self-determination theory highlights types of motivation, or regulation, in terms other than of quantity, level, or amount, as well as differentiates types of behavioural regulation in terms of the degree to which they represent autonomous versus controlled functioning.

Modern age-related and pedagogical psychology of language teaching has a number of researches dealing with various aspects of study motivation: kinds of motives of academic activity [5, 6]; emotional well-being during academic activity [3–5, 23, 32]; and interest as a kind of inner motivation [14, 15, 17, 18].

The studies of Gardner and Lambert reveal the most commonly used framework for understanding motivation that language learners typically have: instrumental and integrative types of motivation [12]. The external needs, coming from outside, reflect the instrumental motivation, learning a language because of practical reasons. Integratively motivated learners want to learn the language so that they can better understand the people who speak that language, and their culture. Integrative motivation has proven to be a strong impetus to successful language learning.

Many scientists note that students in Russia nowadays have an ambiguous attitude to foreign language as a school subject, the level of motivation to learn it being inadequate [3–5, 7, 15, 19, 33]. Certainly it gives rise to a contradiction between the increasing demands of society, when good language skills increase the competitiveness of a specialist in any field, and the traditional system of learning a foreign language organized in most Russian schools. We believe that one of the reasons for this phenomenon is an insufficient level of theoretical and experimental research concerning the problem of motivation in foreign language learning. Consequently, the psycho-pedagogical organization in educational institutions does not meet the educational needs [4, p. 69].

When investigating motivation of language learners one should take into consideration the specific character of a foreign language as a school subject. Thus, following Zimnyaya we shall turn our attention to the analysis of three main features of foreign language as a school subject: objectlessness, infinity and heterogeneity. The essential feature of a foreign language as a subject in comparison with other subjects is that its learning does not give a person immediate knowledge of reality. The foreign language is a means of creating and forming the idea of objective reality. It is only a medium of communication. Foreign language has no separate topic sections, like other subjects. When studying a foreign language, it is impossible to know only vocabulary not knowing grammar. One should know all the grammar and the entire vocabulary required for communication. In this context language as a learning subject is infinite. Heterogeneity is also a significant feature of language. In a broad sense, language includes a number of other phenomena, such as speech activity (i.e. processes of speaking and understanding), and language system (i.e. language defined by a dictionary and grammar), etc. These features of a foreign language as a school subject form a negative attitude to it as a difficult subject [33]. Hence, these factors are one of the reasons why students suffer from a lack of motivation.

1.3 Basic Assumptions

Theanalysis of scientific literature and our own teaching experience show that it is of vital importance for teachers to foster students' enthusiasm and interest in a subject, build a supportive environment that will meet psychological needs, increase motivation and facilitate the process of learning foreign languages.

The goal of our study is to identify the components of motivation, determine the level of each, and find ways of enhancing motivation. The following questions constitute the foundation of our study: (1) What are the reasons why students feel reluctant to learn English as a foreign language? (2) What are the ways to promote students' motivation towards learning it?

2 Materials and Methods

The experimental study was conducted on the basis of Nizhniy Novgorod State Ped-agogical University, the Faculty of Humanities; 30 students majoring in Linguistics took part in the study. To explore the outlined components of motivation, their levels we used qualitative research and content analysis; observation, interview were used to collect data for our study as well as the following questionnaires: (1) the questionnaire "Attitude to studying process and school subjects" [27]; (2) the questionnaire "Emo-tional well-being, activity, mood" [11]; (3) the questionnaire "Motives of studying" [27]. The questionnaires were aimed at distinguishing leading motives of learning English, emotional attitude, factors affecting students' willingness and unwillingness to learn a language.

The following principles were the bedrock of the study: (a) voluntary participation of respondents in the research; (b) avoidance of offensive, discriminatory, or another unacceptable language in the formulation of questionnaire and interview; (c) privacy and anonymity of respondents; (d) maintenance of the highest level of objectivity in discussions and analyses throughout the research [9].

3 Research Data

At the first stage the experimental study was aimed at distinguishing the leading motives of learning a foreign language (see Fig. 1).

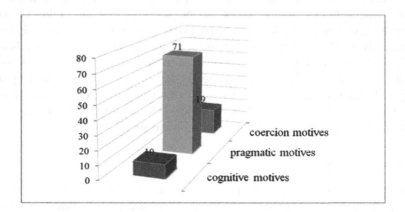

Fig. 1. Leading motives of foreign language learning.

Data received show that only 10% of students have cognitive motives as leading ones. These students are focused on acquiring deep knowledge and feel the need to study independently. This group notes that learning English provides them with the opportunity to know a lot of new, interesting and important things, and show their abilities. Pragmatic motives dominate in 71% of students. They consider learning a foreign language as it gives opportunity to get a prestigious job, stand out among friends, and get a good grade. But still, realizing the importance of a subject studied, more than half of students note that difficulties cause a negative attitude, make the process less attractive. 19% of students chose coercion motives as leading, underlining that language learning is mostly connected with avoiding possible negative consequences. There is no doubt that such motives cannot promote motivation, eagerness to participate in class activities, so students do not take most out of their learning. From our conversation with the students we found out that mostly they have difficulty studying theoretical material, topics concerning phonetics, history of the language and grammar. Half of the students admit that they carry out the tasks, given at lessons, and participate in the discussions only under the teacher's supervision. The presence of the teacher and the possibility of negative consequences in case of failure are the dominant factors when doing the tasks.

The second stage of the experiment is to study students' confidence in a positive result of their activities. The results are shown in Fig. 2.

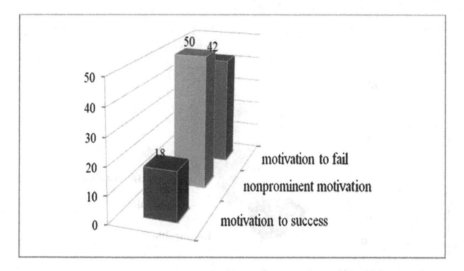

Fig. 2. Motivation to achieve positive results in language learning.

It should be noted that only 18% of the students are set up for success in learning a foreign language. This group of students is focused on achieving positive results. Hope and need for achieving success and positive expectations are the basis of their activity. Motivation to fail is typical for 42% of respondents. Accordingly, when starting the activity they a priori are afraid of a possible failure, and try to find a way to avoid a hypothetical failure. Without desire and will it is impossible to gain fruitful learning results. These students shared that difficult tasks make them panic rather than stimulate or arouse interest in learning. We found out that there is a large number of students (50%) with non-prominent motivation to achieve success and avoid failure.

At the third stage of the experimental study we identified students' interest in lessons, activities and tasks. Interest is one of the main motives when learning any subject, providing a positive impact on the process of learning, thinking, memorizing, and the result of academic activity. Our results show that only 30% of students can be characterized as interested in the content of the subject as well as the forms of teacher's presentation and style. 45% of students note that only some facts are interesting in learning a foreign language, so their interest is episodic. They are attracted by some topics discussed at the lesson, new facts related to culture, history, traditions and customs of the country, forms of work, films and presentations. 25% of students consider the educational tasks boring. They do not believe language to be purposeful and meaningful. They remain indifferent to learning a foreign language, do not strive for gaining new knowledge and achieving considerable results and show no determination and persistence. Certainly, lack of interest, enthusiasm towards learning affects motivation in a negative way, keep motivation levels quite low.

At the final stage we also studied the emotional attitude of students to the process of language learning and students' emotional well-being, the results of which are shown in Fig. 3.

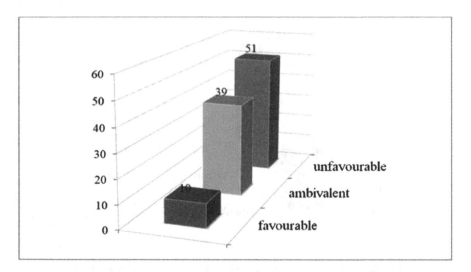

Fig. 3. Students' self-assessment of their emotional state at foreign language lessons.

Based on the obtained results, we can conclude that the minority of students have positive emotional attitude to the process and the result of their activity and feel emotionally comfortable. They are actively involved in discussions, eager to continue their activity and to achieve better results. Negative emotional attitude to the process and the results of activity and emotional discomfort at English lessons are typical of the majority of students. These students point out that in many cases teachers are impatient and unsupportive, have too high expectations of students. Thus, it makes students feel psychologically uncomfortable. The interviews with these students revealed that more than half of them remain indifferent at lessons, often feel nervous when answering, especially when they come to the blackboard, feel nervous about the "ironic smiles of the classmates" and the "teacher's comments". Emotional state of 39% of students is estimated as ambivalent. We can conclude that it is necessary to decrease students' anxiety, build a positive and supportive classroom atmosphere, warm and patient attitude. In this case, such healthy environment will appeal to students and positive emotions determining the level of success will promote students' motivation. Negative emotions of students in the process of studying have a negative impact on motivation, because it is emotions that perform motivation functions in any activity.

The data received demonstrate the significance of finding such methods of teaching English which would best meet psychological needs of students and capture students' interest maintaining motivation. Educators suggest different strategies of doing it, such as creating the situation of success, fostering intercultural communication [15], organizing collective teaching [10], team-work, various projects [7], and using bright hand-out material [29]. We suggest that emotions are motives of any activity. The emotionality of the pedagogical process is one of the most important conditions optimizing the activity of its participants. Feelings and emotions are sure to be paramount, as they provide energy which makes an action possible. A lot of scholars pay attention to music, being of the opinion that music is an art having the greatest power of emotional impact. Music stimulates imagination [25, 30], creative thinking [3, 4], develops speaking skills [20], has stimulative and sedative impact, improves mood and working memory [4]. Thus, we see the prospect of our further studies in the research of musical compositions and studying their impact on the process of education, creating a harmonious, emotionally comfortable and positive classroom atmosphere.

4 Conclusion

Development of society directly depends on the system of education. The aim of educational policy is to change the essence of educational process and organize the teaching process in such a way that it will solve acute problems of modern society and satisfy the modern society demands. Economic globalization has resulted in the ubiquitous use of the English language making it an indispensable part of educational curricula nowadays. However, despite all the efforts, the outcome still does not live up to expectation. Inadequate motivation can be referred to as one of the reasons. Motivation is stated as a significant factor of academic success. The results of our experiment show that students' interest in learning English as a foreign language is accidental, and most of them are indifferent to the process of learning. Striving for success is a characteristic feature of only 18% of students. The rest of students tend to anticipate failure, have a high level of anxiety, are diffident about their achievements and abilities. Students are sensitive to criticism, remain silent being unsure of the correct answer. The results of the study allow us to draw a conclusion that most students have inadequate motivation to learn a language. We consider it significant to find out means and methods of teaching a foreign language that will meet physiological and psychological needs of today's students and enhance motivation to learn a foreign language. More than other activities, learning relies on positive attitude of the student to acquire knowledge. Positive emotions and attitudes towards learning play a key role in increasing motivation and influencing students' performance. Emotions are placed into the forefront of the learning process in terms of enhancing academic achievements. We consider music to be an instrument of inspiration and a tool that can capture students' interest, build a positive classroom atmosphere, deal with negative motivating processes, thus increasing motivation and as a result school success.

References

1. Acat, M.B., Demiral, S.: Sources of motivation in learning foreign language in Turkey. Educ. Adm. Theory Pract. **31**, 312–329 (2002)
2. Aiken, L.A.: Update on attitudes and other affective variables in learning mathematics. Rev. Educ. Res. **61**, 880–915 (1976)
3. Arkhipova, M.V.: Muzykal'noe iskusstvo kak osnova sozdanija uchebnyh jelektron-nyhkursov v obuchenii anglijskomu jazyku studentov nejazykovyh vuzov [Music in e-learning courses of English language at non-linguistic universities]. Hist. Soc.-Educ. Idea **7**(4), 112–114 (2015). (in Russian). doi:10.17748/2075-9908.2015.7.4.112-114
4. Arkhipova, M.V., Shutova, N.V.: Psihologicheskie vozmozhnosti ispol'zovanija muzykal'nogoiskusstva v processe izuchenija inostrannogo jazyka shkol'nikami [Psychological opportunities of using music in the process of learning a foreign language by pupils]. Hist. Soc.-Educ. Idea **1**(17), 68–71 (2013). (in Russian)
5. Aseev, V.G.: Motivacija povedenija i formirovanija lichnosti [Behaviour Motivation and Personality Development]. Mysl', Moscow (1976). (in Russian)
6. Atkinson, J.W.: An Introduction to Motivation. Van Nostrand, Princeton (1964)
7. Belova, E.E.: Metod proektov kak sredstvo intensifikacii obuchenija inostrannomu jazyku [Method of projects as a means of intensifying foreign language teaching]. In: Bikesheva, R. R., Chuvashova O.V. (eds.) Key Problems of Pedagogy, Psychology, Linguistics and Methods of Teaching in Educational Institutions: Digest of Materials of the International Scientific and Practical Conference, pp. 20–24. Astrakhan University Publishing House, Astrakhan (2014). (in Russian)
8. Bolte, A., Goschke, T., Kuhl, J.: The emotion and intuition: effects of positive and negative mood on implicit judgments of semantic coherence. Psychol. Sci. **14**, 416–422 (2003)
9. Bryman, A., Bell, E.: Business Research Methods, 2nd edn. Oxford University Press, Oxford (2007)
10. Dornyei, Z.: The Psychology of Language Learner. Lawrence Erlbaum Associates, Mahwah (2005)
11. Doskin, V.A., Lavrentyeva, N.A., Miroshnikov, M.P., Sharai, V.B.: Test differencirovannoj samoocenki funkcional'nogo sostojanija [Test of differentiated self-assessment of functional state]. Issues Psychol. **6**, 141–145 (1973). (in Russian)
12. Gardner, R.C., Lambert, W.E.: Attitudes and Motivation in Second Language Learning. Newbury House, Rowley (1972)
13. Gökçe, D.: The reasons of lack of motivation from the students' and teachers' voices. J. Acad. Soc. Sci. **1**, 35–45 (2013)
14. Grossman, S.P.: The Biology of Motivation. Annu. Rev. Psychol. **30**, 209–242 (1972)
15. Guseva, L.V.: Principy sozdanija situacij mezhkul'turnogo obshhenija dlja formirovanija mezhkul'turnoj kompetencii studentov po napravleniju "Turizm" (in Russian) [The principles of fostering intercultural communication to form intercultural competence of students majoring in Tourism]. High. Educ. Today **1**, 60–63 (2015)
16. Kauffman, D.F., Humsan, J.: Effects of time perspective on student motivation: introduction to a special issue. Educ. Psychol. Rev. **16**(1), 1–7 (2004)

17. Keller, J.M.: Motivational design of instruction. In: Reigeluth, C.M. (ed.) Instructional-Design Theories and Models: An Overview of Their Current Status, pp. 386–434. Lawrence Erlbaum Associates, Hillsdale (1983)
18. Kesler, J.: Let's Succeed with Our Teenagers. David C. Cook Publishing Co., Elgin (1973)
19. Koroleva, E.V., Kruchinina, G.A.: Teoreticheskie i prakticheskie aspekty lingvistiki, literaturovedenija, metodiki prepodavanija inostrannyh jazykov [Theoretical and practical aspects of linguistics, literature, methodolody of teaching foreign languages]. Bull. Minin Univ. 4(27) (2015). (in Russian). http://vestnik.mininuniver.ru/upload/iblock/1ef/koroleva.pdf. Accessed 16 May 2017
20. Kudravets, O.V.: Muzyka i pesni na urokah nemeckogo jazyka [Music and songs at lessons of the German language]. Foreign Lang. Sch. 2, 45–50 (2001). (in Russian)
21. Lane, J., Lane, A.: Self-efficacy and academic performance. Soc. Behav. Personal. 29(7), 687–694 (2001)
22. Ma, X., Xu, J.: Determining the causal ordering between attitude toward mathematics and achievement in mathematics. Am. J. Educ. 110, 256–280 (2004)
23. Moore, D.G., Burland, K., Davidson, J.W.: The social context of musical success: a developmental account. Br. J. Psychol. 94, 529–550 (2003)
24. Moore, S.C., Oaksford, M.: Some long-term effects of emotion on cognition. Br. J. Psychol. 93, 383–395 (2002)
25. Petrushin, V.I.: Muzykal'naja psihologija [Musical Psychology] Kompozitor, Moscow (1997). (in Russian)
26. Pintrich, P.L., Schunk, D.H.: Motivation in education: theory, research and applications. Prentice Hall Regents, Englewood Cliffs (1996)
27. Rean, A.A.: Psihologija izuchenija lichnosti [Psychology of Personality Research]. Izd-vo Mihajlova V.A, St-Petersburg (1999). (in Russian)
28. Ryan, R.M., Deci, E.L.: Intrinsic and extrinsic motivations: classic definitions and new directions. Contemp. Educ. Psychol. 25, 54–67 (2000)
29. Salanovitch, N.A.: Problemy motivacii i rol' uprazhnenij pri obuchenii francuzskomy jazyku v starshih klassah [The problem of motivation and the means of exercises in teaching the French language at secondary schoo]. Foreign Lang. Sch. 1, 15–18 (1998). (in Russian)
30. Sopchak, A.L.: Individual difference in responses to different types of music in relation to sex, mood and other variables. Psychol. Monogr. 69(11), 369–387 (1955)
31. Utvær, B.K.S., Haugan, G.: The academic motivation scale: dimensionality, reliability, and construct validity among vocational students. Nordic J. Vocat. Educ. Train. 6(2), 17–45 (2016)
32. Zembylas, M.: Emotions and teacher identity: a post structural perspective. Teach. Teach.: Theory Pract. 9, 213–238 (2003)
33. Zimnjaja, I.A.: Psihologija obuchenija inostrannym jazykam v shkole [Psychology of Teaching Foreign Languages at School]. Prosveshhenie, Moscow (1991). (in Russian)

Subject MOOCs as Component of Language Learning Environment

Marina A. Bovtenko[1]([⊠]) [iD] and Galina B. Parshukova[1,2] [iD]

[1] Novosibirsk State Technical University,
Novosibirsk 630073, Russian Federation
bovtenko@is.nstu.ru, g.parchukova@corp.nstu.ru
[2] Novosibirsk State University of Architecture, Design and Arts,
Novosibirsk 630099, Russian Federation

Abstract. The article deals with use of MOOCs as a resource for development of students' professional and foreign language competencies. Based on current trends in teaching foreign languages at tertiary level – a shift from teaching language for specific purposes to content and language integrated learning and use of subject MOOCs in language learning purposes – the authors discuss the ways of acquisition of subject and language knowledge and skills through learning environment which includes MOOCs along with other learning materials, learning management systems and authentic resources both in native and foreign languages. The offered components of bilingual learning environment are described; the examples of content related MOOCs selected for pilot study in frame of Advertising and Public Relations, Psychology, Linguistics and Architecture undergraduate curricula and project, analytical, creative tasks, interactive exercises created in NSTU language learning management system eLang are given; main results of the project and prospect of its further development are presented.

Keywords: Massive online open courses (MOOCs) · Content and language integrated language learning (CLIL) · Language for specific purposes (LSP) · Professional competencies · Learning environment

1 Introduction

1.1 Building a Research Problem

Rapid growth of MOOCs has made available for a wide audience a great number of courses and learning materials on various subjects in different languages. Taking into account that modern education tends to be bilingual or multilingual, MOOCs are widely used in higher education as a resource for teaching and learning subjects, development of proficiency in language for specific purposes, and content and language integrated learning. These practices make topical an exploration of efficient ways of MOOCs integration in a university curriculum. In this article MOOCs are considered as a component of bilingual educational environment which is based on combination of face to face, distance and blended learning and use of variety of traditional and

A. Filchenko and Z. Anikina (eds.), *Linguistic and Cultural Studies:*
Traditions and Innovations, Advances in Intelligent Systems and Computing 677,
DOI 10.1007/978-3-319-67843-6_15

e-learning resources – from learning management systems and authoring tools to information resources and MOOCs both in native and foreign languages.

1.2   Literature Review

The phenomenon of MOOCs is explored in a great number of research works focusing on MOOCs in context of open, continuous, secondary and higher education [1, 10, 18–20] and different educational systems and local practices [7, 9, 18]; target audiences, typology and structure of MOOCs; their advantages and limitations for formal and informal learning, approaches to course development, academic and intercultural communications in MOOCs; students' motivation, collaborative and autonomous learning, conditions for efficient integration of MOOCs in universities curriculum [11, 13, 14, 16], et al. One of the topical research issues is repurposing subject MOOCs for cross-subject and language learning [4, 5, 15, 17]. Studies of MOOCs integration to universities' curriculum show that effectiveness of the integration is provided by transformation of existing courses according to MOOCs content, taking into account results of students' learning through MOOCs in overall course assessment, university support of integrated and hybrid learning based on MOOCs [8, 11, 14, 16].

1.3   Basic Assumptions

The Russian Federal Educational Standards contain general description of students' professional competences and state requirements to curriculum, which can be widen by regional and university components [6]. All Russian Federal Educational Standards of higher education both for undergraduate and postgraduate levels require development of students' competences at least in one foreign language (linguistics, philology, international relations, regional study curricula include two foreign languages). For example, the State Educational Standard of undergraduate programme "Advertising and Public Relations" specifies a graduate's professional area as "communication processes in interpersonal, social, political, economic, cultural, educational, and scientific spheres; technologies and techniques of mass, business and personal communications; technologies and techniques of market positioning and promotion of products, services, commercial companies, non-profit and community non-governmental organizations, state institutions. The professional competences acquired should provide graduates with abilities to take part in development, production, and distribution of promotional products, planning and conducting PR and advertising campaigns, writing projects in a variety of PR and advertising genres such as backgrounders, news, press-, media-, and social media releases, creative briefs, media monitoring reports, etc. Foreign language professional competences acquired should provide graduates with ability to efficiently use of foreign language in frame of the described professional discourse. However, practice shows that an isolated learning of specialization subjects and language for specific purposes may not provide the desired level of both professional and foreign languages competences. In this case subjects MOOCs in foreign languages serve as a resource for content and language integrated learning [2, 5, 12, 15, 17, 19]. Thus, modern university learning environment should offer variety of components and materials (including MOOCs) and learning material should be both in native and in foreign languages. Integration of MOOCs in a university learning environment requires the following steps: selection of MOOCs which are relevant to a specializations and curriculum, comparison of syllabuses of selected courses and MOOCs, analysis of teaching approaches used, making decision about the role and place of the MOOCs in the curriculum, inclusion of learning through MOOCs into specialization courses assessment system. The added value of integrating MOOCs into the curriculum is connected with various ways of training in MOOCs. MOOCs offering self-study and self-paced models can be used as a resource for students' autonomous learning; MOOCs with a strong collaborative focus help students to develop professional and academic communications skills. Use of subject MOOCs both in native and foreign languages would be more efficient if students learn through MOOCs on a regular basis throughout the curriculum and if the training is supported by dedicated language learning materials including materials based on selected subject MOOCs.

2   Materials and Methods

Materials under study can be divided into several groups: (1) Federal State Educational Standards (undergraduate level) of four specializations: "Advertising and Public Relations", "Psychology", "Linguistics" (English and Russian Sign Language), "Architecture"; (2) the specializations curricula, syllabuses of subject and foreign language courses; (3) printed, media, and electronic course materials including materials developed in NSTU program systems – LMS DiSpace and dedicated language LMS (LLMS) eLang; (4) MOOCs relevant to the areas of study; (5) materials developed for the selected courses in frame of developing bilingual learning environment. The methods used include analysis of documentation (Federal State Educational Standards, curricula, syllabuses of selected specialization courses) to determine professional competences (including competences in foreign languages for specific purposes) which are to be formed; selection of MOOCs relevant to specializations courses (syllabus, content, language, communications, implemented teaching approaches), analysis of benefits and limitations of program systems used and selected MOOCs platforms; modeling of bilingual learning environment for target groups of students taking into account their specialization, year of study, foreign language proficiency; project piloting; conducting students surveys. Examples of materials selected for "Advertising and Public Relations" specialization include online data bases of scientific publications, materials of specialists' web-sites and social media accounts, job vacancies and résumés, press-releases; media and social media monitoring tools, reports on media and social media monitoring, media kits, infographics; MOOCs in Russian and English: History and Theory of Media, Modern Communication Technologies, Introduction to Public Relations, Online Advertising, Brand and Product Management, Reputation Management in a Digital World, etc. One of the obligatory components of bilingual learning environment is foreign language modules, containing interactive exercises and quizzes developed in LLMS eLang. The system is designed to provide students' independent work; the authoring tools offer 18 types of interactive tasks with multimedia components – from quizzes, matching, multiple choices, gap filling, jumbling to total text deletion exercises with no limitations in number of exercises and modules [3]. The exercises and quizzes aimed at mastering foreign language skills – knowledge of professional vocabulary, lexical collocations, genres and style of professional communication. The materials are incorporated into curriculum through analytical, project, and creative tasks in Russian and English, such as designing visuals relevant to the type of data presented in a real-life report on monitoring media and social media; writing news based on given newsbreaks for a company's web-site or posting on social media, developing media plans and the media lists for particular companies based on company profiles and analysis of media kits; writing essay on the topic "Gender status varieties" based on translated into Russian of more than 50 gender status offered a user while registering in the social network Facebook; rewriting a web-site news for different social media, comparing social media monitoring tools/plagiarism checkers. The tasks based on MOOCs include group project works on the topics of the courses, preliminary peer review of individual works, and participation in online discussions, recording audio and video reports.

3   Research Data

Proposed approaches to development of learning environment by inclusion of subject MOOCs, authentic materials in native and foreign languages, and interactive tasks for development of professional foreign language competence created in LLMS eLang has been piloting for three years in Novosibirsk State Technical University at Faculty of Humanities and for a year in Novosibirsk State University of Architecture, Design and Arts. 12 groups of undergraduate students majoring in Advertising and Public Relations, Psychology, Linguistics, and Architecture studied general and specializations courses in the described way: Advertising and Publics Relations students took two specialization courses "Technologies of Advertising Products Development" and "Modern Press-Office", Psychology and Linguistics students studied "Theories and Models of Communication", Architecture students studied "Library Science-Bibliography" course. Examples of MOOCs recommended for the subject courses: "Modern Press-Office" – Modern Communications Theories (In Russian) (MGIMO, Uniweb), History and theory of media (In Russian) (HSE, Coursera), PR: Theories and Practices (In Russian) (Pushkin State Russian Language Institute), Introduction to Public Relations (PR Academy), Gathering and Developing the News (Michigan State University, Coursera), Social Media in Public Relations (National University of Singapore, Coursera) Reputation Management in a Digital World (Curtin University, edX); "Technologies of Advertising Products Development" – Advertising and Society (Duke University, Coursera), Online Advertising (OnlineAd), Services Marketing – Selling the Invisible (The University of New South Wales, OpenLearning), Design and Make Infographics (Michigan State University, Coursera); "Theories and Models of Communication" – Introduction to Communication Science (University of Amsterdam, Coursera); Improving Communication Skills (University of Pennsylvania, Coursera), "Library Science-Bibliography" – Copyright for Educators & Librarians (Duke University, Coursera). Students' foreign language proficiency varied from pre-intermediate and intermediate in groups of architecture, psychology, advertising and public relations students to upper intermediate and advanced in linguistics' groups.

The levels determine variations of tasks based on foreign language materials: all groups rendered authentic materials both in Russian and English, but one of the final tasks for linguistics' students was to present carefully edited translation from English to Russian; review of MOOCs materials relevant to the course content was obligatory for all groups, but linguistics students were to fulfill selected tasks or modules in the MOOCs chosen.

4 Conclusion

The main results of piloting bilingual learning environment included authentic materials, subject MOOCs, e-learning tasks, integrated foreign language learning components are the following: empowering specializations learning environment through integration of new components aimed at development of students' professional and foreign language competences; implementation of practice oriented approaches, wide use of authentic materials, project and analytical tasks; development of universities' regulations for procedures of implementation of content and language integrated learning and integration of MOOCs into curriculum; positive dynamic of students' and teachers' attitude to use of MOOCs and learning materials and tasks in foreign language in subject courses; decisions on spreading the experience gained and implementing the approaches offered for engineering and business CLIL syllabus design.

References

1. Billington, P.J., Fronmueller, M.P.: MOOCs and the future of higher education. J. High. Educ. Theory Pract. **13**(3/4), 36–43 (2013)
2. Borshcheva, V.V.: Osobennosti ispol'sovaniya massovuh otkrutuh onlain kursov v obuchenii yazuku dly aspetsialnuh tselei [Special features of using massive open online courses in teaching English for specific purposesl]. Vestnik PNIPU. Problemy âzykoznaniâ I pedagogiki **1**, 86–95 (2017). (in Russian)
3. Bovtenko, M.A.: eLANG: programmnaya sistema dlya razrabotki onlayn kusosov po inostrannym yazukam [eLANG: multiligual online language learning au-thoring system]. Open Distance Educ. **4**(60), 21–26 (2015). (in Russian)
4. Chirtsov, A., Mikushev, V.: MOOK-technologii kak instrumenty sozdaniya transgranichnoy predmetnoy obrazovatelnoy sredy [MOOC-technologies as tools for creating a cross-subject educational environment of the university]. Vestnik PskovGU. Estestvennue i phisiko-matematicheskie nauki. **7**, 33–139 (2015). (in Russian)
5. De Waard, I., Demeulenaere, K.: The MOOC-CLIL project: using MOOCs to increase language, and social and online learning skills for 5th grade K-12 students. In: Kan, Q., Bax, S. (eds.) Beyond the Language Classroom: Researching MOOCs and Other Innovations, pp. 29–42. Research-publishing.net, Dublin (2017)
6. Federalnye Obrzovatelnye Standarty: Reklamaisbyazi s obschestvennostyu. Psikhologiya. Lingvistika. Arkhitectura [State Educational Standards (Bachelor's degree): Advertising and Public Relations, Psychology, Linguistics, Architecture]. (in Russian). http://fgosvo.ru/. Accessed 01 July 2017

7. Godwin-Jones, R.: Global reach and local practice: the promise of MOOCS. Lang. Learn. Technol. **18**(3), 5–15 (2014). http://llt.msu.edu/issues/october2014/emerging.pdf. Accessed 02 July 2017
8. Israel, M.J.: Effectiveness of integrating MOOCs in traditional classrooms for undergraduate students. Int. Rev. Res. Open Distrib. Learn. **16**(5), 102–118 (2016)
9. Jansen, D., Konings, L. MOOCs in Europe. In: Overview of papers, HOME conference Rome. EDTU, Maastricht (2016)
10. Kim, S-W.: MOOCs in higher education. In: Cvetkovic, D. (ed.) Virtual Learning, pp. 121–135 (2016). https://www.intechopen.com/books/virtual-learning/moocs-in-higher-education. Accessed 01 July 2017
11. Kulik, E., Kidimova, K.: Integrating MOOCs in university curriculum. In: Kloos, C.D., Jermann, P., Pérez-Sanagustin, M., Seaton, D.T. White, S. (eds.). Proceedings of Work in Progress Papers of the Experience and Research Tracks and Position Papers of the Policy Track at EMOOCs 2017 co-located with the EMOOCs 2017 (5th European MOOCs Stakeholders Summit, EMOOCs 2017, Madrid, Spain, 22–26 May 2017), pp. 118–127. CEUR Workshop Proceedings, vol. 1841, RWTH Aachen University, Aachen (2017). http://ceur-ws.org/Vol-1841/P05_101.pdf. Accessed 01 July 2017
12. Kuznetsova, D.I., Almazova, N.I., Valieva, F.I., Halyapina L.P.: Innovatsionnye idei i podhody k integrirovannomu obucheniu inostrannym yazykam i professional'nym distsiplinam v sisteme vysshego obrazovaniya. In: Proceedings of International Conference Innovative Ideas and Approaches to Integrated Content and Language Learning at Tertiary Level, 27–30 March 2017. St. Petersburg Polytechnic University Publishing House, St. Petersburg (2017). (in Russian)
13. Lambert, S.R., Alony, I.: Embedding MOOCs in academic programs as a part of curriculum transformation: a pilot case study. In: Li, K.C., Yuen K.S. (eds.) Proceedings of International Conference on Open and Flexible Education, pp. 1–9. Open University of Hong Kong, Hong Kong (2015)
14. Pérez-Sanagustín, M., Hilliger, I., Alario-Hoyos, C., Kloos, C.D., Rayyan, S.: H-MOOC framework: reusing MOOCs for hybrid education. J. Comput. Higher Educ. **29**(1), 47–64 (2017)
15. Rybushkina, S.V., Chuchalin, A.I.: Integrated approach to teaching ESP based on MOOCs. In: Hawwash, K., Léger, C. (eds.) Proceedings of the 43rd SEFI Annual Conference June 29–July 2 2015, Orléans, France, pp. 76–83. SEFI, Brussels (2015)
16. Starodubtsev, V.A.: Personalizirovannie MOOK v smeshannom obuchenii [Personalized MOOCs in blended learning]. Higher Educ. Russ. **10**, 133–144 (2015). (in Russian)
17. Stognieva, O.: Integration of MOOCs in the university ESP course curriculum: beliefs and practices. Prof. Acad. Engl. **48**, 25–33 (2016)
18. Titova, S.V.: MOOK v rossiiskom obrazovanii [MOOCs in Russian Education]. High. Educ. Russ. **12**, 141–145 (2015). (in Russian)
19. Vyushkina, E.G.: Massovie ontkrytye onlain kursy: teoriya, istoriya, permassive open online courses: theory, history, implementation prospects [Massive open online courses: theory, history, implementation]. Izvestiya of Saratov University. New Series. Series: Philosophy. Psychology Pedagogy **2**, 78–83 (2015). (in Russian)
20. Yuan, L., Powell, S.: MOOCs and open education: Implications for higher education. CETIS White Paper (2013). http://www.smarthighered.com/wp-content/uploads/2013/03/MOOCs-and-Open-Education.pdf. Accessed 01 July 2017

Development of Preschool Children Speech Problems

Galiya H. Vahitova[1] and Alexander V. Obskov[2(✉)]

[1] Tomsk State Pedagogical University, Tomsk 634061, Russian Federation
galija2000@mail.ru
[2] Tomsk Polytechnic University, Tomsk 634050, Russian Federation
alexanderobskov@hotmail.com

Abstract. According to the standards of new educational context, this paper highlights children's speech development and pays attention to the aspects connected with communication skills development. The main reasons of pre-school children low speech development have been observed. These reasons cause a lot of difficulties not only for a good grammar development but for a successful social communication. For a full-fledged development, it is important to include each child in a confidential informal communication. It is noted that the methodological methods of activating and developing speech of preschoolers should be thought out and selected the game-based ones. In this case they can cause interest, and the main thing is not to damage neither the physical nor mental health of children.

Keywords: Speech development · Preschool · Development of communication skills · Children's statements · Russian as a foreign language · Word combinations of russian preschool children

1 Introduction

In the present days there are theoretical and practical works in modern science, studying such a great phenomenon like basic nature of speech, its types, functions and mechanisms; and the most important is their role in child development. A lot of researchers and educators study this kind of phenomena, including Herbert Ernest Bates; Maria Montessori; Lev Vygotsky; Alexey Leontiev; Alexander Luriya; Jean William Fritz Piaget; Sergei Rubinstein and other. Today it is necessary to upgrade and bring very important corrections into Russian preschool education system by changing the out of date traditional teaching methods into modern, up to date and effective ones, according to implementation of the Federal State Educational Standards for Preschool Education. This process of preschool education modernization concerns all aspects of educational activity, including organization of a complex section related to development of children's speech. Beginning of upgrading is already included in the structure of general education program, proposed in the standards. The educational area of speech development is aimed at "possession of speech as a means of communication and culture; active vocabulary enrichment; development of coherent, grammatically correct dialogical and monological speech; development of speech creativity; sound

A. Filchenko and Z. Anikina (eds.), *Linguistic and Cultural Studies:*
Traditions and Innovations, Advances in Intelligent Systems and Computing 677,
DOI 10.1007/978-3-319-67843-6_16

and intonation culture development, phonemic hearing; acquaintance with the book culture, children's literature, understanding the texts of various genres of children's literature; formation of sound analytical and synthetic activity as prerequisites for literacy [1]."

2 Literature Review

The research is based on the works of domestic and foreign scholars in such areas as various definitions of development of preschool children speech problem. It allows getting access to researches in the field of preschool education (M.M. Alekseeva, M.L. Kusova, S.V. Plotnikova, S.I. Pozdeeva, O.S. Ushakova, V.I. Yashina, et al.) to point out the decrease in overall level of children speech development due to both objective and subjective reasons.

2.1 Research Objectives

Our theoretical studies allowed us to conduct questionnaires, surveys, observations in the context of polysubject interaction among participants of preschool education [2, 8]. The original position of the authors was based on the need to implement the polysubject approach, which allows reinterpreting the realities of modern preschool education practices like interaction of educators, children, and their parents) and outline ways to overcome the problems of children's speech development.

3 Research Problem

Thus, in comparison with many years the section of existing educational program on speech development, the new content of educational area outlines the range of tasks related, first of all, to form a desire to communicate with peers and adults, and the development of communicative skills of preschoolers. It is no doubt, the solution of these problems requires, above all, the educators' professionalism. However, even the most highly qualified educator in pre-school education cannot fully solve the problems of speech development without taking into account the characteristics of children nowadays. Today's preschooler is completely different from those twenty or thirty years ago. Nevertheless, our knowledge about children is made up of the ideas of those fundamental studies that were conducted more than fifty years ago. It is obvious that culture is changing the situation of modern childhood development [3].

3.1 Preschoolers' Speech Development: Research Focus

Current ideas about children, their psychological, social development, content of preschool education, require a new cultural approach. In the preschool educational environment arises the situation that requires the educators' special efforts to organize the interaction of modern children, easily mastering computer technology. Some educators like [1, 4–6] note a decrease in the overall level of children speech development due to objective as well as subjective reasons. First, this is about development while using computers, radically changing the usual models of communication towards impoverishing live communication. Real communication is replaced by viewing multimedia presentations, cartoons and mostly computer games. In the professional environment of preschool pedagogy educators appeared the expression "Homo-Clicker," instead of "Homo-sapiens." An expression that accurately characterizes the modern preschooler since the child is already developing not only through modeling, drawing, cutting, constructive dialogue, but also as a result of mastering the keyboard of a mobile phone or a computer. It is important to admit that today the child spends much more time in front of a TV screen and computer monitor rather than in direct communication with peers and adults. Such pastime leads to a decrease in the child's speech culture: no diction is formed; the correct sounds pronunciation and words do not develop the ability to clearly, logically express their thoughts, constructively engage in dialogue with peers and adults. As a consequence, the preschooler develops a monosyllabic speech consisting of simple sentences, and sometimes even limited by interjections: "Hey, look!" "Wow! Yes! Cool!" "Hey!" "What?" "Ah, that is cool!" Such dialogues are not a rare phenomenon among preschool children.

As research shows, a five-six-year-old child is very well-oriented in names, the cries of screened characters (often difficult to pronounce even for adults) of foreign cartoons, and cannot explain the meaning of primordially Russian words or words actively used in the Russian vocabulary. For instance, six-year-old Lisa explains the word "great-grandson", comparing it with the word "great-grandfather": if the great-grandfather means a very old grandfather, then the great-grandson means an ancient grandson. At the same time five-year-old Sasha easily and simply finds an explanation to the words "blacksmith" and "violinist". "Blacksmith" means the father of a grasshopper (Translator's Note (TN): words "blacksmith" and "grasshopper" are cognate in the Russian language), and the "violinist" holds the paper together (TN: words "violinist" and "staple" sound similarly), but then he straightens his thoughts and thinks: "No, the "violinist" is a man who creaks (TN: "violinist" and "creak" are cognate words in the Russian language). Similarly, by establishing the connection between the meaning of the word and the sound, other children explain their meanings: the machinist drives the machines, the turner conducts the current, and the joiner makes the tables. And if the horses live in the horsehouse, so, a cow lives in the cowhouse, and a bull lives in the bullhouse, a goose lives in the goosehouse, and a duck lives in the duckhouse. And there are many more similar examples, and they cannot but cause concern for preschool educators.

At present, figurative, rich in synonyms, additions and descriptions, speech of pre-school children is a very rare phenomenon. Our research that has been done in different preschool groups in the city of Tomsk and Kemerovo region as well, confirms the vocabulary poverty in terms of reducing the knowledge necessary for full communication and socialization in the future. Within our research framework, the lexical dictionary of preschool children was analyzed. One hundred and seven children (39 five-year-old and 68 six-year old children) were asked to explain the meaning and give the definition to twenty words, which were divided into two groups, ten words each. The first group included names, nicknames of foreign cartoon heroes like Ben Ten, Batman, Goofy, Dumbo, Sponge Bob, Ninjago, Scrooge McDuck, Tweety, Shrek,

Skipper, and the second group included Russian words like great-grandson, great-grandfather, stable, violinist, turner, carpenter, weaver, confectioner, veterinarian and forester. It turned out that 5 to 6 characters in the first group are well-known among preschoolers. 86% explained 8 to 10 words, 7% explained 4 to 7 words and 3 or less words were explained by 8% of preschoolers. As for the second group, preschoolers faced some serious difficulties. Among five-year-old children only 5% recognized all the words, 7% didn't know the meaning of any word at all, 26% were familiar with half or more words, 23% explained the meaning of less than five words. Among 68 seven-year-old children, it was found that 3% cannot explain a single word from the list, 13% cannot explain three or fewer words, 41% cannot explain four or five words, 30% know more than half of the words, and the entire group of words was recognized by 13% of preschoolers. The obtained results prove that today, in conditions of mass intellectualization of preschool children in preschools and daycares, etc., children successfully master computer technologies, get involved in project activities, but at the same time have a poor speech characteristics. In addition, as our long-term research of children in pre-school education groups show, a child is not always ready to cooperate, does not know how to correct their own mistakes, does not know how to follow the plan and instructions, he/she does not always hear and understand the requests of adults and peers, their functional illiteracy. These circumstances oblige educators and parents to take due care of the timely formation of children's speech, its purity and correctness, warning and correcting various violations that consider any deviations from the generally accepted norms of the Russian language. Another, no less relevant reason, impoverishing the speech development of the preschoolers, is cause by the disruption of ties between parents and children. Young parents tell their children stories, epics and fables less than before. In many families, tradition of reading together followed by a discussion of what has been read, has been lost. Communication between children and adult family members, due to busy life of the latter, is often superficial and limited to everyday topics or reduced to instructions on what and how to raise a child.

Table 1 shows the results of the survey among 92 parents with preschool children, of whom only 49% read fairy tales to their children; they only introduce their children to the tales, nursery rhymes 11–12% (apparently this genre leaves the practice of raising children), only 53% mothers/grandmothers sing lullabies, 36% recite the stories of children's writers. 85% of dads and moms do not read with the children together. Parents consider the main task is to feed and to dress a child, i.e. to provide for financially, and specialists - teachers, educators, methodologists, speech therapists, etc., are for all other aspects of the child's psychological and pedagogical development, in their opinion. Of course, such narrow understanding of education goals by parents can

Table 1. Research results among parents.

Parents activity	Responders
Providing financial protection	85%
Singing lullabies	53%
Reading fairy tales	49%
Recite the stories of children's writers	36%
Introducing the Russian tales	12%

lead to artificial slowdown in the overall development of children; verbal aspect is no exception in this case. To these quite serious reasons, leading to a decrease in the level of speech development of modern preschoolers, it is worth adding the objective complexity of the Russian language system itself, and the poor quality of speech environment surrounding the child, and various abnormalities of the child's development, in which speech suffers inevitably. In general, all these reasons for reducing the overall level of speech development of preschool children cause certain difficulties not only for formation of a coherent, grammatically correct dialogical and monological speech, but for their successful communication and socialization with peers and adults. Certainly, this circumstance should not be left out of attention of adults raising a child. To achieve a communicative competence at the stage of pre-school education, the educator must contribute in every way to its development by solving the problems of forming different aspects of the child's speech: developing coherent speech, enriching the vocabulary, mastering grammatically correct speech and its sound culture. The full development of children, including speech, can be carried out only in a live direct communication, important for each subject. Child inclusion in a personally significant and trusting informal communication is a very important task, which can only be solved by joint efforts of educators and parents. In this regard, I would like to note that the methodological methods of activating and developing the speech and speech activity of preschool children must be carefully thought out and only those selected that are game-based, i.e. accessible for their perception and understanding. Experts in the field of pre-school education noted that the speech development methods for preschoolers are diverse [1, 4–6]. The choice of these methods and forms of work is carried out by the educator and depends on many reasons: cultural and regional characteristics, the specifics and equipment of the institution, teacher's experience and creativity, the type of exercise, and finally, the level of children development, their activity and independence.

Thus, organizing games for speech development, we proceed from the needs of children in competition. For example, the games "Say the Opposite", "Call the Cub", "Who Lives Where", "Guess the Profession" - are held in the form of competition. Mastering the rules for constructing meaningful speech segments (phrases, sentences, and small texts), not only subject pictures are used, but also cubes of Nikitin, geometric blocks of Gyenes, Montessori's inserts, the games "Fold square", "Tangram", etc. Preschool children are involved in activity of making proposals by means of a cut book or a cube of actions and show interest in it. We stress that whatever methods are used, they should be accessible because interest and, most importantly, not damage the physical or mental health of children. Therefore, the educator needs to remember to create the environment that fully corresponds to the age and individual needs of pre-school children. At the same time, the training should take place in such a way that children do not realize that educational process is taking place here and now. And such organization of training activities will proceed easily and freely if a trusting relationship has developed between the child and the teacher, if the teacher does not perform a supervisor and dictator functions, but more of a tutor and facilitator, to whom one can turn for help at any time.

4 Conclusion

Timely and complete formation of speech at the preschool age is one of the main conditions for normal development of a child and their future successful schooling. The more enriched and correct the child's speech, the easier it is for them to express their thoughts, the better their relationship with adults and peers. Construction of educational process in pre-school education should be based on methods and forms of work with children that are appropriate to the age and focus not only on selection of appropriate methods for the children's speech development, but create comfortable conditions for teachers in which children will not be afraid of committing errors.

References

1. Alekseeva, M.M., Yashina, V.I.: Rechevoe razvitie doshkolnikov [Speech development of preschoolers]. Vlados, Moscow (2004). (in Russian)
2. Federal State Educational Standard for Preschool Education (FSESPE). http://www.rg.ru/2013/11/25/doshkstandartdok.html. Accessed 05 Feb 2017
3. Vygotsky, L.S.: Problema kulturnogo razvitiya rebenka [The problem of cultural development of the child]. Bulletin of the Moscow University, Moscow (1928). (in Russian)
4. Kusova, M.L.: Slovo kak obyekt refleksii v detskoi rechi [The word as an object of reflection in children's speech]. Chelyabinsk State Pedagogical Univ. Bull. 8, 68–77 (2014). (in Russian)
5. Plotnikova, S.V.: Razvitie leksikona rebenka [Development of the Child's Lexicon], 2nd edn. Phlinta, Moscow (2013). (in Russian)
6. Pozdeeva, S.I.: K problem obucheniya pervonachalnomu chteniyu rebenka starshego doshkolnogo vozrasta [To the problem of teaching the initial reading of a child of senior preschool age]. Kindergarten A to Z. J. Teachers Parents 5(41), 27–37 (2009). (in Russian)
7. Ushakova, O.S., Strunina, E.M.: Metodika razvitiya rechi detej doshkolnogo vozrasta [Methods of Speech Development for Preschool Children]. Vlados, Moscow (2010). (in Russian)
8. Vakhitova, G.H.: Polisubektnoe vzaimodejstvie v kontekste predshkolnogo obrazovaniya [Polysubject interaction in the context of preschool education]. Tomsk State Pedagogical Univ. Bull. 13, 71–75 (2013). (in Russian)

Complex-Incomplete and Elliptical Foreign Language Sentences Translation Training of Original Non-literary Texts by Technical University Students

Tatiana A. Dakukina[1,2(✉)] and Alexey V. Balastov[1]

[1] Tomsk Polytechnic University, 634050 Tomsk, Russian Federation
sharda@yandex.ru, lexiusb@gmail.com
[2] Tomsk State Pedagogical University, 634061 Tomsk, Russian Federation

Abstract. Modern translation training approaches are not aimed at the development of the future engineers' necessary translation skills. Therefore there is an increasing need to teach engineering students how to translate original foreign language non-literary texts. We suggest to develop more efficient and up-to-date translation training model and experimentally prove its reliability. The training model will be based on original non-literary text translation of complex-incomplete and elliptical sentences. The proposed model feature is that it takes into account these sentenses perception and recognition peculiarities.

We have analyzed scientific literature, foreign and local textbooks, learning and teaching sets and performed the approbation of the experimental model for the engineering students training. As a result, we concluded that complex-incomplete and elliptic sentences complexity affects their perception and recognition. Perception during these texts translation training is not always equal to complete recognition of the content. Firstly, the known grammatical phenomenon recognition occurs. Secondly, the process of the new information perception starts followed by the subsequent updating of knowledge in the transfer process – this phenomenon is understudied scientific literature. Complex-incomplete sentence structure and elliptical models are various. Consequently, it is necessary to apply different methods to reach the optimal translation.

The performed comparative analysis of the engineering students teaching to original non-literary text translation before and after the experimental training has shown positive results. Correctness of the complex-incomplete and elliptic sentences translations has doubled. Theoretically, we have proposed a new translation teaching approach concerning the relationship between two concepts - perception and recognition. The developed translation training model of the original non-literary texts takes into account their peculiarities and serves as a starting point for the further research.

Keywords: Original non-literary texts · Complex-incomplete sentences · Models of elliptical sentences · Structural analysis · Pre-translation and translation exercises · Scheme of the sentences' analysis

© Springer International Publishing AG 2018
A. Filchenko and Z. Anikina (eds.), *Linguistic and Cultural Studies:*
Traditions and Innovations, Advances in Intelligent Systems and Computing 677,
DOI 10.1007/978-3-319-67843-6_17

1 Introduction

Nowadays, higher educational institutions provide training process, taking into consideration modern labor market requirements. Our society needs professionally flexible specialists capable of fast adapting to the constantly changing labour market conditions [6, 9, 12]. At the same time the specialists with the foreign language knowledge are in high demand. Hence, the goal of the educational institution is to train mobile students who are able to move from one professional position to another.

The important components characterizing specialist mobility are their readiness to adaptation in production conditions as well as the professional competence that is the intelligence with good command of the foreign language, creative capabilities and professionalism [4].

Professional competence formation and the foreign language knowledge enhancement relying on the "modern concept of higher education" [3] could be possibly based on the academic mobility programs. An important issue at the same time is the language of academic mobility. When students arrives in a foreign university for mobility programs training they should be sufficiently proficient in English or the host country language (Dalinger [6]). The foreign language lectures and practical training assume working with foreign educational literature. This literature reading and understanding skills directly depend on the trained students' translation abilities but the current level of such skills is still unsatisfying. Our students experienced some difficulties during teaching process when dealing with the foreign literature. Nevertheless within the limits of only one scientific research it is impossible to solve all the raised issues. Taking into consideration the limited amount of time devoted to the foreign language learning in a technical university, we decided to pay attention to the most complex phenomena: complex-incomplete and elliptical foreign language sentences. As a result of the study we developed translation training model based on complex-incomplete and elliptic sentences of original non-literary texts and took into account the peculiarities of their perception and recognition. The developed training model based on complex-incomplete and elliptic sentence translation of the original non-literary texts sufficiently improved the students' translation skills. Thus, this article represents an attempt to theoretically and practically prove the feasibility and the need of the suggested model application.

2 Materials and Methods

The paper outlines the methodology for the study, including pedagogical aim, task design, the training model development on the basis of complex-incomplete and elliptical foreign language sentences translation and its experimental verification. According to the aim of this study, the following pedagogical objectives were formulated:

- to prove the necessity to teach engineering students to the original non-literary foreign language text translation;
- to study the content of the Tomsk Polytechnic University foreign engineering students curricula;

- to analyze the original non-literary literature;
- to develop original non-literary texts' translation training model, theoretically and experimentally prove its efficiency.

When solving the above stated tasks both theoretical and empirical methods were applied: the analysis, the literature review, the experimental training with the subsequent data processing.

When dealing with the first task, we considered the interview with Filippov, Rectop of the Peoples' Friendship University of Russia, concerning the engineering students training. He confirmed the relevance of this issue and stated the necessity to provide engineering students with the opportunity to obtain overprofessional competencies, both within the basic educational programs and additional training programs framework.

Internship is one of the forms contributing to professional competency developing. Engineering students can and should do an internship not only in the Russian enterprises but also abroad. Here we mean IAESTE (International Association for the Exchange of Students for Technical Experience) universities and enterprises with the support of the scholarship exchange programs, especially in our case, the German language summer university courses for foreign students in Germany.

This requires a good knowledge of the foreign language, the ability to work with and translate foreign educational literature. Such engineering students can work on joint projects with foreign partners in the engineering, energy, heat and power engineering fields, etc. [9].

When working under the second task we have studied a number of technical university programs that suggest the following: the students involved in scientific activity starting from the very first year of training: making presentations during so-called "conference weeks", scientific articles writing, participation in competitions, presenting scientific reports at conferences. These activities demand study of a certain quantity of scientific literature, including foreign language sources, i.e. original non-literary texts. For the engineering students it is impossible to succeed in the above mentioned activities if they lack non-literary texts translating skills.

In the scientific literature there are a lot of studies on the original literature translation problems [2, 10, 11, 14, 15, 17]. Having analyzed the information on the issue we were able to draw the following conclusions:

- to achieve the positive results, it is necessary to know psycholinguistic bases of the original foreign-language texts in general and ceratain grammatical structures in particular [10, 14, 17];
- the translation essence, in our opinion, is defined not only as an activity result [10], but also as the semantic analysis;
- visual text material perception is required as well as its recognition to evaluate the information given in the text [17];
- the process of teaching such texts translation is aimed at mastering students' new skills and their subsequent activation, thus, it is important to know how to manage the available knowledge and to acquire a new one [2, 11, 15].

All the above mentioned conclusions play an important role in the original foreign language texts translation teaching.

The performed original non-literary texts analysis allowed us to distinguish the following: these texts are complex both in terms of grammar and vocabulary, they contain a large number of cultural and regional concepts and realities, they have complex syntactic constructions, particularly complex-incomplete and elliptical sentences.

The term "complex-incomplete" sentence, from our personal point of view, emphasizes its relation to a complex sentence. Nevertheless, for the complex-incomplete sentence a poly predicativeness is characteristic with the omission of the subsequent subject or the presence of a number of subjects with the absence of one of the predicates. Such sentences, unlike the complex ones, are less cumbersome in the syntactic sense. At the same time, the taxis relations are expressed in the structure of the complex-incomplete sentence, that is, the temporal correlation of actions denoted by predicates: strict/weak simultaneity, complete/partial simultaneity. The difference in the types of the taxis relations depends on the lexical-semantic composition of the predicates, but can be supported by such adverbial specifiers as "sometimes", "simultaneously", "while", etc. Shishkova, Nord distinguish the following structures of the complex-incomplete sentences:

(1) two or three subjects with an incomplete predicates. For example: Wir standen vor dem Tor, *sie im Hof.*

(2) Several subjects + one complete predicate + several incomplete predicates: Der Knabe *schien ruhig*, das Mädchen *erregt*. Sehr *hell war* der Himmel, *silbern* sein Licht [15, 16].

At present, along with the full sentence information, there is a language economy tendency that is expressed in the use of the incomplete structure sentences: the parts of a sentence which meaning is understandable from the context can be omitted and this is usual for both the principal and subordinane parts of the sentence. In this case we are talking about an elliptical sentence – syntactic unity, where there is no part of the structural sentence scheme, which semantic completeness is clear from the previous context.

The complex-incomplete sentences are functioning in the form of the following models:

Model 1: the subject + the nominal part of the compound predicate. For example: Viele Köpfe, viele Sinne. Both parts are expressed by a noun.

Model 2: the subject + the nominal part of the compound predicate but both parts are expressed by participles. Gesagt – getan.

Model 3: the subject + the nominal part of the compound predicate. The first part is expressed by a noun or pronoun, the second – by an adjective, an adverb, a participle. For example: (Sprenggranaten krachten in die Keller.) Flammen. Einstürzende Häuser. Klirrende Panzerketten überall. Panik. Regellose Flucht.

Model 4: subject + part of the predicative group (adverbial, part of the verbal predicate). Keine Seele weit und breit. Dort an der Ecke ein Auto. Wer da? Was dann?

Model 5: subject + negation from the predicative group. "Ich weiß es schon, erwiderte sie. "So? Ich nicht!"
Model 6: The question-response unity:

(a) in the form of interrogative words:

− Der vierte August. Mein Lieber, Sie haben über ein Jahr geschlafen.
− Was?

(b) in the form of verbal expression: − Sie sollten nicht so viel reden. Also: Was hatten Sie da zu suchen? Reden Sie!

− Nichts Schlimmes, sagte ich mühsam. − Eine Zaubervorstellung.
− Was bitte?
− Eine Zaubervorstellung.

(c) one subordinate part of the sentence that asks the question:

− Ein berühmter Zauberer, vielleicht kennen Sie ihn. Jan van Rode.
− Sicher.

(d) verbal predicate + subordinate part of the sentence (subject is omitted): Suche Kleinkühlschrank. Kaufe Schrankwand. Biete Zweizimmerwohnung.
(e) sentence division into smaller parts: Sie verlassen den amerikanischen Sektor. Das geschlossene Tor des jüdischen Friedhofs in Ostberlin. Ein Lokal mit Musik aus den vierziger Jahren.
(f) greeting/farewell formulas: Auf baldiges Wiedersehen! Auf Ihr Wohl! Herzlichsten Glückwunsch!

In the process of teaching to complex-incomplete and elliptical sentences traslation, in our opinion, it is necessary to train engineering students to determine the above-mentioned models of these sentences as it will allow them to reach the optimal translation solution. Student translational skills will be formed from the certain actions and operations.

Such actions and operations nature is mainly hypothetical, it consists of the results of comparing the source texts and the translation. The translator's mental actions occur beyond the threshold of his or her consciousness, they are not available for direct observation [14], just as the processes of recognition, perception and understanding. These aspects are identified rather conditionally [4, 10, 17].

Speech processes are not separable in time. They are realized at each moment simultaneously and are interrelated [4, 8, 13, 17]. However, in our opinion, when translating original foreign language texts, the perception is primary related to understanding, because under the translation activity conditions the decoding processes of the text subject content and the language semantics turn to be divided in time.

In this case, the engineering students learn to use the new language means in order to refer to the facts already known to them and to the objective world phenomena.

We analyzed the psychological investigations on the general theory of recognition and perception, some of which were further used as a basic material for our study [4, 13, 17].

Text perception is subjected essentially to the same rules as other kinds of perception and by that we mean the process of retrieving the transmitted content by the sender in the mind of the recipient [5]. The communicative norms of the recipient are strongly influenced by this process. The recognition occurs during the perception process if the facts are known to the translator.

Recognition of grammatical phenomenon in the text is subject to the basic laws of reality objects and phenomenon recognition. However it has some specific features associated with both the written speech perception in general and with the writing perception in the foreign language in particular [1].

Being a mental process, recognition involves: identifying the necessary distinctive features in the perceived object, updating the reference in the students' memory, comparing the perceived object with the actualized standards and assigning it to a certain category. Recognizing the perceived word is the process of comparing it to the standard that is stored in the students' long-term memory, and finally choosing the appropriate meaning. The result of recognition is the determination of the word meaning in the connection with the context where this word is perceived [4, 10, 17].

Grammatical phenomenon recognition in a translated foreign language text can proceed in different ways conditioned by its application. Recognition can be simultaneous and discursive depending on the student' formed skill.

When perceiving grammatical material for the first time, engineering students first actualize their knowledge about one particular grammatical category and then, according to this knowledge, the grammatical fact recognition or non-recognition could occur. In the secondary perception, engineering students search for peculiarities of the grammatical fact that are characteristic of this grammar category. During the initial grammatical material perception there could be a "presumptive\supposed" recognition (the term is given by Dodonov [7]). At this stage, the students appeal to their experience, with no theoretical knowledge involved, which manifests itself in the form of so called "language intuition." At the next stage, the grammatical analysis methods are directly based on the students' theoretical knowledge. Secondary grammatical phenomena perception, as a consequence, turns into so-called "certifying" recognition. This type of grammatical phenomenon recognition is possible when the translator's memory has a combination of morphological and syntactic levels. The grammatical material recognition is accompanied by a comparison of the incoming information with the predicted one on the basis of the accumulated variants of the combination schemes (phrase stereotypes).

In this regard we mean teaching grammar combined with teaching vocabulary.

Grammatical phenomena are not artificially filled structures, they, of course, are lexico-grammatical blocks. In the process of translating original foreign non-literary texts, lexical and grammatical phenomena operations recognition occur as an interrelated process. Each operation for translator is a special purposeful process with its own aim. In this case, we provide the translator's training by fulfilling each operation concerned with the unique corresponding task. These are the following operations:

- the phenomenon is determined out of other lexical and grammatical phenomena;
- distinctive features of the perceived phenomenon are defined;
- grammatical phenomenon memory actualization;
- comparison of the perceived signs with the actualized ones;
- grammatical homonyms or grammatical phenomena having similar components are discerned;
- identification.

If lexical and grammatical unit recognition has taken place, then it is possible to provide the following foreign text semantic understanding and translation.

In our work we propose to apply the developed model by taking into account the peculiarities of perception and recognition of the elliptical and complex-incomplete foreign language sentences. This model consists of the translation activity stages, including certain actions and operations.

Step 1. Theoretical positions synthesis of the studied subject.

Activity 1.1. Terminology knowledge generalization. The goal is to consolidate the studied theoretical material.

Activity 1.2. Fill in the gaps in the text with the words from the list.

The goal is to draw students' attention to the complex-incomplete sentences and elliptical model structure. The first activity consists of operations on the restoring information, translating sentences according to the given samples from the foreign language into Russian.

Activity 1.3. Test tasks accomplishment. This activity includes choosing the right variant and translating sentences from the foreign language into Russian.

Step 2. Foreign sentence translation analysis.

Activity 2.1. Work with original non-literary text. The goal is a consistent complex-incomplete and elliptical sentence translation, taking into account the features of their perception and recognition. Operations: divide sentences into simple, complex-incomplete and elliptical ones, determine their features.

Activity 2.2. Determination of the studied sentences structure. The goal is to teach students to apply their theoretical knowledge directly when working with original non-literary text. Operations: definition of the complex sentence and the elliptic model structure with the subsequent translation.

Activity 2.3. Sentence analysis according to the proposed scheme. The goal is to systemize previously studied activities. Operations: conducting the given sentence schematic analysis, comparing engineering students translations with the professional ones followed by mistakes analysis.

Step 3: Unassisted translation of the original non-literary texts from Russian into the foreign language.

The goal is to translate the text on the already studied topics. Operations: consciously controlled written translation operations, the result is the student's translated text.

3 Results and Discussion

During the 2016–2017, experimental training was performed, the purpose was to prove the training methods and selected experimental materials. Experiment pre-requisites:

- purpose: methodical;
- content: search-and-test;
- organization: natural, because it was conducted with the Tomsk Polytechnic University engineering students under normal teaching conditions.

The experiment was performed by comparing the results of the pre- and post-experimental testing in the experimental group.

The purpose of the initial testing was to determine the level of translating skills among the first year engineering students. The students were given the original text in the foreign language that had to be translated into Russian. The text consisted of 15 sentences, containing complex syntactic constructions. Students' works were analyzed in terms of quantitative and qualitative criteria. Quantitative criterion: the volume of the grammatically correct sentences. Qualitative criterion: the completeness of the original non-literary text information (determined by the teacher) (Table 1).

Table 1. Skills level.

TS	Skills development level
$5 < TS \leq 15$	High
$1,5 \leq TS \leq 5$	Medium
$1 \leq TS < 1,5$	Low

The translation skills level (quantitative criterion) was determined by the following formula:

$$TS = \frac{15}{n},$$

where
TS – translation skills
15 – number of the original text sentences
n – number of correctly translated sentences

The students initial testing results showed their low translation skills level: the certain vocabulary absence, the culturally-labeled vocabulary inadequate translating methods and skills, etc. However the biggest difficulty was the translation of syntactically complex sentences, in particularly complex-incomplete and elliptical ones.

Having assumed that the level of the translating skills of the complex-incomplete and elliptical sentences among engineering students would increase, we applied the developed model with further experimental training. During the students training we took into account the features of the complex-incomplete and elliptical sentences, their structure and models, the sequence of the structural-semantic and semantic connection between the individual parts of the sentence, their unification into one syntactic unity.

The exercises were based on original non-literary texts that clearly illustrated the basic syntactic features of the German language, in accordance with the developed stages. These are the following exercises:

- synthesis of the theoretical positions on the separately studied material. All grammatical material is divided into separate groups; it is necessary to choose the appropriate title;
- terms revision. Students were given the glossary of syntactic concepts, they had to study and complete it with their own examples, arrange in alphabetical order and finally add the missing terms (Table 2).

Table 2. Sample task.

Topic: complex-incomplete and elliptical sentences	Examples	Additional relevant terms
• Der elliptische Satz (eng.: the elliptical sentence)		
• Das Subjekt-Prädikativ Verhältnis (eng.: the subject-predicate relation)		
• Der zusammengezogene Satz (eng.: the complex-incomplete sentence)		

- Making up their own glossary on a new topic: it's necessary to study the grammatical material and compose the glossary of necessary terms;
- grammatical terms translation (group work);
- game "Guess the term";
- gap filling (Table 3).

Table 3. Gap filling example.

Grundlagen des theoretischen Stoffs zu dem Thema "Zusammengezogene und elliptische Sätze" (eng.: The main theoretical positions on the topic «Complex-incomplete and elliptical foreign language sentences». Then there is the task in the target language.)
➢　　　entsteht durch __(1)_ zweier einfacher Sätze;
➢　　　sind zwei oder mehrere __(2)_ vorhanden;
➢　　　nur __(3)_. Hier sind zwei Fälle:
a) das Prädikat wird für alle Teile des Satzes __(4)_ gegeben (z.B. Wir standen vor dem Tor, *sie im Hof.*);
b) das Prädikat wird nur einmal vollständig gegeben, in den anderen Teilen fehlt __(5)_ (z.B. Der Knabe *schien ruhig*, das Mädchen *erregt*. Sehr *hell war* der Himmel, *silbern* sein Licht.)...

– Test tasks (*I step*) (Table 4).

Table 4. Test task example.

This table shows a part of the test material in the target language.
Test 1.
1. "Das weiß man nicht. Kanada, Karlsruhe, Kirgisien, Köln. Ich habe keine Ahnung." Bestimmen Sie das Modell von Sätzen.
    a) die Parzellierung   b) verblose Sätze
    c) subjektlose Sätze
    2. "Viertel vor elf, genau." Bestimmen Sie die Unterart einer Frage-Antwort-Einheit.
    a) die Ellipse in der Form eines Fragewortes
    b) die Ellipse in der Form von Formeln
    c) die Ellipse in der Form eines erfragten Satzgliedes
    3. "Die Gewerkschaftsfunktionäre. Sollen hereinkommen!" Das sind ___.
    a) zwei subjektlose Sätze
    b) zwei parzellierte Sätze
    c) Subjekt - Prädikativ – Sätze ...
    usw. (etc.)

The second step is textual, that is, all the given grammar exercises are text-oriented. In our case, it is the original non-literary text. Practical exercises are the following:

– read the text, choose simple sentences, complex-incomplete and elliptical ones, arrange them into groups (worksheet 1) (Table 5).

Table 5. Worksheet sample 1.

Sätze, №. (eng.: Sentences, №.)	Der Satz ist ... (The sentence is ...)		
	einfach (eng.: simple)	elliptisch (elliptical)	zusammengezogen (complex-incomplete)
1			
2			
3			
usw. (etc.)			

– Analyze the complex sentence structure. Translate them into Russian (worksheet 2) (Table 6).
– Determine the elliptical sentence model, identify the relevant features. Translate them into Russian (Table 7).

Table 6. Worksheet sample 2.

Sentence	Structure	Structure analysis and translation
1. Das weiß man nicht. Kanada, Karlsruhe, Kirgisien, Köln. Ich habe keine Ahnung. 2. ...usw. (etc.)		

Table 7. Worksheet sample 3.

Model №.	Sentence	Translation
1		
2		
3		
4		
5		
6		
7		
8		

Table 8. The sentence analysis scheme № 1.

Das Schema "Analyse von einfachen, elliptischen und zusammengezogenen Sätzen".

– Bestimmen Sie eine Satzart:

a) nach der Struktur (einfach / zusammengezogen / elliptisch);

b) nach der Zieleinstellung (ein Aussage-, Frage-, Aufforderungs- oder Ausrufesatz) und eine Unterart bei dem Fragesatz (eine Entscheidungs-, Ergänzungs-, Doppel-, Bestätigungsfrage oder eine rhetorische Frage);

c) nach dem Bau (erweitert oder unerweitert), eingliedrig / zweigliedrig;

d) nach der Art des Subjekts (persönlich / unbestimmt-persönlich / unpersönlich);

– Bestimmen Sie, ob der Satz (die Sätze) bejahend oder verneinend (wodurch ist die Negation ausgedrückt) ist (sind).

– Bestimmen Sie, wodurch die Modalität des Satzes ausgedrückt ist (Modalverben; Verben mit modaler Schattierung; Modalwörter; drei Modusformen des Verbes).

– Bestimmen Sie das Satzmodell (für elliptische Sätze).

– Geben Sie die Charakteristik der **Hauptglieder** des Satzes (der Sätze):

a)   des Subjekts (wodurch ist das Subjekt ausgedrückt);

b)   des Prädikats (bestimmen Sie die Art des Prädikats: das einfache verbale, das zusammengesetzte verbale, das nominale, das verbal-nominale, das phraseologische; die Bestandteile des Prädikats).

Eng.: A scheme for analyzing simple, elliptical and complex-incomplete sentence.

- Determine the sentence type:
a) by the structure (simple, elliptic and complex-incomplete);
b) according to the purpose (interrogative sentence sub-type);
c) by the sentence construction (extended / unextended, one-member / two member sentences);
d) by the subject type (personal / impersonal / indefinitely personal).

- Determine whether the sentence is affirmative or negative (what expresses negation).
- Determine the sentence modality.
- Determine the sentence model (elliptical).
- Descrbe the main members of the sentence:
a) subject (what it is expressed by);
b) predicate (the predicate type, the predicate part).

etc.

- Study the elliptical sentence scheme. Make up the sentence scheme. Translate into Russian (Table 8).
- Translate the text passage, compare it with the original one, identify mistakes and analyze them.

The **third step** exercises are translational. Their implementation requires students to know the previous material. Typically, this is a translation of a small original non-literary text, a comparison of the students' translated texts, their analysis and finally the choice of the correct optimal variant.

In order to analyze the students' results after the experimental training, a post-experimental test was performed, that was done the same way as the pre-experimental one. The quantitative criteria analysis of the two groups, i.e. post- and pre-experimental testing, showed the positive trend. There were also qualitative changes in the students' works: the text content was translated better than before the experimental training.

4 Conclusion

Thus, according to the obtained results we could draw a positive conclusion about the suggested model feasibility and applicability. The results comparison of two experimental group testings has confirmed the proposed assumption and exercises efficiency. The goal of the experimental work was achieved.

The content of the student's translated text becomes more accurate if they are gradually taught to perform translational activities, in our case, complex-incomplete and elliptical sentences, taking into account the features of their recognition and perception. Analysis of the non-literary sentences translation allowed us to see and analyze mistakes in the translation process. The developed model and the obtained data application can significantly increase the translation skills level of the engineering students and, as a consequence, encourage them to work with the foreign scientific literature in the future.

Acknowledgements. We issued two educational and methodical papers on the results of the conducted experiment. They included the author's developed pre-translation and translation practical exercises. The authors express gratitude to the foreign language department of the Institute of Power Engineering, Tomsk Polytechnical University, for the help in performing our scientific experiment and also Tomsk State Pedagogical University publishing house for the free edition of the specified materials. Today these materials are used in the translation teaching to both humanitarian and engineering students.

References

1. Balastov, A.V., Sokolova, E.Y.: Adult learners' communicative foreign language competence development in highernschool via information technology and multimedia implementation. Mediterr. J. Soc. Sci. 6(1), 537–543 (2016)

2. Bernardo, A.M.: Perevodcheskaya kompetentsiya - razrabotka kontseptsii perevoda v lejptsigskoj shkole (in German) [Translation competence - the concept development in the Leipzig translation scientific school]. In: Fleischmann, E., Schmitt, P.A., Wotjak, G. (eds.) Conference 2001, LICTRA, pp. 43–49. Stauffenburg Verlag, Tübingen (2004)

3. Buran, A.L.: How to use blogs in creating special opportunities for language learning. Mediterr. J. Soc. Sci. **6**(1), 532–536 (2015)

4. Dakukina, T.A.: Formirovanie professional'noj mobil'nosti studentov v ramkakh «Evropejskogo» obrazovaniya» [Formation of students professional mobility in the framework of the "European" education]. In: International scientific video conference "Student youth in the context of globalization", pp. 37–39. TOV "Innovation", Dnepropetrovsk (2013). (in Russian)

5. Dakukina, T.A: CHtenie original'nykh tekstov s polnym i tochnym ponimaniem uchashhimisya starshej stupeni obucheniya srednikh shkol [Reading of original texts with full and exact understanding by students of the secondary schools senior stage]. Tomsk State Univ. Pedagogical Univ. Bull. **9**(137), 171–174. TSPU Publishers, Tomsk (2013). (in Russian)

6. Dalinger, V.A.: Problemy povysheniya kachestva podgotovki inzhenernykh kadrov [Problems of improving the engineering personnel training quality]. Fundam. Res. **9**, 55–56. Academy of Natural Science Publishing House, Penza (2005). (in Russian)

7. Dodonov, B.I.: Protsess kategorial'nogo uznavaniya grammaticheskogo materiala [The process of grammatical material categorical recognition]. Quest. Psychol. **2**, 157–168 (1959). (in Russian)

8. Ehlers, S.: Lesen als Verstehen [Reading as Understanding]. Langenscheidt, Berlin (1992). (in German)

9. Filippov, V.: Sovremennyye inzhenery dolzhny obladat' nadprofessional'nymi kompetentsiyami [Modern engineers should have overprofessional competencies]. (in Russian). https://www.ucheba.ru/article/2669

10. Gubareva, T.Y.: Psikholingvisticheskij analiz ponimaniya pis'mennogo teksta [Psycholinguistic analysis of the written text understanding]. Unpublished theses (1997). (in Russian)

11. Kade, O.: Perevod kak sotsial'nyj fenomen i predmet nauchnogo issledovaniya [Translation as a Social Phenomenon and Scientific Research Subject]. VEB Verlag Enzyklopadie, Leipzig (1980). (in German)

12. Kaplina, S.E.: Rol' inostrannogo yazyka v formirovanii professionally mobil'nosti budushchego spetsialista [The role of a foreign language in the future specialist professional mobility formation]. In: Foreign Languages in the Educational Space of a Technical College, pp. 146–150. YURSTU (NPI), Novocherkassk (2007). (in Russian)

13. Kobenko, Y.V., Tarasova, E.S.: Peculiarities of translation realionyms into German. Procedia – Soc. Behav. Sci. **206**, 3–7 (2015)

14. Latyshev, L.K.: Tekhnologiya perevoda [Translation technology]. Center "Academy", Moscow (2008). (in Russian)

15. Nord, C.: Text Analysis in Translator Training Teaching translation and Interpreting. Jukius Groos Verlag, Amsterdam (1992)

16. Shishkova, L.V.: Sintaksis sovremennogo nemetskogo yazyka [Modern German syntax]. Center "Academy", Moscow (2003). (in Russian)

17. Zimnyaya, I.A.: Psikhologiya obucheniya inostrannym yazykam v shkole [Psychology of teaching foreign languages in school]. Education, Moscow (1991). (in Russian)

Pedagogical and Testological Optimization Methods in University Foreign Language Training of Future Engineers

Maria V. Druzhinina$^{(\boxtimes)}$ (ID) and Natalya V. Stolyarova (ID)

Northern (Arctic) Federal University named after M.V. Lomonosov,
Archangelsk 1634002, Russian Federation
{m.druzhinina, n.stolyarova}@narfu.ru

Abstract. This article investigates the issue of the university language education optimization for future engineers. The pedagogical ideas that were chosen
for the study are the following: the educational potential of control means,
applied character of university language education and students'
self-development. These ideas are considered to be the methods that impact the
quality of language education and, therefore, professional education. The
necessity of the following testological ideas for successful educational practice
is proved: monitoring of educational process, a validity of test tasks, the training
and developing potential of tests and autonomy of students' work. The experimental data about influence of these interconnected methods on quality of the
university language education for future engineers is analyzed. The investigation
of pedagogical and testological methods confirms that harmonious interaction
between pedagogical and testological ideas provides a positive effect. This
outcome can be observed in transformation of testing from the controlling
procedure to the training and developing tool. The specific results obtained
during the experimental work performed with the engineering students are
presented: improved oral and written communication; increased quality and
quantity of creative products; formed ability to work independently and to
transfer the knowledge in new professional and language situations.

Keywords: Optimization methods · Language education of future engineers ·
Pedagogical and testological ideas · Professional communication · Creative
outcomes

1 Introduction

1.1 Problem Statement

The higher education system in universities of Russia undergoes the process of mod-
ernisation that is characterized by high requirements to the quality of teaching and
learning foreign languages for professional purposes. The large-scale internationalization of all scientific and educational processes of universities in the Russian Federation
makes optimization of university foreign language education extremely significant
especially regarding professional training of future engineers [11, 15, 17, 27].

© Springer International Publishing AG 2018
A. Filchenko and Z. Anikina (eds.), *Linguistic and Cultural Studies:*
Traditions and Innovations, Advances in Intelligent Systems and Computing 677,
DOI 10.1007/978-3-319-67843-6_18

1.2   Literature Review

We understand pedagogical methods as pedagogical ideas about education, control and development of personality in the process of teaching foreign languages. Under testological methods we mean the ideas of testological science about monitoring of educational process, about validity of the test tasks and about influence of testing on students' personality development. Pedagogical and testological methods are closely connected with each other and they are both improve and optimize language education.

In Russian pedagogics "optimization" is a scientific theory which allows to use pedagogical ideas and methods in the course of training. They are applied in the most efficient way for real situations to solve such problems as poor students' progress, irrational use of students' and teachers' time, intensification of educational process, overcoming formalism in assessment [2, 7, 28].

The following progressive ideas are offered in pedagogical and partially in psychological studies on university education: ensuring quality of educational process [32, 34]; self-realization of students personality [5, 26]; competence developing [19, 29]; necessity of close connection between educational process and future professional activity [8, 13].

The pedagogical ideas about university language education are determined by applied character of university language education in the forefront [12, 13, 18, 19, 35].

Pedagogical ideas on self-development are of vital importance for education at universities [5, 38].

There is a significant amount of research on such phenomena as self-realization of the personality, self-formation, self-updating, self-implementation, self-improvement, self-development in the field of psychology [1, 3, 23, 36]. In Russian pedagogical science these issues are also being developed, mainly through the processes of personality image formation, mature personality development, personal growth and self-development [5, 7, 20, 21, 35]. The studies on self-development are fully covered by Russian psychologists [4, 31].

Three groups of pedagogical ideas should be considered as the ones having impact on the quality of language education: (1) the educational potential of controlling means; (2) applied character of university language education; (3) students' self-development. As the majority of assessment is provided by tests, scientific investigations on problems of pedagogical measurements are performed [6, 10, 22, 24, 39].

Applied nature of university language education demands that testing tasks should correspond to the conditions of real daily and professional communication [25]. The pedagogical theory on self-development and self-realization of the personality correlates with the idea of using tests not only for control, but also for training and also for development of the students [5, 24]. In order to reveal the training character and develop the personality, testing should be organically included in naturally organized educational process [30]. It is highlighted that testing of students should be voluntary and independent [24]. Thus, for successful educational process the following testological ideas are significant: (1) monitoring of educational process; (2) testing tasks validity; (3) opportunity to work autonomously and developing students' personality through testing.

It should be noted that the systemized pedagogical and testological methods comply with the following ideas discussed by foreign scientists-testologists: the training testing held directly during educational process [37]; testing as means able to reveal students cognitive opportunities [33]; testing tasks simulating and appealing to the real communication and the use of original, authentic texts in testing tasks [9]; testing as assistance to students personal growth [16].

1.3 Basic Assumptions

In compliance with the research purpose and hypothesis of language testing optimization in university engineering programmes, the following issues are to be solved in this article: requirements to optimization methods and the analysis of experimental data.

The solution of these problems will confirm or disprove the hypothesis about the impact of the selected methods on quality of university language education and give an answer to the question, whether pedagogical and testological ideas are capable to influence the quality of future engineers' language education.

2 Materials and Methods

In 2009–2015 we conducted an experiment on integration of testing into educational process and use of tests for students' training and personal development was carried out.

To verify the hypothesis on the efficiency of testing in certain pedagogical conditions, we used the following research instruments: tests in foreign language for professional purposes, analysis of students' creative works.

We provided the experiment in university language education optimization by testing students during the period of 12/13, 13/14, 14/15 academic years. The main stage of experiment performed in 4 groups of students, where 2 groups were control groups (30 students), and 2 groups – experimental ones (30 students). The purpose of the experiment was to verify the hypothesis that pedagogical and testological methods could increase the quality of university language education of future engineers. The requirements to the testing materials applied in the process of study were the following:

(1) tests were used systematically for training with the obligatory subsequent analysis of students' performance results;
(2) tests were valid according to the professional contents, they corresponded to problems of engineering;
(3) tests allowed students to work independently and creatively, revealing their cognitive potential.

3 Research Data

Results of experiment showed that the level of language proficiency in the control group increased from A1 to A2, and in the experimental group – from A1 to B1 respectively (see Fig. 1). Here the scale of All-European competences of foreign language skills is used [14].

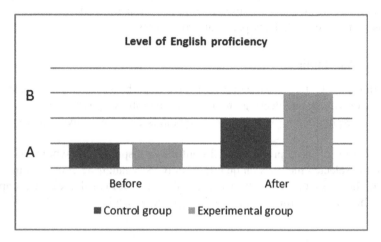

Fig. 1. Changing the level of students language proficiency after the experiment.

Quality of creative works (volume, content, literacy) in the control and experimental groups increased by 30% and 60% correspondingly (see Fig. 2).

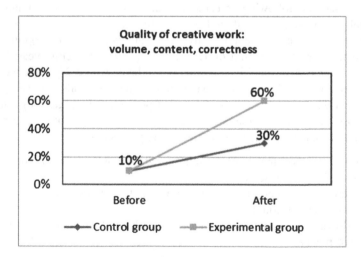

Fig. 2. Growth in the quality of students creative works after the experiment: volume, content, correctness.

While assessing self-realization of the students we considered different types of scientific and educational activities of students in the process of foreign languages learning: participation in the academic competitions in the English language; participation in debates and discussions, conferences; creative performances in concerts, preparing reports and public speaking, professional and language projects, reviews, presentations, etc. The level of students' self-realization in the control groups raised by 5 activities, i.e. 10%, and in the experimental one – by 20 activities, i.e. 40% (see Figs. 3 and 4). Figure 3 shows that students' self-realization in the control group grew by 10% after the experiment. Figure 4 highlights that students' self-realization in the experimental group grew by 40% after the experiment.

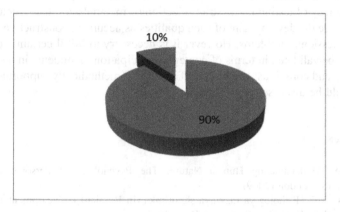

Fig. 3. 10% growth of students' self-realization in the control group.

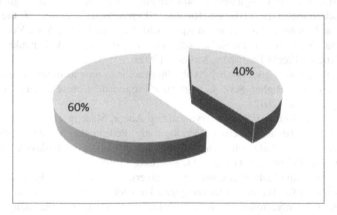

Fig. 4. 40% growth of students' self-realization in the experimental group.

Thus, during an experimental study it became clear that the interrelation and complementarity of the pedagogical and the testological methods are traced in testing. Their interaction and application in total becomes an optimization method of university foreign language education for engineering students.

4 Conclusion

The following pedagogical methods are applied in educational process at the university: pedagogical control of students educational achievements in learning foreign languages; attributing the applied and training character to language education for the professional purposes; personality self-realization in language education of future engineers. These pedagogical methods should be obligatory complemented by following testological methods: measurement of proficiency level at all stages of educational process; taking into account a substantial and statistical validity of testing tasks for engineering programmes of study; use of testing as means of learning autonomy and realization of future engineers' cognitive potential.

The accuracy of testing tasks, strict selection of materials and the structuring of the test can provide the development of such qualities as accuracy, constructiveness, ability to solve professional problems. However it is necessary to fulfill certain requirements: tests should be validated in terms of content; participation of students in testing should be voluntary and conscious; teaching staff should be methodically supported, results of testing should be discussed with students.

References

1. Adler, A.: Understanding Human Nature: The Psychology of Personality. Oneworld Publications, London (2009)
2. Alfyorov, A.D.: Psihologiya razvitiya shkol'nika (in Russian) [Developmental Psychology of a Schoolchild]. Feniks, Rostov-on-Don (2001)
3. Allport, G.: Pattern and growth in personality. Holt, Rinehart & Winston, New York (1961)
4. Ancyferova, L.I.: Psihologicheskoe soderzhanie fenomena sub"ekt i granicy sub"ekt-no-deyatel'nostnogo podhoda (in Russian) [Psychological Matter of the Phenomena 'Subject' and Limits of Subjecttaction Approach]. In: Brushlinskij, A.V., Volovikova, M.I., Druzhinina, V.N. (eds.) Problema sub"ekta v psihologicheskoj nauke, pp. 29–30. Akademicheskij Proekt Publishers, Moscow (2000)
5. Andreev, V. I.: Pedagogika vysshej shkoly. Innovacionno-prognosticheskij kurs (in Russian) [Pedagogy of the Higher School. Innovative-Prognostic Course]. Centr innovacionnyh tekhnologij, Kazan' (2012)
6. Avanesov, V.S.: Kompoziciya testovyh zadanij. Adept, Moscow (1998)
7. Babanskij, Y.K.: Izbrannye pedagogicheskie trudy. Pedagogika, Moscow (1989)
8. Balakireva, E.V.: Professiologicheskij podhod k pedagogicheskomu obrazovaniyu. Izdatel'stvo RGPU im. A.I. Gercena, St. Peterburg (2008)
9. Buck, G.: Assessing Listening. Cambridge University Press, Cambridge (2005)
10. Chelyshkova, M.B.: Teoretiko-metodologicheskie i tekhnologicheskie osnovy adaptivnogo testirovaniya v obrazovanii. Dis … dokt. ped. nauk (in Russian) [Theoretical-Methodological and Technological Basics of Adaptive Assessments in Education. Theses abstract]. Institut stali i splavov, Moskva (2001)
11. Chuchalin, A., Boev, O.: Trebovaniya k kompetenciyam vypusknikov inzhenernyh programm. Vysshee obrazovanie v Rossii 9, 25–29 (2007)
12. Dmitrenko, T.A.: Professional'no-orientirovannye tekhnologii obucheniya. (in Russian) [Professionally-oriented learning technologies]. Prometej, Moscow (2009)

13. Druzhinina, M.V.: Formirovanie yazykovoj obrazovatel'noj politiki universiteta kak faktora obespecheniya kachestva professional'noj podgotovki sovremennyh specialistov (in Russian) [Formation of University Language Educational Policy as a Factor of Quality Assurance of Modern Specialists' Professional Training]. Pomorskij universitet, Arhangel'sk (2007)

14. European Language Portfolio (2002). http://www.coe.int/t/dg4/education/elp/elpreg/Source/assessement_grid/assessment_grid_russian.pdf. Accessed 24 Nov 2015

15. Fedorov, I., Medvedev, V.: Tradicii i innovacii v podgotovke inzhenernyh kadrov (in Russian) [Traditions and Innovations in Training of Engineering Personnel]. Vysshee obrazovanie v Rossii **6**, 30–35 (2008)

16. Hughes, A.: Testing for Language Teachers. Cambridge University Press, Cambridge (2003)

17. Kirsanov, A., Ivanov, V., Kondrat'ev, V., Gur'e, L.: Inzhenernoe obrazovanie, inzhenernaya pedagogika, inzhenernaya deyatel'nost' (in Russian) [Engineering Education, Engineering Pedagogy, Engineering Activity]. Vysshee obrazovanie v Rossii **6**, 37–40 (2008)

18. Krupchenko, A.K., Kuznetsov, A.N.: Genezis i principy professional'noj lingvodidaktiki (in Russian) [Genesis and Principles of Professional Language Pedagogy]. APKiPPRO, Moscow (2011)

19. Kuznetzov Y.E.: Professional'no-yazykovaya kompetentnost' kak faktor uspeshnosti uchebnoj deyatel'nosti studentov tekhnicheskogo vuza. Dis. …kand. ped. nauk (in Russian) [Professional-Language Competence as a Factor of Training Activity Success of Technical Institution of Higher Education Students. Theses abstract]. Sibirskij gosudarstvennyj tekhnologicheskij universitet, Krasnoyarsk (2005)

20. Lednev, V.S.: Soderzhanie obrazovaniya: sushchnost', struktura, perspektivy. (in Russian) [Content of Education: Essence, Structure, Perspectives]. Vysshaya shkola, Moscow (1991)

21. Lerner, I.Y.: Razvivayushchee obuchenie s didakticheskih pozicij (in Russian) [Developing Training from the Point of Didactics]. Pedagogika **2**, 7–11 (1999)

22. Majorov, A.N.: Teoriya i praktika sozdaniya testov dlya sistemy obrazovaniya (in Russian) [Theory and Practice of Production of Tests for Educational System]. Intellekt-centr, Moscow (2001)

23. Maslow, A.H.: Towards a psychology of being. D. Van Nostrand Company, Princeton (1962)

24. Mil'rud, R.P., Matienko, A.V.: Yazykovoj test: problemy pedagogicheskih izmerenij (in Russian) [Language Test: Problems of Pedagogical Investigations]. Inostrannye yazyki v shkole **5**, 7–13 (2006)

25. Norejko, L.N.: Sovremennoe lingvodidakticheskoe testirovanie v svete kommunikativnoj teorii (in Russian) [Modern Language Pedagogical Testing in View of Communicative Theory] (1998). www.testor.ru/files/Conferens/probl_ur…/JLH.doc. Accessed 04 Oct 2015

26. Petrovskij, A.V.: Lichnost' v psihologii (in Russian) [Personality in Psychology]. Feniks, Rostov-on-Don (1996)

27. Poholkov, Y.P., Agranovich, B.L.: Podhody k formirovaniyu nacional'noj doktriny inzhenernogo obrazovaniya Rossii v usloviyah novoj industrializacii: problemy, celi, vyzovy (in Russian) [Approaches to Formation of National Doctrine of Engineering Training in Russia in the Conditions of New Industrialization: Problems, Goals, Challenges]. Inzhenernoe obrazovanie **9**, 5–11 (2012)

28. Polyakova, T.S.: Analiz zatrudnenij v pedagogicheskoj deyatel'nosti nachinayushchih uchitelej (in Russian) [Analysis of Challenges in Pedagogical Activity of Teacher-Beginners]. Pedagogika, Moscow (1983)

29. Radionova, N.F.: Razvivayushchee i razvivayushcheesya vzaimodejstvie prepodavatelej i studentov v usloviyah sovremennogo pedagogicheskogo vuza (in Russian) [Developmental and Developing Interaction of Teachers and Students in the Conditions of Modern Pedagogical Institute of Higher Education]. In: Druzhinina, M.V., Sokolova, M.L. (eds.) Mirovidenie, pp. 64–73. Pomorskij universitet, Arhangel'sk (2004)

30. Richards, J.C., Renandya, W.A.: Methodology in Language Teaching An Anthology of Current Practice. Cambridge University Press, Cambridge (2002)

31. Rubinshtejn, S.L.: Bytie i soznanie. Chelovek i mir (in Russian) [Being and Consciousness. Human Being and World]. Piter, St. Peterburg (2003)

32. Selezneva, N.A.: Kachestvo vysshego obrazovaniya kak ob"ekt sistemnogo issledovaniya (in Russian) [Quality of Higher Education as an Object of Systematical Investigation]. Issledovatel'skij centr problem kachestva podgotovki specialistov, Moscow (2003)

33. Shohamy, E.: Alternative assessment in language testing: applying a multiplism approach. In: Li, E., James, G. (eds.) Testing and Evaluation in Second Language Education, pp. 9–11. Language Center, Hong Kong University of Science and Technology, Hong Kong (1998)

34. Subetto, A.I.: Ot kvalimetrii cheloveka – k kvalimetrii obrazovaniya (genezis) (in Russian) [From Qualimetry of a Person to Qualimetry of Education (genesis)]. Issledovatel'skij centr problem kachestva podgotovki specialistov, Moscow (1993)

35. Verbickij, A.A., Dubovickaya, T.D.: Konteksty soderzhaniya obrazovaniya (in Russian) [Contexts of Education Content]. Al'fa, Moscow (2003)

36. White, R.W.: Motivation reconsidered: The concept of competence. Psychol. Rev. **66**(5), 297–333 (1959)

37. Wiggins, G.: The case for authentic assessment. Pract. Assess. Res. Eval. **2**(2) (1990). http://pareonline.net/getvn.asp?v=2&n=2. Accessed 30 May 2017

38. Zagvyazinskij, V.I.: Strategicheskie orientiry i real'naya politika razvitiya obrazovaniya (in Russian) [Strategical Guidelines and Real Politics of Education Development]. Pedagogika **6**, 10–14 (2005)

39. Zvonnikov, V.I.: Izmereniya i shkalirovanie v obrazovanii (in Russian) [Measurements and Scaling in Education]. Universitetskaya kniga, Mocow (2006)

Cloud Platform SmartCAT in Teaching Future Translators

Marina A. Ivleva[✉] and Elena A. Melekhina

Novosibirsk State Technical University,
Novosibirsk 630072, Russian Federation
{m.ivleva,melexina}@corp.nstu.ru

Abstract. Modern translators face the necessity to cope with substantial amount of translation done at a shorter time than ever. Computer assisted translation (CAT) tools have been successfully used for this purpose. Therefore future translators should possess the skills of CAT tools application. We argue that translation internship, which students have during their final year, can serve a platform to master these tools. We compare the students' performance without and with the use of SmartCAT during the internship. The advantages of this tool application are evident, since the number of terminological inaccuracies, lexical and grammar mistakes have significantly decreased, the marking-up process became easier and more transparent. Students aptly used the theoretical and practical knowledge in fulfilling the task and self-analysis. Both students and supervisors report positive attitude towards implementation of SmartCAT. Considering the prevailing beneficial reviews we conclude that the skills in CAT tools should become an obligatory part in future translators' curriculum.

Keywords: CAT tools · Translators training · Translation internship · Cloud platform SmartCAT

1 Introduction

Teaching translation is a complex process involving instructing learners how to master a foreign language, expand cultural background and develop special skills of a translator. These skills are diversified and according to an experienced translator and author of books on translation [7, p. 18] include 'cultural understanding, project management, language and literacy, making decisions, communication and information technology'. The acquisition of all skills except the last one have traditionally been in the focus of translators' instruction, but the rapid development of information technologies, which have become inherent to digitally native students, stimulate teachers to widely use them in practice. The information technologies provide quick access to electronic dictionaries and encyclopedias as well as to specialized software for computer-assisted translation (CAT), which according to Jaworski [5] 'is a term used to describe computer techniques used to facilitate the process of translation'. That is why future translators need to learn about the possibilities CAT offers for their profession during their study at the university. The necessity of CAT tools involvement has been argued by many authors, among them Bowker [2], Jekat and Massey [6] and Christensen and

© Springer International Publishing AG 2018
A. Filchenko and Z. Anikina (eds.), *Linguistic and Cultural Studies:*
Traditions and Innovations, Advances in Intelligent Systems and Computing 677,
DOI 10.1007/978-3-319-67843-6_19

Schjoldager [3]. There is a report of a successful CAT tool application in the context of higher education by Serpil [8].

2 Research Question

2.1 Educational Context

The educational context for this research is the translator training program at Novosibirsk State Technical University. The program provides among other courses a substantial amount of supervised translation internship, which is in line with the opinion of Gurding-Salas [4] saying that translator should be exposed to continuous training because the quality of translation can become the matter of life and death. The students are assumed to choose between oral and written specialization in translation and the majority prefers to practice in written translation because oral translation is rarely demanded, especially that of undergraduates. In 2015, 20 students had their internship in several departments of our university and local translation agencies.

As translation internship supervisors we received immediate feedback from the students learning about theirs strengths and drawbacks. One of such lacunas in their training pointed out by the translation agency was the inability to confidently use CAT tools. Indeed, students mostly preferred to rely heavily on open access machine translation (MT) resources such as Google Translate, with little or no post-editing. The evidence of this was received during the final stage of the internship when they had to present the results of their translation and a reflective essay about the experience. Students with low motivation or those who combined studies with part-time jobs to earn money for education did not even bother to post-edit MT, which required overall proofreading of the translated material. Another group, that completed the assignment more successfully, still was reluctant to understand in detail the terminology (Sociology) they were translating which resulted in certain amount of lexical mistakes in the translation. Only a few students managed to cope with the translation tasks without serious mistakes.

Analyzing the feedback from students and their supervisors we decided to change the format of the internship and ensure that our students are better positioned in translation market by introducing obligatory use of CAT tools during such practice and allowing them to select the texts in field of knowledge they want to translate or to join in the translation of the suggested texts. The final outcome of this internship should contain among other documents a part of translation presented in a form of a table with source text (ST), target text (TT) and commentary on the techniques employed during translation.

Our hypothesis was that the use of CAT tools during internship would yield better results not only from the point of view of translation quality but would also encourage students to actively use the theoretical and practical knowledge they received during classes and critically analyze their performance. So, this research sought to answer the following research question:

How could the use of CAT tools during internship contribute to developing written translation skills of future translators?

2.2   Methodology

The participants were 30 fourth year students majoring in translation. They all received a theoretical and practical training in written translation and were instructed by teachers how to use CAT tools before internship in accordance with suggestions made by Bowker [2]. The feedback data were collected through semi-structured interviews of the participants, reflective essays and internship reports' analysis.

As a CAT tool we suggested cloud platform SmartCAT which was chosen because it is free for both freelancers and administrators and because it combines Translation Memory (TM), MT and terminology management systems (TMS). Though SmartCAT is not widely accepted among language service providers, the basic principles of CAT tools can be easily understood and implemented in the further work of our graduates.

SmartCAT, being a product of Russia based company ABBYY Language Services, incorporates the TM, MT and TMS with one of the most reliable dictionaries ABBYY Lingvo and concordance search within ST and TT. It also allows working with different text formats and provides OCR services which enable processing such files as PDF, JPEG and the like. The portal contains numerous other services for freelance translators – webinars, blogs and job proposals.

During the classes students were given full information about the possibilities to use this cloud platform and told that it is most efficient for the translation of non-fiction texts, which might contain certain repetitive fragments. SmartCAT with all its applications speeds up the translation process since the translator actually conducts post-editing and MT errors correction. Another advantage of this platform is that translator does not need to bother about text formatting – all the peculiar text features such as color, hyperlinks, types or fonts are automatically converted with the help of special tags.

As soon as a new project is started and the file for translation is chosen, the platform splits it into fragments, usually ended either by punctuation marks (except comma) or end of paragraph mark.

The internship required certain preparation connected with the registration of the involved students as Assignees and internship supervisors as Administrators. This allowed the supervisors to assign certain tasks to the students, control their Glossary suggestions and monitor the time and amount of the work done. The transparency of this tool motivated students interaction while working under the same text (chatroom is another helpful option of the portal) and necessitated timely completion of the assigned task in order not to let down the translators' team.

Coming across the terms students suggested their inclusion into the Glossary, however this has never occurred automatically because the Administrator was to approve of the term and its translation. Though it slowed the process to a certain extent, only the validated terminology was consistently used throughout the final translation variant.

3   Results and Discussion

The internship results showed that the adjusted structure improved the translation quality, since the form of the table containing ST, TT and commentary enables the students and supervisors to clearly see if the main task of translation, i.e. equivalence between the ST and TT was achieved. The necessary commentary part made students reflect upon their professional actions and analyze them according to the theoretical and practical issues discussed during classes of translation. Even students with low motivation to study managed to achieve acceptable results. Generally, the number and quality of grammar and lexical mistakes became lower and the majority of students managed to do all the assigned tasks in time.

Moreover, the quality of the translated text in terms of structure and cohesion became higher, which demonstrated that students' text editing skills improved. The insufficient development of editing skills is often reported by supervisors from translation agencies. That is why, we instructed the students on how CAT can assist in text editing, because we share the idea of Bell [1, p. 36] that a translator should possess 'the decoding skills of reading and encoding skills of writing' among other skills of a good communicator.

In the final reflective essay students positively assessed the experience of using SmartCAT stating that it gave them the possibility to do the work promptly and with enhanced quality, because the immediate correlation of the fragment and its MT highlights the mistakes of the latter. They also noted that the possibility to download the results of the translation in two forms (translation only and bilingual table) made the reporting procedure easier. The following excerpts from their essays may illustrate the general pattern:

Student A

"Generally I assess the translation internship very positively and believe that I gained useful experience which will make me a better professional in the future".

Student B

"After the completion of all the practical tasks I feel confident that I will be able to meet the requirements of the profession".

Student C

"The intuitive interface of SmartCAT alongside other helpful resources of the platform enabled me to fulfil the assigned task timely and on a high level".

The major difficulties reported by the students were the terms. This issue is noted nearly in every essay. Finding and validating the terminology is one of the most challenging translation tasks, so this problem was the expected one.

We conducted a discussion session after the internship where students were encouraged to state advantages and disadvantages of the CAT tool. The evaluation ranged from "Fun" to "OK" or "Acceptable". The advantages stated were the clear interface, enhancing the speed of translation, useful additional options which enabled quick correction of misused terminology and convenient conditions for team work on one text.

The results of semi-structured students' interviews analysis are represented in Fig. 1. We asked them the question: "How did SmartCAT help you in becoming a professional translator?"

The most often answers include gaining numerous skills in handling the CAT tool which students found useful in their future career. They also noted certain level of automation in MT translation post-editing they acquired in the process. Almost half of the students remarked acquiring practical understanding of the correlation between the SL and TL needed for the achievement of translation equivalence.

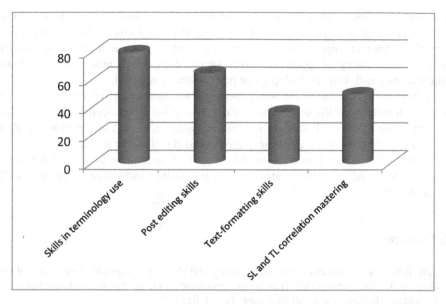

Fig. 1. The results of semi-structured students' interviews analysis.

The most heated discussion was about the fragmentation imperfection. Another questionable point was the interface of the platform: some students believed it could be more wisely organized.

The fragmentation issue, in fact, can easily be eliminated by the ST layout and text check. This option was not fully discussed prior to the internship which is another lesson for the supervisors to learn.

The discussion of the internship results among supervisors confirmed the positive assessment of the students. The use of CAT tools during the internship facilitated performance, command of translation tools and growing confidence among the students. Reporting procedure became more transparent. Among the advantages of the reformed internship the supervisors noted:

- easier marking process;
- students' mistakes shown more evidently;
- the decreased number of mistakes;
- properly applied syntactic transformations of the TT;
- terminological consistency;
- students' self-analysis and critique.

4   Conclusion

There are several implications of the findings in this study. First, the use of CAT tools in teaching translation enhances students' understanding of future profession and raises motivation through active involvement into the process of electronic text processing.

Moreover, it helps to improve students' editing and text-formatting skills by implementing grammar and spell check and providing the text format which requires little correction. Another implication is that by means of CAT tools learners managed to build up the glossary of specialized texts' terminology which made their translation more accurate and their internship more professionally relevant.

Although the results of the study demonstrate some positive dynamics in our students' translation skills development, there is one issue that needs to be taken into account while using CAT tools in teaching translation. First and foremost, teachers must be expert users of the tools they recommend the students to use.

There is also some limitation to the study that should be addressed for future research. We did not use statistics for measuring individual student's skills improvement.

References

1. Bell, R.T.: Translation and Translation: Theory and Practice. Longman, New York (1993)
2. Bowker, L.: Computer-Aided Translation: Translator Training. Routledge Encyclopedia of Translation Technology. Routledge, New York (2015)
3. Christensen, T.P., Schjoldager, A.: Computer-aided translation tools–the uptake and use by Danish translation service providers. J. Spec. Trans. 25, 89–105 (2016)
4. Geding-Salas, C.: Teaching translations – problems and solutions. Trans. J. 4 (2000). http://translationjournal.net/journal/13educ.htm. Accessed 11 May 2017
5. Jaworski, R.: Anubis - speeding up computer-aided translation. Comput. Linguist. – Appl. Stud. Comput. Intell. 458, 263–280 (2013)
6. Jekat, S., Massey. G.: The puzzle of translation skills. Towards an integration of e-learning and special concepts of computational linguistics into the training of future translators. Linguistik online 17(5) (2013). https://bop.unibe.ch/linguistik-online/article/view/785/1350. Accessed 18 May 2017
7. Samuelson-Brown, G.F.: A Practical Guide for Translators, 5th edn. Multilingual Matters Press, Clevedon (2010)
8. Serpil, H.: Employing computer-assisted translation tools to achieve terminology standardization in institutional translation: making a case for higher education. Procedia – Soc. Behav. Sci. 231, 76–78 (2016)

Main Features of Students' Research Competency Formation in a Liberal Arts Curriculum

Lidia A. Kazarina[1] , Irina A. Itsenko[1] , Olga I. Kachalova[2] ,
and Alexander V. Obskov[1(✉)]

[1] Tomsk Polytechnic University, Tomsk 634050, Russian Federation
{lidiak2003, i.icenko}@mail.ru,
alexanderobskov@hotmail.com
[2] The Russian Presidential Academy of National Economy and Public
Administration, Tomsk 634050, Russian Federation
kachoi@mail.ru

Abstract. There are main features of students' research competency formation in a Liberal Arts Curriculum. Researchers highlight the main features of that process: firstly, this research competency formation is based on the students' individual and collaborative research activities; secondly, the development of a reflexive attitude to research activity and its results; thirdly, results of positioning their own research experience in the educational and academic community. Basic forms of research activity organization at each stage are examined: students' creative associations, teachers' and students' problem-creative group, scientific and practical conferences. It is distinguished that focused efforts in forming of students' research competency in the classroom and after-hours activities contribute to students' successful implementation in professional self-determination. As a result, students can try their hand at applying research skills for various educational projects and their future professional activity.

Keywords: Subject-oriented curriculum · Research competency · Liberal education · Liberal arts curriculum · Pedagogical model

1 Introduction

1.1 Building a Research Problem

The introduction of subject-oriented instruction at the senior high school is considered as comprehensive means of improving the education quality and effectiveness. At the training stage, students take their course in professional activity, making a choice of further instruction to develop their personality [5, 6, 10, 14, 22, 23]. Students, as individuals of educational process, whose development and self-realization are considered to be the priority of pedagogical process, have the opportunity to acquire some special subjects in accordance with their training programs in selected subjects like Physics and Mathematics, Natural Sciences (Physics, Chemistry, Biology, and

A. Filchenko and Z. Anikina (eds.), *Linguistic and Cultural Studies:*
Traditions and Innovations, Advances in Intelligent Systems and Computing 677,
DOI 10.1007/978-3-319-67843-6_20

Geography), Socio-Economic, Liberal Arts, Philological, Information Technology, Agriculture Technology, industrial-technological, artistic and aesthetic, military courses and sports. Subject-specific training contributes to a conscious mastery of the necessary knowledge, skills and abilities; activates the students' creative potential; encourages active cognitive activity, and helps to acquire the profound knowledge and sustainable research skills.

It is traditionally believed that natural science areas in such subject-specific training like Physics, Chemistry, Biology, etc., are mostly oriented toward the organization of students' research competency. It allows organizing research procedures like observation, training experiments, laboratory work etc., in class and after-hour class activities. As for the Humanities, it is not always applicable to the student research fields. The specific study in liberal education is determined by that research objectives in Liberal Arts are different from those in natural scientific research. Thus, a special organization of student research activities is required. The distinction between the research characteristics in Natural Sciences and Liberal education is in the liberal knowledge specificity; makes it advisable to concentrate scientifically on the main features of students' research competency formation within the Liberal Arts curriculum. The research presents the main features of students' research competency formation in the Liberal Arts curriculum.

1.2 Literature Review

As for foreign researchers, certain aspects of this topic got covered in the works of the following researchers who studied the problems of students' independence and self-organization in high schools [2, 3]; modern educational concepts, including subject-specific training [7, 13, 21, 26, 28] competence approach to the organization of education in high school [9, 27]. Within the main features of students' research competency formation in a liberal arts curriculum framework, such subject-specific training aspects were extracted from such Russian researchers as [5, 6, 10, 14, 22, 23], organization of students' research competency [9, 12, 15, 24]; problems of students' research competency development [11, 18–20, 25, 29]; students' research competency formation [1, 25].

1.3 Basic Assumptions

The students' research competency in the Liberal Arts curriculum is understood as the integrating individuals' quality, which assumes readiness and ability to conduct productive research activities in the Liberal Arts. The structural and component composition of the students' research competency in the Liberal Arts curriculum in high school is defined as the aggregate of personal, cognitive and activity components represented by the groups of personal, cognitive, and activity-accented competences. Competencies, their totality provides the students' research competence in the Liberal Arts curriculum in high school, are defined as research competencies. Research competence is formed by mastering of research competencies [8]. The most effective research competency is formed within the framework of the respective pedagogical model of students' research competency formation in the Liberal Arts curriculum in high school; it includes the structured-systemic components like goals, tasks, approaches and principles, and functional-organized like research and development environment and conditions of organizing and functioning, researchers' scientific society and characteristics. The research functioning and development environment includes a set of stages and methods of the research competency forming, and forms of students' research activities organization.

2 Materials and Methods

According to the problem raised, the methodological base was a set of theoretical methods, included in pedagogical and psychological literature analysis, dissertational studies of the problem under discussion, modeling, comparison, generalization, and empirical ones included questionnaires, activity analysis like working in group or individual projects, peer reviews method; the analysis of the summary portfolio and essay. Any pedagogical model is characterized by periodicity, which is an expression of the model duration, i.e. its quantitative parameter, and the students' progress, their growth as a researcher. Following Kikot, the students' research competence formation is carried out in three stages [9]. The first stage is to force (form) the competencies necessary for students' individual and collaborative research activities. The second one is to force (form) a reflective attitude to one's own research activity and its results. The third one is to present the research experience results. Each stage is characterized by the purpose, content, and forms of collaborative activities organization. Selected stages in the research competency formation was based on high school students' psychological characteristics like the formation of stable cognitive and professional interests, the increased need for research activity; availability of content and forms of research activities organization for this age category; presence of personal changes, development of activity and operational abilities of students. The stages of students' research competency formation predetermine the collective system (students' creative associations, subject-specific courses, problem-creative group, and individual one (taking part in scientific conferences in schools and universities) forms of research activities organization. These forms are interconnected by the students' collaborative activity along with professors [16] and are integrated into the overall organizational structure - the Scientific Society of Researchers (SSR). The collaborative activity of students and professors is aimed at final result of the students' research competence formation in the Liberal Art curriculum in high school.

Each form of research activities organization is applied in certain steps. The most effective way occurs in conditions of interrelation of organizational forms used in strict sequence from the collective to individual. For instance, in a creative association or elective courses, students are involved in collective research activities, where each student initially receives an individual task to address a common problem that is solved collectively. In future, the individual work can be generalized as an abstract or report, and used to speak at school or university scientific and practical conferences, which enhances cognitive motivation and forms a positive attitude towards the chosen field of scientific knowledge. The key thing is that each student is included in the sphere of creative research. At this stage, a system of knowledge and ideas about the students' research competence are being formed. Inclusion of students in research activities in the problem-creative groups like role-playing, business games or debates, involves the mastery of collective methods and methods of carrying out research activity; facilitates the acquisition of experience of interpersonal and group interaction for collective decision-making. Participation in student scientific and practical conferences at school or university is the highest point of research methods implementation of, and is an individual organizational form mostly; however, collective forms of participation in conferences are also allowed. For example, group reports, group management of roundtables, and discussion of projects, etc. Undoubtedly, the forms of research activities are fixed at certain stages of students' research competency formation. At the first stage, the method of collective search is implemented, ensuring the student's involvement in research activities through interaction with other participants and acquisition of experience in its implementation (a group, then an individual research project) [17]. Some research "samples" are used in the collaborative activities (research and mini-research projects). A form of creative association, represented by a club or a subject circle, is organized. The number of organized clubs is not limited and depends on the range of students' interests in the Liberal Arts. At the same time, a compulsory modular course (twice a week) is carried out, supplementing and deepening the knowledge in the subject areas for which the circles/clubs work, consisting of four elective courses of Liberal Arts ("Research Activity Basics", "Linguistics", "Culture of Self-Knowledge", "Fundamentals of Design"). The first stage results are students developing interest in research activities and increasing motivation, and the formation of competencies necessary for individual and collaborative research activities. At the second stage, students often resort to different types and forms of research, it becomes important to reflect on their activities: "What has been done?", "What method of research has been used?", "What conclusions can be drawn?", "What has been confirmed, and has the hypothesis of the study been confirmed?", "What conclusions have been formulated?". At this stage, the form of problem-creative group is used, suggesting that students take part in the competition of written research, role-playing and business games, debates, try to apply the knowledge gained in various situations of research communication. Within the framework of the problem-creative group, the experience of interpersonal and group interaction is formed to solve common research problems. The result of interaction between subjects of the scientific community at this stage is the reflective attitude formed by students to their own research activities and its results. At the third stage, the method of individual research is used, involving the acquisition of students' own research experiences, the development of individual activity and initiative of beginner researchers. The students' research activity is organized in the form of scientific and practical conferences where a dialogical scheme of interaction between students as members of scientific community assumes the active position of a young researcher who tests the results of research activities in the form of report. The result of interaction among the scientific community researchers at this stage is positioning of their own research experience results. Thus, these stages during students' research competency formation reveal the logic of its development from collaborative research activities and collective research projects to independent research activities and results presentation in the academic and scientific community. The gradual formation of research competency is done using the system of collective and individual forms of research activity organization, which include (1) creative unification of students, (2) special courses, (3) problem-creative group, (4) scientific and practical conferences conducted as inside the school, and beyond its borders (in the university).

Since conditions of research organization and functioning, and development environment is the integration of lessons (research tasks, search terms, study experiment, and special course) and extracurricular activities like creative association, problem-creative group, conference and participation in SSR can integrate these forms of organization for a successful implementation of research activities.

3 Research Data

Pedagogical model approbation was carried out in High School No. 30 in Tomsk from 2010 to 2014. There were 96 students who took part in the research, and there were 11 high school teachers. In the process of organizing the pilot experimental verification of the model, two groups were formed: the control group (CG) and the experimental group (EG). There were 48 participants in CG who took part in the educational process where training was conducted in accordance with traditional programs, and there were 48 participants in EG where training was conducted in accordance with research competency formation. Experimental run for students' research competency formation included three stages: ascertaining, forming, and generalizing. Assessment of the pedagogical model effectiveness was carried out with the help of criteria, the level of students' competencies formation, included in the structure of research competency; the level of research activity of students; the focus on research activity in the context of the perspective of education in the university; and indicators of its effectiveness. There is an estimate of the model effectiveness based on the criterion "Level of students' research activity", for the results of the summary analysis of students' portfolio in EG and CG were used (see Table 1). The level of students' research activity (analysis of summary portfolio).

Table 1. The level of students' research activity (analysis of summary portfolio).

Index	Experimental group EG included 48 students	Control group CG included 48 students
Availability of certified documents that confirm individual educational achievements of students	83%	44%
Availability of works (creative, design, research)	88%	54%
Participation in conferences, competitions, elective courses, practices, etc.	100%	33%
Availability of positive feedback and written assessment of students' accomplishments	85%	48%

As can be seen from Table 1, a high level of research activity is observed mainly among students in the EG. From 2011 to 2014 students took part in 6 events of international and local level, 10 events of regional and municipal level, 3 school events, where they showed good results, were given awards and diplomas. Thus, the data of experimental testing of the pedagogical model according to the criterion "The level of students' research activity" shows an increase in the level of students' research activity at the summarizing-generalization stage.

4   Conclusion

The following can be attributed to the results of students' research competency formation in Liberal Arts curriculum: (1) positive dynamics of the level of research competencies development by students in the experimental group; (2) increasing the level of student research activity at the final generalizing stage; (3) development of internal motivation of students for research activities in the context of university education perspective. These results indicate that research activity becomes one of the means of implementing further educational and career steps for high school students. The presented main features allow effective organization of research activities and students' research competency forming within the liberal education.

References

1. Aleksandrova, N.A.: Razvitie issledovatelskikh kompetentsiy - uchashchikhsya sredstvami istorikorodoslovnogo kraevedeniya (in Russian) [Development of research competencies of students by means of historical and genealogical local lore]. Publishing house of the Institute of Social Pedagogy of RAS, Moscow (2011)
2. Baer, M., Guldimann, T., Kocher, M., Wyss, C.: Kognitive Aktivierung als Ausbildungsziel der Lehrer/innenbildung. Was zeigt den Blick in den Unterricht? Schulpraktika in der Lehrerbildung: Theoretische Grundlagen, Konzeptionen, Prozesse und Effekte. Waxmann, Münster (2014)
3. Helmke, A.: Forschung zur Lernwirksamkeit des Lehrerhandelns. Handbuch der Forschung zum Lehrerberuf. Waxmann, Münster (2014)
4. Helmke, A.: Unterrichtsqualität und Lehrerprofessionalität: Evaluation und Verbesserung des Unterrichts. Klett-Kallmeyer, Seelze (2009)
5. Khoraskina, R.I.: Profilnoe obuchenie kak sredstvo socializatsii uchashchikhsya gumani-tarnykh klassov obshcheobrazovatelnoyj shkoly (in Russian) [Subject training as a means of socializing students in the humanitarian classes of the general education school]. IGMA, Kazan (2010)
6. Idilova, I.S.: Profilnoe obuchenie kak faktor podgotovki shkolnikov k prodolzheniyu obrazovaniya v vuze na primere predmeta «Inostrannyi yazyk» (in Russian) [Profile education as a factor in the preparation of schoolchildren to continue their education in higher education: on the subject example of the "Foreign Language"]. Intermet, Ryazan (2007)
7. Jürgens, E.: Was ist «guter» Unterricht. Namhafte Experten und Expertinnen geben Antwort. Klinkhardt, Standop, Leipzig (2010)

8. Kazarina, L.A., Khasanshin, Y.R., Smyshlyaeva, L.G.: Teaching model of pupils' research competence formation in the context of humanitarian subject-oriented classes of general education school: functional and organizational characteristics. In: Kachalov, N. (eds.) Proceedings of the 15th International Conference Linguistic and Cultural Studies: Traditions and Innovations, pp. 241–246. TPU, Tomsk (2015)

9. Kikot, E.N.: Teoreticheskie osnovy razvitiya issledovatel'skoy deyatel'nosti uchashchikhsya v uchebnom komplekse "litsey-vuz" (in Russian) [Theoretical basis of research activity development of pupils in educational complex lyceum-higher educational institution]. BGARF, Kaliningrad (2002)

10. Kravtsov, S.S.: Teoriya I praktika organizacii profilnogo obucheniya v shkolah ross federacii (in Russian) [Theory and practice of organization of subject-oriented education at Russian schools. Publishing house Institute of content and methods of RAS, Moscow (2007)

11. Krivenko, Y.V.: Formirovanie issledovatel'skoy kompetentnosti starsheklassnikov v usloviyakh profil'noy shkoly kraevedeniya (in Russian) [Formation of research competence of senior grade pupils in conditions of local history subject orientation]. OGPU, Omsk (2006)

12. Leontovich, A.V.: Proektirovanie issledovatelskoj deyatelnosti- uchashchihsya (in Russian) [Designing research activities of students]. MSPU, Moscow (2003)

13. Mietzel, G.: Pädagogische Psychologie des Lernens und Lehrens. Hogrefe, Göttingen (2007)

14. Novozhilova, M.M.: Formirovanie kultury starsheklassnikov v usloviyakh profilnogo obucheniya (in Russian) [Formation of the culture of senior pupils in the conditions of profile training]. MPGU, Moscow (2008)

15. Podiyakov, A.N.: Issledovatelskoe povedenie. Strategii poznaniya, pomoshch, protivodeystvie, konflikt (in Russian) [Research behavior. Strategies of cognition, help, opposition, conflict]. Erebus, Moscow (2000)

16. Pozdeeva, S.I.: Kontseptsiya razvitiya otkrytogo sovmestnogo deistviya pedagoga i rebenka v nachalnoy shkole (in Russian) [The concept of development of an open joint action of a teacher and a child in an elementary school]. Deltaplan, Tomsk (2005)

17. Pozdeeva, S.I., Obskov, A.V.: Justification of the main pedagogical conditions of interactive teaching a foreign language in high school. Procedia – Soc. Behav. Sci. 207, 166–172 (2015)

18. Phedotova, N.A.: Razvitie issledovatelskoy kompetentnosti starsheklassnikov v usloviyakh profilnogo obucheniya (in Russian) [Development of research competence of senior pupils in conditions of profile training]. BSU, Ulan-Ude (2010)

19. Pheskova, E.V.: Stanovlenie issledovatel'skoy kompetentnosti uchashchikhsya v dopolnitel'nom obrazovanii i profil'nom obuchenii (in Russian) [Formation of research competence of pupils at subject-oriented training in secondary education]. KGPU, Krasnoyarsk (2005)

20. Phorkunova, L.V.: Metodika formirovaniya issledovatelskoj kompetentnosti shkolnikov v oblasti prilozheniya matematiki pri vzaimodejstvii shkoly i vuza (in Russian) [The method of forming research competence of schoolchildren in the field of application of mathematics in the interaction of school and university]. Orel, OGU, Arkhangelsk (2010)

21. Precht, D.R.: Anna, die Schule und der liebe Gott. Der Verrat des Bildungssystems an unseren Kindern. Wilhelm Goldmann Verlag, München (2013)

22. Ramazanova, V.N.: Profilnoe obuchenie starsheklassnikov pri setevom vzaimodeystvii obrazovatelnyh uchrezhdeniy i organizatsiy (in Russian) [Specialized training of senior pupils at network interaction of educational institutions and organizations]. Penza State Pedagogical University named after V.G. Belinsky, Penza (2011)

23. Savchina, T.G.: Profilnoe obuchenie v nepreryvnom obrazovanii "shkola vuz" na primere prepodavaniya istorii (in Russian) [Specialized education in continuous education "school-university": the example of teaching history]. Sputnik, Moscow (2009)

24. Savenkov, A.I.: Put k odarennosti issledovatelskoe povedenie shkolnikov (in Russian) [The path to giftedness: the research behavior of schoolchildren]. Piter, St. Petersburg (2004)
25. Scarbich, S.N.: Formirovanie issledovatelskikh kompetentsiy uchashchikhsya v protsesse obucheniya resheniyu planimetricheskikh zadach v usloviyakh lichnostno orientirovannogo podhoda obucheniya (in Russian) [Formation of the research competences of students in the process of teaching the solution of planimetric problems in conditions of the person-oriented approach of training]. OSPU, Omsk (2006)
26. Scheibe, W.: Die reformpädagogische Bewegung. Eine einführende Darstellung. Juventa, Berlin (2005)
27. Schiersmann, C., Thiel, H.-U.: Organisationsentwicklung: Prinzipien und Strategien von Veränderungsprozessen. Springer, Lëtzebuerg (2014)
28. Schlemminger, G.: Dem Lernen einen Sinn geben. Fragen und Versuche **37**, 18–23 (2013)
29. Ushakov, A.A.: Razvitie issledovatel'skoy kompetentnosti uchashchikhsya obshcheobrazo-vatel'noy shkoly v usloviyakh profil'nogo obucheniya (in Russian) [Development of research competence of pupils of general education school in conditions of subject-oriented education]. AGU, Maykop (2008)

Signs and Symbols in Teaching English Vocabulary to Primary-School Children

Galina A. Kruchinina$^{(\boxtimes)}$ ⓘ, Elena V. Koroleva ⓘ,
and Anastasia A. Oladyshkina ⓘ

Minin University, Nizhny Novgorod 603005, Russian Federation
galina-kru@yandex.ru, e-koroleva@mail.ru,
a.a.oladyshkina@gmail.com

Abstract. In this article the experimental work of forming lexical competence by the signs and symbolical means is described. Signs provide the form of lexical units and symbols depict their meaning. Primary school children's cognitive abilities are not high, though they are to learn a lot of new words and collocations in a very short period of time. It was proved, and showed the effect of creating symbolical means for memorizing as many lexical units as possible. Working with younger students on the use of lexical material, it's useful to rely on the semiotic and symbolic functions of the language. We have described the components of the offered model.

Keywords: Primary school children · Vocabulary skills · Signs and symbols · Model of teaching · Learning strategies

1 Introduction

1.1 Building a Research Problem

Teaching a foreign language at elementary school has been widely adopted now. The subject "Foreign language" assumes not only development of foreign language lexics and grammar, but also acquaintance of the younger school students with a new social experience: acquaintance with the world of foreign peers, foreign children's folklore [1] and available samples of fiction. Large volume of lexical units and their use in speech are required, despite the limited cognitive abilities of students.

Having analyzed the researches on speech activity [3, 8, 18], we have come to a conclusion about the importance of the language functions and consciousness in formation of speech activity. A special value among all functions, in our opinion, belongs to semiotics and sign-symbolical functions.

Educational activities for skills developing in foreign language assume the use of certain algorithms of actions for goals achievement. According to the remark of I. Bim, a Russian methodologist, the purpose, contents and methods of training in a foreign language are in close interrelation with each other [2]. They allow simulating and anticipating the course of educational process.

© Springer International Publishing AG 2018
A. Filchenko and Z. Anikina (eds.), *Linguistic and Cultural Studies:*
Traditions and Innovations, Advances in Intelligent Systems and Computing 677,
DOI 10.1007/978-3-319-67843-6_21

1.2 Literature Review

Certain aspects of this topic got covered in the works of the following researchers who studied the problems of the place of language in psychology, according to the cognitive approach in education [9, 16, 19, 20]; the theory of early learning of foreign languages [10, 12, 17], the communicative method for teaching English [2, 5]; and the problem of signs and symbols usage [4, 11, 13, 14].

1.3 Basic Assumptions

The specific feature of a foreign language is its practical orientation. It is closely connected with the most attractive kinds of activities for younger school students: art and aesthetic, and different games. The named kinds of activities have sign and symbolical nature and demand formation of skills and abilities for work with them. There is a set of classifications of sign and symbolical means, such as: philosophical, linguistic, semiotic, psychological, semantic, etc. Sign and symbolical means for a foreign language learning are subdivided into verbal and nonverbal. Pictures, diagrams, symbols, geometrical figures, charts, sketches and maps belong to nonverbal means. The verbal include appliances for creation of a sound, visual and symbolical image of the word. A specific place in the course of training, in our opinion, is taken by situation modeling as a link between verbal and nonverbal means. Designation of visualization by language signs, mastering the internal relations between the designated and the designating; fixation of speech experience by modeling tools promote formation and development of mental vocabulary.

2 Materials and Methods

According to the raised problem, the methodological basis was a set of theoretical methods, including pedagogical and psychological literature analysis, theses researches on the problem. Modeling, comparison, generalization, and empirical methods include questionnaires, activity analysis like working in group and individual tasks.

Analysis of theoretical studies of modeling process and in a particular the model of lexical skills and competences formation, has allowed us to create a new model of speech training in the lexical part, based on sign and symbolical means. This model assumes the use of such means as effective way of necessary lexical units storing and extraction from memory in a certain context. With account for psychological features of younger school students, determining of strategy for lexical skills development, use of a complex of sign and symbolical means have allowed simulating the lexical skills training.

The model of training consists of the following components: (a) target; (b) substantial; (c) conceptual; (d) technological; (e) productive. All components are connected with each other and aimed at results achieving. The target component includes the statement of purpose and defines tasks for its achievement. The substantial component is determined by approaches and the content of foreign language training at elementary school. The all-methodological, all-didactic and special principles form a conceptual component of the offered model. The strategies of lexicon training, developed on the basis of principles, methods of training, and exercises for assimilation of lexical units, belong to the technological component [3]. The result of the offered model is the usage of sign and symbolical means for formation of different lexical skills in a foreign language training, according to requirements of the Federal state educational standard and the programme for students of 2–4 grades.

The model assumes interaction of teacher and students for positive results and goal achievement. The teacher plays a role not only as an information source. Gradually he becomes the leader and the assistant in the process of students interaction in class. The particular feature of the offered model is the teacher's invaluable role in creation of the student's educational style, i.e. individual way of learning, and development of educational activity. The main objective of the teacher is to teach to study.

Successful mastering is in provision of lexical materials, first of all, with a semantisation and systematisation of lexical units (use of presentation, diagrams, and symbols). The action with the sign and symbolical means (replacement, coding, decoding, and modeling) promotes strong mastering of vocabulary and creation of own models of lexical material organization.

So, for example, students are offered to decode the name of products which they will take to the picnic (See Fig. 1).

	A	E	I	O	Y
P	A	G	M	S	Y
T	B	H	N	T	Z
G	C	I	O	U	
S	D	J	P	V	
M	E	K	Q	W	
B	F	L	R	X	

You and your classmates decide to have a picnic. You have a list of products to buy but you must decode it.
TA PA TI PA TI PA PO
GI BI PA TI PE MA PO
SE GO GE GA MA
TA BI MA PA SA
MO PA TO MA BI
TO GI PI PA TO GI MAPO
GA GO GA GO PI TA MA BI PO

Fig. 1. Example of a task for school students.

The task increases motivation, promotes use of the alphabet, correct writing of words, teaches one of ways to code/decode information. Students are offered to create the code system and to code the place of a picnic.

Using different types of signs and symbols in work facilitates a process of theoretical thinking development by students. During work with the sign and symbolical means younger school students learn to apply such means in different types of activities (education, game, art and aesthetic).

Exercises on generalization and revision of lexicon are the necessary condition of learning lexical units and their use for doing communicative tasks.

For example, having mastered the word meanings for basketball, hockey, football, and skating, students refer them to the word "sport", as shown (See Fig. 2).

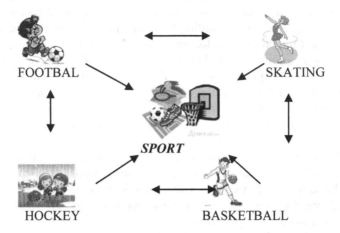

Fig. 2. Illustration of a task for school students.

Guided by the drawing, students remember not only separate words, but also can reproduce sentences on the studied topic and make a monological statement.

Further work with the word "sport" assumes its use as a sign for making sentences and complete statements on the topic.

3   Research Data

Pedagogical model approbation was carried out in primary schools No. 21 and 115 in Nizhniy Novgorod from 2011 to 2014. There were 140 students who took part in the research, and there were 8 school teachers. In the process of organizing the pilot experimental verification of the model, two groups were formed: the control group (CG) and the experimental group (EG). There were 70 participants in CG who took part in the educational process where training was conducted in accordance with traditional programs, and there were 70 participants in EG where training was conducted using signs and symbols for formation of lexical competence. The results of experimental training have confirmed the offered method efficiency in formation of lexical skills of younger school students, based on the sign and symbolical means. A high level of performance of operations in the structure of different lexical skills proves it.

As can be seen from Table 1, strategies of using certain operations in forming lexical competence are observed mainly among students in the EG. Experimental training has proved efficiency of the developed strategies for younger school students training in the lexical part of foreign language speech on the basis of the sign and symbolical means.

Table 1. The usage of some operations in forming lexical competence.

Index	Experimental group EG included 70 students	Control group CG included 70 students
Operation of reminding lexics	84%	53%
Operation of using lexics in a speech pattern	87%	56%
Operation of reminding the audio form of lexics	86%	38%
Operation of visual images of lexics	83%	49%

4   Conclusion

Thus, it is possible to draw a conclusion that the language sign is both a means of designation (fixing of value to a certain sign), and an instrument for speech process ensuring. In its turn, the model of lexical skills training on the basis of the sign and symbolical means promotes thorough learning of lexical material and its use in speech. It also increases motivation and develops theoretical thinking of students. The skills in work with models, diagrams and symbols, which are formed at foreign language classes, verbalization of signs and symbols can be actively used in further educational and cognitive activity. The technique developed for younger school students for foreign language lexical skills formation on the basis of the sign and symbolical means provides an effective training of lexicon, creates the basis for a successful education at high school, and allows using the learned foreign language in its different functions.

References

1. Arhipova, M.V., Shutova, N.V.: The problem of using specific musical compositions in educational process. Hist. Soc. Educ. Ideas **7**(3), 171–174 (2015). doi:10.17748/2075-9908. 2015.7.3.171-174
2. Bim, I.L.: Methods of Teaching Foreign Languages as a Science and the Problems of School Textbook. Russkii yazyk Publishers, Moscow (1977)
3. Bozhivich, E.D.: Research of Students' Mental Activity while Doing Grammar Exercises. MSU Publishing House, Moscow (1972)
4. Gamezo, M.V.: Psychological aspect of methodology and general theory of signs and sign system. Psychological problem of sign data processing. Publishing House "Nauka", Moscow (1977)
5. Haleeva, I.I.: Basic Theory of Teaching of Foreign Speech Understanding. Publishing House "High School", Moscow (1989)
6. Koroleva, E.V., Kruchinina, G.A.: Theoretical and practical linguistics, literature, methodology of foreign languages teaching. Bull. Minin Univ. **4**(12), 27 (2015)
7. Kruchinina, G.A.: Strategies of teaching vocabulary of primary school children on the basis of sign-symbolic means. The Emissia.Offline Letters: Digit. Sci. J. (2014). http://www.emissia.org/offline/2014/2213.htm. Accessed 16 June 2017

8. Luriya, A.R.: The Language and the Consciousness. Publishing House "Feniks", Moscow (1998)
9. Miller, G.A.: The Place of Language in a Scientific Psychology. Psychological Science, Moscow (1990)
10. Nikitenko, Z.N.: Technology of teaching vocabulary in course of the English language of 6-year-old children at school. Foreign Lang. Sch. **4**, 52–59 (1991)
11. Paivio, A.: Mental Representations: A Dual Coding Approach. Oxford University Press, New York (1986)
12. Rogova, G.V.: Metodics of Teaching Foreign Language at Primary School. Publishing House "Prosveschenie", Moscow (1988)
13. Salmina, N.G.: Sign and Symbol in Training. MSU Publishing House, Moscow (1988)
14. Salmina, N.G.: Sign and Symbol in Training. MSU Publishing House, Moscow (1989)
15. Shamov, A.N.: Modeling the learning process, focusing on the lexical side of foreign language speech. Foreign Lang. Second. Sch. **2**, 2–10 (2014)
16. Solso, R.L.: Cognitive Psychology. Publishing House "Trivola", Moscow (1996)
17. Vereschagin, E.M.: The Language and the Culture. Publishing House "The Russian language", Moscow (1990)
18. Vygotsky, L.S.: Intellect and the Speech. Publishing House "Labirint", Moscow (1999)
19. Wickens, D.D.: Imaginary and abstractness in shorten memory. J. Exp. Psychol. **84**, 272–286 (1970)
20. Williams, M., Burden, R.: Psychology for Language Teachers. Cambridge University Press, Cambridge (1998)

Language Learners Communication in MOOCs

Artyom D. Zubkov$^{(\boxtimes)}$ ⓘ and Maya A. Morozova ⓘ

Novosibirsk State Technical University,
Novosibirsk 630073, Russian Federation
{zubkov_nstu, majamorozova}@mail.ru

Abstract. The article explores linguistic features of communication in massive open online courses (MOOCs), the possibilities of their usage in professionally oriented foreign language teaching at tertiary level. A number of linguistic features of various types of texts created by MOOCs participants are revealed, possibilities of this type of foreign language resources in teaching specific language are analyzed, the structure of foreign language competencies in professional and scientific spheres and foreign language competencies related to the usage of information and communication technologies (ICT) are determined.

Keywords: Massive open online course (MOOC) · Foreign language · Network interaction · Discussion forum · Online communication · Foreign language competence · Professionally-oriented teaching

1 Introduction

1.1 Building a Research Problem

Professionally oriented foreign language teaching makes it necessary to combine foreign language courses and special disciplines. Currently, a foreign language teacher has to use modern teaching technologies and methods which allow forming professional foreign language competence of future specialists. In this regard, it is currently important to study the possibilities of using massive open online courses (MOOCs) in foreign language teaching process at tertiary level.

Currently, MOOCs is actively used in non-formal education, as well as in educational process at universities as a component of higher education programs, programs for additional vocational education and qualification upgrading courses.

Due to its features, subject MOOCs in foreign languages can act as a means for language training of students in various study fields. This article explores various aspects of online communication of massive open online courses participants as a tool for professional foreign language communicative competence development.

1.2 Literature Review

As for researchers, certain aspects of this issue are covered in the works of the following researchers who studied the massive open online courses teaching experiences

A. Filchenko and Z. Anikina (eds.), *Linguistic and Cultural Studies: Traditions and Innovations*, Advances in Intelligent Systems and Computing 677, DOI 10.1007/978-3-319-67843-6_22

[1], the usage of MOOCs discussion forums [2], content analysis of MOOCs discussion forums and participants behavior [3], MOOCs participants interaction [4–6], communication patterns in massive open online courses [7], the features of self-presentation [8, 9], evolution of MOOCs communication means [10], motivation in asynchronous online discussions with MOOC mode [3, 11], features of self-presentation and self-awareness in computer-mediated communication [12, 13], and online discussion in educational process in tertiary organizations [14].

1.3 Basic Assumptions

Massive open online courses are understood as courses provided online and developed by the lead scientific, educational and cultural organizations all over the world. These courses are characterized by a potentially unlimited number of participants and open authorized access.

The authors of the study believe that the communication opportunities and features of the MOOCs discussion forums can be an effective tool for students language training in higher education. The possibilities of massive open online courses in integrated learning of special disciplines and foreign languages can be realized in the framework of such approaches as Language for Specific Purposes (LSP), Content and Language Integrated Learning (CLIL), English Mediated Instruction (EMI).

2   Materials and Methods

While conducting this research, the following methods were implemented: linguistic analysis of the features of different types of texts (self-presentation, comments, request/clarification of information) in MOOCs, quantitative data processing, and synthesis.

In the framework of this study an analysis of the texts of students' self-presentations, requests for information and comments in 8 different English and German MOOCs on technical, liberal and economic sciences, academic writing on the web platforms OpenHPI (Hasso Platner Institute, Potsdam), iversity (various European Universities, Berlin), Mooin (Lübeck University, Germany), FutureLearn (University of Liverpool, UK), Coursera (University of Pennsylvania, University of Illinois at Urbana-Champaign) was carried out.

Research materials included 400 texts of self-presentation, 200 comments and 200 information requests. The courses analyzed cover the disciplines in the field of information and communication technologies (OpenHPI), energy (Mooin), economics and business (iversity, Coursera), English for specific purposes (Coursera), electronics and electrical engineering (FutureLearn).

While analyzing the texts of self-presentations, comments and requests of information, the following criteria were used: content, structure, coherence, politeness and realization of communicative intention.

3   Research Data

3.1   Analysis of Self-presentation

Content components included nickname (presence/absence), indication of country and/or city, interests, hobbies, age, speciality(acquired)/profession, information about the available work experience in the relevant field, motivation for learning on the course chosen, expectations associated with the course, presence of greeting/farewell phrases, wishes for course developers and participants, course feedback provided by participants.

The criterion concerning structure and coherence included a consistent and logical exposition, compliance with the structure, adequacy of information reported to communicative situation. In general, the texts of self-presentations were characterized by the preservation of structure and coherence: 42% of them fully met the requirements for structure and coherence, and 55% met them partially. There were occurrences of inconsistent presentation, for example, the participant expressed some value judgment and only then presented or described his experience in detail while forgetting to mention motivation in the course chosen and expectations associated with the course. The authors identified only 4% of self-presentations that had serious structure and coherence violations. Most of these texts included incomplete sentences, the absence of proposals and the formulation of a statement in the form of listing some information that is characterized by lack of concord. There was also information that appeared as a rationale for choosing a course and information that was not directly related to a course. A violation of the adequacy of the information reported in the communicative situation, for example, an indication of the non-substantiated choice of course interests was noticed.

As for content, some participants did not specify such significant issues as specialty (acquired)/profession, occupation, existing work experience in the relevant field, and motivation. There were also self-presentations in which participants did not specify their name (with a significant percentage of fictitious names while registering - 21%), place of residence (country and/or city), and status. At the same time the majority of participants wrote their expectations related to the course and motivation to learn on the course. As a rule, some participants gave a positive feedback about the course. Despite the fact that participants were asked to fill out a questionnaire at the beginning of the course, where they have already answered these questions, many people considered it necessary in their presentations to write about their motivation and expectations related to the course again and in more detail.

A number of self-presentations expressed wishes to developers and other course participants. Participants of the previous version of the course indicated the reasons for re-registration for the course. Limitations and recommendations regarding the volume of self-presentations in the courses analyzed were absent. The volume of self-presentations varied from one sentence to several paragraphs. The most common volume of self-presentation text presented 60–80 words. The detailed degree of information provided by the course participants about themselves in most cases could be considered sufficient. Generally, name, place of residence, motivation for the course, goals and plans for completing the course were indicated.

The criterion of politeness included the greeting/farewell phrases, signature at the end of a message, various means of lexical, grammatical, syntactic and stylistic levels.

The analysis showed that in almost 50% of self-presentations the norms of politeness were observed, in 45% of self-presentations the norms of politeness were partially observed and only 6% of self-presentations pointed out the lack of politeness, manifested in ignoring the courtesy formulas, usage of greeting forms that are common among young people and in a special interests group, for instance, the exclamation "Huhu" (au!, hey!), the usage as a greeting word of Hawaiian origin, denoting at the same time "hello", "goodbye", "welcome", and just a wish for peace and joy "Aloha", the usage of little-known and consequently poorly used by many users dialectal expressions, for example: "Grüzi mitenand" (greeting in the eastern Swiss dialect of German language).

The criterion of speech intention realization included such components as the providing of necessary and sufficient information about themselves, nickname (presence of a full/incomplete real name or feigned name). In general, most of course participants indicated necessary and sufficient information about themselves. Realization of speech intention was noted in 37% of self-presentations. 61% of self-presentation texts had a partial realization of speech intention. In a number of instances, information on the place of residence, occupation, interests and hobbies, information on available work experience in the relevant field, expectations associated with the course were not provided. Only 3% of self-presentations reflected an unrealized speech intention expressed in indicating unnecessary information that was not relevant to the course and not important to other participants and using little-known reductions or indicating insufficient information about themselves. The study of self-presentation texts showed that, in general, the components relating to content, structure, coherence, politeness and speech intention were being realized. The percentage of violation in the first three components was low, and did not exceed 6%. At the same time, the value of unrealized speech intentions in this genre of network communication was only 3% (Fig. 1).

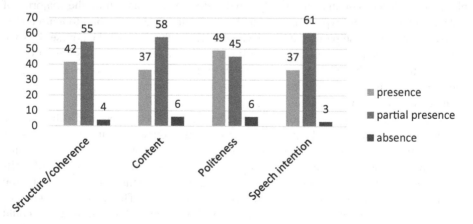

Fig. 1. The features of learners' self-presentation in MOOCs.

3.2 Analysis of Comments

While analyzing the texts of comments (Fig. 2), the content criteria included the following: nickname (presence/absence), expression of gratitude (if necessary), usage of greeting/farewell phrases. The structure and coherence criteria included a consistent and logical presentation and compliance with the structure. The politeness criteria included the same components as in the previous genre of network communication. While analyzing this and subsequent types of texts (requests for information), such a component of the realization of speech intention criterion as presence of comments/ requests of information remarks, reflecting conversation partner verbal response, was used.

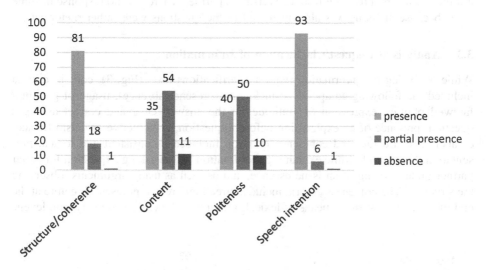

Fig. 2. Features of learners commenting in the MOOCs.

The analysis of comment texts content showed that 35% of the material examined revealed all the necessary components, 54% of the texts lacked some content components, 11% lacked such significant components as gratitude, exhaustive explanation where this would be necessary. In rare cases, non-constructive criticism was noticed. In this case, nickname was presented for all participants. 81% of message texts retained structure and coherence, 18% of them generally retained structure and coherence but had some violations, concerning incorrect choice of communicative strategy for speech intention expression. At the same time errors in statements structure were noted: incorrect word choice, large volume and complexity of the message, illogicality, absence or insufficiently convincing arguing. A number of errors can be explained by spontaneity which is a characteristic of informal speech. Comments with significant violations in the structure and coherence (1%) were almost not found.

The analysis of courtesy phrases usage and politeness strategies showed that courtesy was present in 40% of comments and in 50% of materials it was present partially. In 10% of comments neither polite phrases nor politeness strategies were used. It should also be noted that the usage of greeting/farewell phrases was taken into account in the analysis only at the beginning of the discussion. Then the participants focused on discussing certain problems and issues in which the courtesy phrases were superfluous. Despite the absence of courtesy phrases in some cases, participants used communicative strategies of politeness [15], for example, the expression of approval, sympathy for conversation partner, offer of help, an expression of apology and gratitude, encouraging wishes, etc.

Almost all participants' remarks received a verbal response from the interlocutors (93%). 6% of comments also followed a reaction but in some cases due to incorrect, incomplete or inaccurate information clarification was required because the interlocutor did not quite understand any details of the discussion issue. Only 1% of comments did not receive a verbal reaction from conversation partner. There was no response in some cases because the issue was already solved or the questions were rather rhetorical.

3.3   Analysis of Request/Clarification of Information

While analyzing the information requests/clarifications texts (Fig. 3), content criteria included the following components: nickname (presence/absence), usage of greeting/farewell phrases, expression of gratitude for the answer, clear and correct asking a question, presence of an explanation before a question, constructive criticism (if necessary). The structure and coherence criterion included a consistent and logical presentation, for example while explaining the situation before asking a question for better partner understanding of an issue essence, and as well as using arguments, observing the structure. The politeness criteria included greeting/farewell phrases, signature at the end of a message, various means of lexical, grammatical, syntactic and stylistic levels.

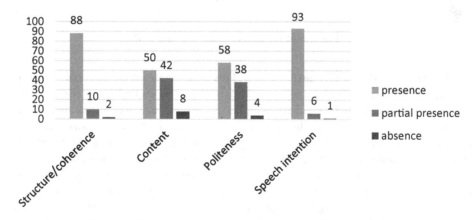

Fig. 3. Features of the learners request/clarification of information in MOOCs.

The analysis conducted showed that in most requests/clarification of information texts (88%) the structure and coherence were observed. 10% of the texts had deviations in relation to this criterion, they concerned inaccuracy in the formulation of the request.

Only 1% of the materials had significant errors that made their understanding difficult. 50% of the texts revealed all the necessary content components, in 42% of materials examined some content components were missing, there were no significant components such as clear/correct formulation of the question, explanation before a question (where it would be necessary) in 8% of instances.

As for the politeness criterion, 58% of the texts fully complied with all the requirements determined by the criterion. 38% of the materials had deviations, for example, an expression of gratitude for the answer. Only 4% of the texts did not contain all the criterion components. However, only cases where this was really necessary were taken into account.

Upon that, 93% of the texts had a speech intention, 6% of materials explored had a partially realized speech intention (it was required to clarify the interlocutors due to an incorrectly formulated request), and only 1% of the texts did not show the speech intention realized.

While analyzing the usage of nicknames (Fig. 4), it was found that the percentage of using full and incomplete genuine names is higher in self-presentations (51% and 35%) than while commenting (28% and 36%) and requesting for information (21% and 29%). This can be explained by the fact that there was a discussion in the last two types of texts, it was characterized by the speed of messages exchange, the concentration on joint solution of important issues and more free communication.

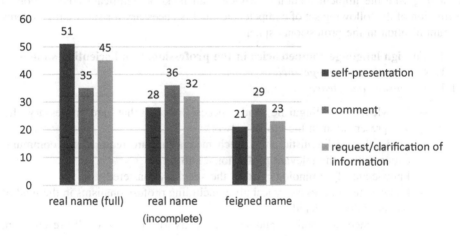

Fig. 4. Usage of nicknames in MOOCs.

Within the framework of the courses analyzed, two types of communication were offered: academic discussion and free (spontaneous) discussion of course participants. Therefore, an opportunity for implementation of formal and unofficial foreign language communication in professional area (area of knowledge studied) and an opportunity for implementation of speech intentions in different communication situations were created.

3.4   Linguistic Features of Communication in MOOCs

Linguistic analysis of online communication in MOOCs was carried out on phonetic-graphic, lexical, morphological and syntactic language levels [16–18]. The following features of communication in MOOCs were revealed:

1. Phonetic-graphic (reduction of sounds, word merging and other cases, emoticons, cases of multiple spelling of punctuation marks, use of capital letters instead of lowercase letters, multiple repetition of the same letter, use of graphic means in messages),
2. Lexical (use of alphabetic-syllable and alphabetic-digital words writing, typical vocabulary for the subject of the course is found),
3. Morphological (active voice was used more often than passive voice; there were no significant deviations in the usage of grammatical categories),
4. Syntactic (the presence of complete and incomplete sentences and also sentences with inverted word order).

Thus, communication in MOOCs has the evidence of oral-written form of speech with a great emphasis on written speech.

3.5   Foreign Language Competencies

Teaching with the implementation of MOOC can make a significant contribution to formation of the following set of competencies that is necessary for successful students communication in the professional sphere.

I. **Foreign language competencies in the professional and scientific sphere**
1. *Communicative competencies*
1.1 *Linguistic competence*

- knowledge of linguistic and speech means that are necessary for self-presentation in the global network;
- knowledge of linguistic and speech means that are required for communication within the relevant professional subject;
- knowledge of terminology within the specialty mastered;
- knowledge of necessary vocabulary including professionalisms in the field of the specialty mastered;
- competence in linguistic and speech means for creating a self-presentation, writing a request, clarifying, etc.;
- mastery of the language means of expressing politeness.

1.2 *Sociolinguistic competence*

- usage of colloquial clichés in constructing dialogue/polylogue in situations of verbal communication (when addressing, greeting/farewell, requesting information, expressing a wish, gratitude, consent or disagreement, etc., taking into account the status of the addressee, communication situation, relationship nature between communicants);
- ability to complete the necessary documents (application form, registration form, etc.).

1.3 *Pragmatic competence*
13.1 *Discursive competence:*

- ability to discourse on a topic given, get into touch with foreign mates;
- oral and written language skills in a language acquired, coherence observance in terms of content and expression, the logical structure of a message, rhetorical effectiveness, reasoning and completeness of an utterance;
- ability to prove personal point of view reasonably.

1.3.2 *Functional competence:*

- ability to get into touch with the interlocutor;
- ability to ask for interesting information.

2. **Sociocultural competence**

- knowledge of cultural traditions of the target language countries and their manifestations in communication situations and significance in the context of corporate and internal communication;
- understanding of dialectal differences and importance of using dialectisms in communication with different age groups;
- taking into account the social characteristics, status and interests of an addressee while establishing and maintaining contact;
- following the rules of speaking etiquette in the framework of academic and professional communication;
- adequate reaction to critical remarks.

2.1 *Intercultural competence*

- knowledge of conflict situations communication features of target language speakers, taking into account the preservation of the positive/negative behavior of communication partner;
- tolerance to national and cultural features (to students from different countries);
- overcoming of other cultures stereotypes that impede effective communication.

2.2 *Existential competence*

- open, friendly tone of communication in relation to an interlocutor;
- readiness for ideas exchange and cooperation;
- respectful and courteous attitude to an interlocutor;
- tolerance to the others opinion;
- readiness to acquire new experience, share experiences and acquire new knowledge;
- readiness for collaboration;
- readiness for foreign language communication;
- ability to reflect and sum up the results of personal activities;
- ability to self-evaluation and evaluation of course mates.

II. **Foreign language competencies related to the usage of ICT**
1. *Communicative Competencies*
1.1. *Linguistic Competencies*

- knowledge of the course interface terminology;
- knowledge of the genre and style features of communication in MOOCs;
- knowledge of the techniques for dialogue linguistic expression and dialogue construction.

1.2 *Sociolinguistic competence*

- knowledge of the netiquette rules (taking into account the status of an addressee, communication situation, the nature of relations between communicants), self-positioning in the network;
- adequate reflection in subject and text of a message of an addressee's speech intention;
- knowledge of the foreign punctuation rules;
- knowledge of greeting and farewell formulas;
- knowledge related to the architectonics of a message text (the possibilities of varying the structure of a message, using or ignoring greeting and farewell, etc.);
- ability to use style of communication in accordance with a type of a text in foreign language.

1.3 *Pragmatic competence*
1.3.1 *Discursive competence*

- ability to predict the communicative relevance of the speech means in accordance with the style/genre of communication in different types of online resources;
- ability to carry on a dialogue;
- usage of communication strategies (prompt reaction, argumentation, error recognition, expression of gratitude, request, etc.);
- discussion of professional problems.

1.3.2 *Functional competence*

- skills in the usage of functional capabilities and tools in MOOCs;
- knowledge of technical and functional capabilities and limitations in MOOCs.

2. *Sociocultural competence*

- observance of the rules of educational and professional etiquette in the framework of oral and written communication (rules are universal and adopted in a foreign culture);
- observance of the rule of obligatory reply to a message;
- adequate reflection of a speech intention in the subject of addressee message;
- sending messages (tasks, documents) on time.

2.1 Intercultural competence

- understanding of the role of using national customs and traditions in communication;
- knowledge of traditions, customs, norms and rules of foreign language culture.

2.2 Existential competence

- readiness for collaborative work in MOOCs;
- readiness for foreign language communication in professional area with the usage of different types of electronic resources.

4 Conclusion

This teaching and learning format can organically combine the possibility of mastering an impressive amount of complex materials (placement of a large number of learning materials) with the possibility of implementing collaborative work and realization communication with mates and course developers. In addition, professional foreign language communication is realized in the form of official communication (while performing assignments, official appeals) and in the form of informal communication (while discussing the issues that arise on forums). Thus, students acquire professional communication experience at different levels. In official communication the most important special terminology (of area studied) is mastered, and while communicating informally, acquaintance with professionalisms and professional slang occurs, i.e. acquaintance with the other important side of the specialists communication, which makes it possible to master the volume of terms necessary for professional communication, it also gives an opportunity to gain an impression of "live" communication that is outside the official communication but also appears as an integral part of specialists communication. Such courses unite absolutely different people, when they solve problems jointly they learn to work together, come to mutual understanding and tolerant attitude to another culture. The opportunities for communication offered by massive open online courses contribute to development of two blocks of important competencies: foreign language competencies in the professional and scientific spheres and foreign language competencies related to the usage of information and communication technologies (ICT).

References

1. Kop, R.: The challenges to connectivist learning on open online networks: learning experiences during a massive open online course. Int. Rev. Res. Open Distance Learn. **12**(3), 20–38 (2011)
2. Onah, D., Sinclair, J., Boyatt, R.: Exploring the use of MOOC discussion forums. In: London International Conference on Education 2014, pp. 1–4. Infonomics Society, London (2014)

3. Yang, D., Piergallini, M., Howley, I., Rose, C.: Forum thread recommendation for massive open online courses. In: Stamper, J., Pardos, Z., Mavrikis, M., McLaren, B.M. (eds.) 7th International Conference on Educational Data Mining 2014, London. LNCS, pp. 257–260 (2014)
4. Sinha, T., Li N., Jermann, P., Dillenbourg, P.: Capturing "attrition intensifying" structural traits from didactic interaction sequences of MOOC learners. In: Conference on Empirical Methods in Natural Language Processing 2014, pp. 42–49. Taberg Media Group AB, Taberg (2014)
5. Wong, J., Pursel, B., Divinsky, A., Jansen, J.B.: An analysis of MOOC discussion forum interactions from the most active users. In: Social Computing, Behavioral-Cultural Modeling, and Prediction 2015, pp. 452–457. Springer, Heidelberg (2015)
6. Zhang, Q., Peck, K.L., Hristova, A., Jablokow, K.W., Hoffman, V., Park, E., Bayeck, R.Y.: Exploring the communication preferences of MOOC learners and the value of preference-based groups: is grouping enough? Educ. Tech. Res. Dev. **64**, 1361 (2011)
7. Gillani, N., Eynon, R.: Communication patterns in massively open online courses. Internet High. Educ. **23**, 18–26 (2014)
8. Tu, C.: From presentation to interaction: new goals for online learning technologies. Educ. Media Int. **3**, 189–206 (2005)
9. Hogan, B.: The presentation of self in the age of social media: distinguishing performances and exhibitions online. Bull. Sci. Technol. Soc. **30**(6), 377–386 (2010)
10. García-Peñalvo, F.J., Cruz-Benito, J., Borrás-Gené, O., Blanco, A.F.: Evolution of the conversation and knowledge acquisition in social networks related to a MOOC course. In: Panayiotis, Z., Andri, I. (eds.) Learning and Collaboration Technologies, pp. 470–481. Springer International Publishing, Cham (2015)
11. Yang, Q.: Students motivation in asynchronous online discussions with MOOC mode. Am. J. Educ. Res. **5**, 325–330 (2014)
12. Joinson, A.N.: Self-disclosure in computer-mediated communication: the role of self-awareness and visual anonymity. Eur. J. Soc. Psychol. **31**, 177–192 (2001)
13. Walther, J.B.: Selective self-presentation in computer-mediated communication: hyperpersonal dimensions of technology, language, and cognition. Comput. Hum. Behav. **23**, 2538–2557 (2007)
14. Hammond, M.: A review of recent papers on online discussion in teaching and learning in higher education. J. Asynchronous Learn. **3**, 9–23 (2005)
15. Gazifov, R.A.: Sposoby funkcionirovaniya implicitnoj vezhlivosti v ehkspressivnyh rechevyh aktah (na materiale nemeckoj lingvokul'tury) (in Russian) [Methods of functioning implicit politeness in expressive speech acts (on the basis of German linguistic culture)]. Vestnik CHelyabinskogo gosudarstvennogo universiteta **29**(320), 24–26 (2013)
16. Shhipicina, L.J.: Zhanry komp'juterno-oposredovannoj kommunikacii: monografija (in Russian) [Genres of computer-mediated communication: monograph]. Pomorskij universitet, Arkhangelsk (2009)
17. Shhipicina, L.J.: Komp'juterno-oposredovannaja kommunikacija. Lingvisticheskij aspekt analiza (in Russian) [Computer-mediated communication. Linguistic Aspect of Analysis]. Krasand, Moscow (2010)
18. Goroshko, E.I., Polyakova, T.L.: Lingvisticheskie osobennosti angloyazychnogo tvittera (in Russian) [Linguistic features of Anglophone twitter]. Uchenye zapiski Tavricheskogo nacional'nogo universiteta im. V.I. Vernadskogo. Seriya «Filologiya. Social'nye kommunikacii» **24**(63), 53–58 (2011)

The Activation Model of University - Employer Interaction in the Field of Master's Students' Foreign Language Proficiency

Elena M. Pokrovskaya[(✉)] (iD), Ludmila Ye. Lychkovskaya(iD),
and Olesya A. Smirnova(iD)

Tomsk State University of Control Systems and Radioelectronics,
Tomsk 634050, Russian Federation
{pemod,lef2001}@yandex.ru, smirnova_alisee@mail.ru

Abstract. The research work is dedicated to the problem of gaining useful learning benefits from providing the "University – Employer" interaction system in the field of foreign language training. What is important in this respect is to have a clear idea of what skills students will need to carry out the tasks set by the Master's degree programs.

The article presents some ideas for computer-based language learning activities, namely, using on-line training courses for master's students. Attention is drawn to the importance of methodological issues which relate both to the Internet and language teaching. It is vital to be sure that E-learning, a popular educational trend nowadays, could provide ways of following the trends of global information and communication space, checking tightly the specified competences which students are asked to demonstrate in practice. In fact, such approach presents the practical value of the project.

The article gives valuable information about the existing on-line platforms both in Russia and abroad. Their advantages and disadvantages are specially noted. The data are given about the advantageous use of modern training on-line platform Moodle that challenges students to trace their own progress and create the training independently.

By way of conclusion the authors offer the list of chief functions and characteristics of the "University - Employer" dyad and the training course structure revealing different methods providing interrelated formation and development of almost all the components of communication.

Keywords: E-learning · Master's students' training course · "University – Employer" interaction

1 Introduction

The questions of science, education and production integration have actively been discussed in Russia recently, and new forms of their potential interconnection have been searched. Scientific knowledge and experience, that is necessary for understanding the processes of development of mutually beneficial cooperation, allow us to conceptualize a model of activation of the "University-Employer" interaction system in

© Springer International Publishing AG 2018
A. Filchenko and Z. Anikina (eds.), *Linguistic and Cultural Studies:*
Traditions and Innovations, Advances in Intelligent Systems and Computing 677,
DOI 10.1007/978-3-319-67843-6_23

the field of language training in target graduate programs. Under the target graduate program the authors mean a form of education aimed at engineering staff of companies working in the field of high-tech industries. Currently network coordination of clusters, education and production is implemented; however, there is the locality of its operation due to, among other factors, insufficiency to ensure a unified methodological and conceptual base in the field of language training for high-tech enterprises and high-tech industries [5]. To ensure the successful formation of innovative infrastructure of language training, you should perform a set of scientific and organizational-theoretical measures, the main of which is creation of areas of development of educational content, coordinating the work of the "University-Employer" dyad [3]. Note that learning a foreign language is relevant to all master's degree programs that are in demand by employers of specialized cluster in the region and the country as a whole. Currently, most high-tech enterprises in the city and region, enterprises of the military-industrial complex focus on language skills of the master's degree program graduates during selection of candidates for vacancies. Thus, the demand for graduates depends on the system of language training in the master's degree program.

2   Theory and Methods

The laws and principles of system analysis (integrity, complexity and emergence; systemic and integrative), management science and organization theory are used as theoretical and methodological basis for the research. As the main research tool, we used the methods of systemic and expert analysis. World practice shows that E-learning today is a popular educational trend, providing the opportunity to follow the trends of global information and communication space, namely to individualize learning, to differentiate tasks, depending on levels of difficulty and psycho-individual characteristics of the student; personalize the learning environment by selecting the required student information (answer the question: "Why do I need this?").

Today, there are several sites of educational content development: Coursera (www.coursera.org), EdX (www.edx.org), Uda-city (www.udacity.com), MIT Open-CourseWare (www.ocw.mit.edu), Academic Earth (www.aca-demicearth.org), Khan Academy (www.khan-academy.org), TED (www.ted.com) and etc. Each of them brings together many of the world's leading universities that are developing courses, presents them for public access and is engaged in teaching support courses. One of the most significant examples is the EdX platform, founded in 2012 with the partnership of such advanced and well-known universities as Harvard University and the Massachusetts Institute of Technology. It aims at creating a global community of on-line training by other universities in any part of the world. EdX should support all the modern technology remote on-line learning (video segments, memorable quiz prompt feedback, student rating of questions and answers, orientation to student paced learning, on-line labs and more).

Most known Russian online platforms are universarium.org, lektorium.tv, intuit.ru, while coursera.org, edx.org, eduson.tv, khanacademy.org, which differ in the amount of educational service materials applied, and refer to the most notable ones.

All of this allows us saying that today we are dealing with the practical implementation of a distributed (or network) training, run by leading universities in the world.

For comparative analysis let's consider the following.

General characteristics of foreign online platforms.

1. Most vital and attractive subjects are provided nowadays at the following courses: Management, Economics, Electronics, Finance, Computer Programming, and Foreign Languages.
2. The leading universities of the world are involved in the process of online platforms development: Boston, Harvard, London, Yale, Michigan, Berkley, and also leading Business schools: INSED, HEC Paris, Wharton Graduate School of Business, and London Business School.
3. The training forms of educational online platforms are similar; they are mainly videolectures with the set hometasks and tests to them.
4. Courses duration is different, starting with minicourses (5 min) to several hours (www.eduson.tv), some of them lasting from 4 to 15 weeks (www.edx.org).
5. A notable feature of online platforms is their integrating character. By way of illustration it is worth mentioning that the online platform coursera.org provides courses in 23 languages, the platform khanacademy.org – in 20 global languages. Besides, all of them have a great variety of training subjects.

The disadvantages of foreign educational online platforms:

1. Small number of courses available for training free of charge.
2. A special registration is required.
3. Graduation certificates are either payable, or they are not delivered at all.
4. Access to the online course is time-limited (every test or hometask must be done only within the set period).

With regard to Russian online platforms, generally, they coincide with the foreign ones in duration of the courses, training forms and subjects. However, most of Russian online courses are free of charge. They are also remarkable for valuable video archives. Many online platforms are developed by the specialists from the Intel and Microsoft Academies; there is also a wide choice of specialization courses, in the humanities, in particular.

Upon analysis [1, 2, 8–10], it is necessary to point out that nowadays the existing educational platforms and learning management systems allow one to use scientific and educational potential of leading institutes and universities more actively, to involve the best teachers into the creation of the courses for distant training to expand the number of trainees.

Nevertheless, implementation of these ideas and concepts in the Russian practice articulates a number of difficulties, namely:

• The absence of consistent interaction between education and production establishments, and no mechanism of cluster system interaction is defined.
• Although application of the latest means of information technology in various spheres of human activity, including information, gets the increasing urgency and

computerization of educational process is considered as one of the vital factors for organization of training in this or that subject, the content of this training process, its validity and verification are still rather problematic, in contrast to "primitive" social networks, such as *Facebook* and *VKontakte*.

- Consistent work in the sphere of educational technology development is insufficient, especially the one based on university interaction, which in its turn guarantees the quality of educational process.
- Traditions of reflective awareness in the field of language training are lacking.

3 The Results and Discussion

Let's present the model of university and employer interaction in the field of language training, coming from the outlined problems.

Figure 1 presents the related functions of interaction of the model activation elements. Foreign language training of master's students is the basic one in this model. It is done by the means of (educational course on foreign language) online courses "EdConForLang", the subject "Foreign Languages" for Master's degree programme on the basis of network dual platforms of the university. Namely, such kind of awareness offoreign language special training functions as local integrator for the master's students in the field of professional foreign language and university technical partners' cooperation. It is also necessary to develop a new management methodology in respect to the "University - Employer" concept and methodological ground for selecting special training exercises for the professional subject, skill requirements for both sides: university and employer.

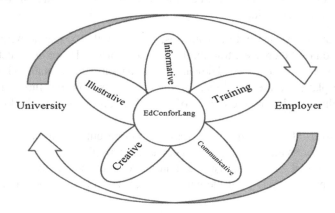

Fig. 1. The main functions of the "University-Employer" dyad.

The main functions of the "University - Employer" dyad are the following:

- informative (publication of training package);
- training (preparation for the tests);
- illustrative (multimedia presentation of the studied material; attached pdf-files);
- communicative (informing the students and tutorials):
- creative (self-actualization, autonomy and critical thinking development).

Let's consider the given conceptional model implementation on the example of the educational project developed by TUSUR. The specific feature of the project is defined by its transdisciplinarity.

The novelty of the project is teaching Science subjects in English which is supposed to be done by online training course for the master's students in "EdConFor-Lang" subject. This can be conducted with due regard to the requirements for participants of network platforms, thus doing a joint teacher's work.

Practical value of the project is stressing the formation of personality through language learning but not for the sake of learning a foreign language properly. The project developers a are guided by the social-constructionist approach and methodological foundations, suggested by Vygotsky [7] considers the environment as a development source of the lofty psychical functions, which are notable for obliquity, awareness, arbitrariness and consistency. The given project actualizes the students' potential concerning online training by means of mastering special exercises. The logic of the course lays stress on the level of generalization and synthesis, evaluation and way of thinking, to which the structure of the training modules is contributing: it comprises the basic textbooks and courseware in the "Foreign Language" course; a set of training tasks to form, develop and improve the corresponding competences and also control-and-measuring tasks to evaluate the level of the corresponding competence formation at the stage of formative and summative assessment.

The given project contributes to development of a new type of thinking due to consistent work of students within the semester (as a result of grade-rating system); due to weekly planning of work, but allocating one's time independently; self-evaluation of knowledge; the students' interest in using new technologies. In such a way a personal mobile educational system is formed. Obviously, the advantageous use of modern training online platform Moodle makes it possible for students to create the training on their own, to trace their own progress and achievements with the help of automation system of evaluating means and control-and-assessing tasks [4, 6].

The basis for interaction management is the principle of sustaining consistent work alongside with constant adaptation to changeable conditions (actual environment conditions), which provides the necessary accuracy, stability and quality of management process.

Considering the "University - Employer" interaction as a basic element in the field of language proficiency, it is necessary to define the functionality of the "University - Employer" dyad and distinguish the main structured blocks, comprising a set of minimum requirement management components of the educational path.

The course structure

- **Organization block**
  - Course abstract
  - Course scheme and work programme
  - Information about the author
  - News
  - Consultations (forum)
- **Reference block**
  - Glossary
  - Literature (course and module)
  - Participants
  - Grades
  - Course materials (for downloading)
- **Teaching units**
  - Calendar schedule
  - Lectures, presentations, references to the training resources and so on
  - Organization instructions, concerning students' individual work and practical studies
  - Check-yourself tests
  - Credit and examination tests

Testing is the most frequently used function of e-learning course. Each section of the unit comprises tests for checking the students' progress. At the end of each section there are communicative tasks (based on the video selected in accordance with the studied language material) that are aimed at developing the speaking skills and listening comprehension skills.

One of advantages of the given online course is, indisputably, giving the teacher a possibility to trace and control the work of trainees. All information about the students' activity, that is the date and time, number of students getting access to the web-site, results of exercises – everything will be available to the teacher.

On the whole, the online course was developed in accordance with the new requirements to the level of language proficiency for master's students, taking into account the process of intensification and informatization of education and specific language training. Thus, implementation of the online course in LMS Moodle allows us not only to involve multi- and hypermedia resources of the modern information technologies but also to provide their systematic using.

In conjunction with traditional means of foreign language teaching, video- and audio materials, in turn, provide interrelated formation and development of almost all components of the communication competence. Besides, the use of innovative means is an important requirement of our time which blends seamlessly into the modern concept of education modernization and training of future specialists.

4   Conclusion

To conclude, we note the following:

- Proposed units as the reference model elements for intensification of "University – Employer" interaction in the field of language training provide a complex system functioning;
- Complex and consistent use of innovative methods and technical training resources makes it possible to optimize and intensify, to a large extent, the process of language training of the master's students;
- Developed mechanism of the University interaction with the Employer, depending on the market, may change the content of assumptions, spatial positioning, *etc.* It allows us to justify and build a system for successful functioning of the "University – Employer" dyad or the purpose of sustainable development of Russia, the transfer of its innovative ideas and developments in the international market;
- It is necessary to implement the system monitoring and feedback between teacher and student, teacher and employer, employer and student;
- The proposed mechanism of interaction and training in the discipline "Foreign language" via on-line course "EdConForLang" can be extrapolated to other regions of Siberia, the Far East, and further – the CIS countries;
- The main obstacle in the "University – Employer" interaction is the lack of methodological framework for managing learning outcomes interrelated within the competence approach in the context of innovative interdisciplinary projects.

Acknowledgements. The work is performed at support of the Vladimir Potanin Foundation, project no. 170001293.

References

1. Costello, E.: Opening up to open source: looking at how Moodle was adopted in higher education. J. Open Distance e-Learn. **28**(3), 187–200 (2013)
2. Gavin, W.P.: Free choice of learning management systems: do student habits override inherent system quality? Interact. Tech Smart Ed **10**(2), 84–94 (2013)
3. Lychkovskaya, L.E., Mengardt, E.R.: Ispol'zovanie tekhnologii MOODLE pri razrabotke uchebno-metodicheskogo kompleksa po distsipline «Angliyskiy yazyk» (in Russian) [Use of Moodle resources in development of E-learning course "English for students of engineering faculties"]. Materialy Mezhdunarodnoi nauchno-prakticheskoi konferentsii. Sovremennoe obrazovanie: problemy vzaimosvyazi obrazovatel'nykh i professional'nykh standartov. TUSUR Publishers, Tomsk (2016)
4. Protsenko, E.A., Zhivokina, M.A., Kirgintseva, N.S.: Sovremennye infokommunikatsionnye tekhnologii v prepodavanii angliyskogo yazyka (in Russian) [Modern inforcommunicative techniques in the English language teaching]. Vestnik Voronezhskogo instituta MVD Rossii **2**, 225–233 (2013)

5. Federal'nyi zakon "O vnesenii izmenenii v Trudovoi kodeks Rossiiskoi Federatsii i stat'i 11 i 73 Federal'nogo zakona "Ob obrazovanii v Rossiiskoi Federatsii" (in Russian) [The federal law "About modification of the Russian Federation labour code" and Article no. 11 and 73 of the federal law "About education in the Russian Federation"]. http://www.consultant.ru/document/cons_doc_LAW_178864/. Accessed 7 Dec 2016
6. Vonog, V.V., Prokhorova, O.A.: Ispol'zovanie LMS MOODLE pri obuchenii inostrannomu yazyku v aspiranture v ramkakh smeshannogo i distantsionnogo obrazovaniya (in Russian) [The use of LMS Moodle in the process of teaching a foreign language in the post-graduate courses in the context of mixed and distance education]. Vestnik Kemerovskogo gosudarstvennogo universiteta **3**, 27–30 (2015)
7. Vygotsky, L.S.: Pedagogicheskaya psikhologiya (in Russian) [Pedagogical psychology]. Astrel', Moscow (2008)
8. www.englex.ru, http://englex.ru/educational-platforms/
9. www.open.edu, http://www.open.edu/openlearn/languages
10. www.ru.wikipedia.org, https://ru.wikipedia.org/wiki/Coursera#.D0.9A.D1.83.D1.80.D1.81.D1.8B

Teaching Tasks in Self-directed Reading as a Part of Foreign Language Course

Elena A. Sysa$^{(\boxtimes)}$ ⓘ, Liubov A. Sobinova ⓘ,
and Elena K. Prokhorets ⓘ

Tomsk Polytechnic University, Tomsk 634050, Russian Federation
jella-lumen@yandex.ru, sla_19.82@mail.ru,
lenpro@tpu.ru

Abstract. Higher education teaching encompasses more than merely giving students knowledge and developing skills. It entails engaging students in their own self-development in order to be able to grow in both ways personal and professional. Under the conditions of time constraints in technical universities of Russia, the emphasis on student independent work motivates researchers to look for new teaching methods that would make students responsible for their learning results and collected skills. In this regard, the roles of higher education faculty are constantly modified in response to the educational paradigm change. The task of a teacher is to help a student to realize him/herself as a subject of educational process, to teach them how to search, systematize and integrate knowledge from different areas of global experience. Another core task is to provide students with learning strategies that could be widely used in their further independent study of a foreign language and professional fields. From this perspective, the authors' aim is to investigate the teacher's role and formulate the main tasks for teacher aimed at development of skills in foreign language reading and self-manage learning activity at technical universities.

Keywords: Self-directed reading · Learning activity · Autonomy · Role of higher education teachers · Foreign language reading · Teaching task · Foreign language reading · Self-directed learning activity

1 Introduction

The development of skills to be a self-directed learner is currently one of important directions in the foreign languages teaching methodology; its main tenet is to focus the learning process on the student's personality. We agree with the opinion of many modern researchers such as Bimmel [5], Dickinson [7], Little [13], Nosacheva [15], Oxford [16], Prochorets [18], Rampillon [19], Solovova [20], Tambovkina [22] and others that a modern teacher changes her/his role from a lecturer who just delivers lectures to a consultant or a moderator in the educational process: who informs, advises and encourages learners how to be independent in their studying. Moreover, the teacher can work with students together as a groupmate or coworker doing the same tasks, research projects or taking part in the student-led discussions. The following role and function of a university teacher is unusual both for Russian teachers and students that is why we indicate the necessity to examine this research field.

© Springer International Publishing AG 2018
A. Filchenko and Z. Anikina (eds.), *Linguistic and Cultural Studies:*
Traditions and Innovations, Advances in Intelligent Systems and Computing 677,
DOI 10.1007/978-3-319-67843-6_24

Changes of pedagogical paradigm are reflected in the Federal State Educational Standard of Higher Education (2015). According to it, the university graduate should strive for self-development, enhance his/her skills, have the ability to evaluate advantages and disadvantages of the learning process critically, choose the proper tools for self-development and eliminate the gaps, use the basic methods, ways and means of obtaining, storing and processing information [8].

One of the methods that helps implement the result-oriented paradigm in educational practice can be the development of *self-directed reading* [7, 9, 12, 14, 21], because the identified key aim of the 21st century education is not only to teach students how to learn, but also self-manage their learning, that is, to be able to plan, organize educational activities, to implement self-reflection and self-correction [2, 5, 12, 13, 16, 17, 19, 21]. Awareness of their own actions importance for achievement of the educational purpose, acceptance by students of responsibility for result of learning will contribute to independent organization of self-directed learning activity and further acquisition of a foreign language, intensifies the process of foreign language teaching and learning, and also will increase competitiveness of graduates of technical universities [16].

In the process of learning a foreign language the independent educational student activity occurs with the development of skills in various types of activities, both in written and oral forms. In a technical university, to our mind, the most suitable activity for this purpose is reading, because it provides the basis for processing of scientific and professional information in the studied field, search for materials in the field of grants and projects within the disciplines of the professional cycle, and therefore, it supports practical expression in future professional activities.

The objective of this article is to investigate the teacher role and formulate the main teaching tasks aimed at developing of skills in self-directed reading as a part of foreign language course and self-manage learning activity at technical universities. Speaking about a great importance of a teacher in formation of self-directed reading during foreign language course, we find it necessary to define the term "self-directed reading". So, we specify it as *the learning activity where a learner develops the ability to self-manage his/her reading process (without teacher's help or the textassignments), not only apply reading strategies to understand a specific foreign text, but to gain the reader experience, which is required for further self-study of a foreign language.*

We define the reading strategy as the way to achieve the purpose of information perception being chosen consciously and based on the individual plan of certain mental actions.

2 Literature Review

It goes without saying that students' ability for self-directed reading largely depends on the teacher. According to Bagdziuniene and Kazlauskiene, analyzing the factors of improving quality of education, researchers have repeatedly provided evidence that teachers' professionalism has a greater impact on students learning in comparison with curricula, learning environment, funding, and number of students in the classroom [3, 12].

As it was already mentioned, the concept of self-directed learning and reading contributed immensely to the shift from the teacher-centered to the learner-centered approach in foreign language teaching. "This shift prompted to move from a traditional teacher's role of knowledge deliverer to one of a facilitator. The learner, who at the times of the grammar-translation and audio-lingual methods, was an empty vessel, is now expected to actively participate in the teaching/learning process inside and outside the language classroom. This immediately implies that the foreign language teaching/learning process presupposes an equal participation of both teachers and learners leading to shared responsibility" [4, p. 36].

According to Kazlauskiene, "the concept of competence is based on holistic and future oriented thinking; in practice, it should underpin decision-making structures, especially such competencies as critical evaluation of viewpoints and possibilities, clarification of values and commitment to engage and undertake risks" [11, p. 118] and to take responsibility for learning outcome. All these activities move the role of a teacher to a new level opposite to the traditional one where teachers only gives knowledge.

The process of students' abilities formation up to self-directed reading demands introducing learners to strategies of an effective work on foreign language texts from teachers supporting organization of educational process; strengthening the student faith in his/her own potential, promoting the awareness of students in managing their own learning and being responsible for the results. Thus, the teacher's role is changing. The teacher functions as a consultant, informs, advises and encourages students to independent actions. Cubukcu underlines that "the teacher's role is central to the development of learner capacity for self-directed learning. The teacher will need to support student independence in learning. This may involve a teacher first addressing learners' past educational experience, then slowly raising their awareness of increased independence benefits in their learning. Thus, learning helps develop the ability to think and act interdependently" [6, p. 17].

Stronge states, the essence of pedagogical counseling is "to support the learner in the educational process and provide advice in resolving problems related to educational activity and personal development in general" [21]. The teacher advises the student, promotes the maximum student independence in problem situations, encourages the manifestation of independence, observes and analyzes the learning activity of students, anticipates possible difficulties, if necessary, assists in the correction of learning activities – accompanies the learner in the educational process.

Boyadzhieva indicates in her research, "self-directed reading, as it is mentioned above, intrinsically implies freedom of choice. However, too many choices, which seemingly give more freedom, may in fact lead to obstacles due to low self-esteem and fear of failure and to the effect of blocking motivation, especially if a wrong choice causing dissatisfaction is made. This psychological phenomenon is known as the paradox of choice. What is more, some cultures are more sensitive to decision making than others, and the paradox of choice is very likely to result in blocking motivation, thus, affecting the whole teaching/learning process" [4, p. 37].

Everybody should always respect the opinion of the elderly. "It is interesting that questioning of "authority" is a supposed mainstay of science. In the end striking a balance between the two tensions of preserving our cultural identity and helping students become free and conscious thinkers, and tolerant to diversity of viewpoints and opinions could be probably better. Why can't we do both simultaneously?" [14, p. 696].

We agree with F. Cubukcu that "learners and teachers are partners in the learning process" [6, p. 17]. According to Nosacheva [15], the ability to be a consultant includes:

- Desire to perceive the student as the subject of educational activity and to approve the chosen ways of learning a foreign language;
- Ability to describe the possible ways of working on a foreign language visually and clearly.

In our opinion, the process of developing abilities is a gradual experience. It is a process where both a teacher and a learner are involved: the teacher does not control the learning process, but *gradually involves the student in co-management* of learning process. *Educational counseling and support* will gradually help to achieve the objectives of the above educational process.

Analysis of psychological and pedagogical literature allows making a conclusion that many researchers [1, 3–7, 9, 10, 12–21] believe that the teacher's role is delicate and requires a high degree of such professional skills' development as:

- Ability to work with students;
- Ability to analyze learning and cognitive activities;
- Ability to anticipate possible difficulties that students may have;
- Ability to create the conditions, ensuring development of student autonomy, the desire for self-development and self-education while learning and using a foreign language.

3   Research Methods

During examining the problem the authors have chosen the following research methods: analysis of psychological and pedagogical literature, monitoring of the academic process at the technical university, studying the experience of foreign language learning at the technical university, surveying of students and teachers. These methods allowed us to conclude that recently the position of a teacher in the traditional lesson is dominant. Students are not quite ready for self-directed learning activity, are not aware of their role in the management of educational and cognitive activity, students do not want to be responsible for their results. Cognitive independence of students is reduced, students do not believe in their abilities.

In order to make reliable conclusions about the level of student autonomy in the process of reading foreign texts and to identify the difficulties encountered by students working on foreign-language texts, a questionnaire survey of students and teachers of a foreign language was conducted.

3.1 Participants and Procedure

The questionnaire encompasses 7 German language teachers working at Tomsk Polytechnic University and 56 freshmen of technical specialization at the same university. Students were asked to make a list of skills from the suggested table in the process of reading a foreign-language text. The crucial condition for this work was to act independently. To obtain more objective results, teachers were asked to accompany the students work on the text, to evaluate their level of skills in the course of foreign text reading. Fifteen main parameters of self-directed learning activity were evaluated and included in the table (Table 1).

Table 1. The parameters of (self-) evaluation.

The parameters of (self-) evaluations	Self-directed reading	Points (0–2)
The psychological attitude to learning activities	1. Self-control/self-evaluation – getting the goal of reading 2. Self-motivation for successful completion of the work	
Self-management of learning activities	3. Self-planning of learning activities/steps in the context of the task/the text genre 4. Self-determination of the reading goal 5. Choosing the most effective ways to achieve the goal/the optimal task execution 6. Defining of reading types in accordance with the purpose	
Social interaction	7. Discussion of the obtained results in groups 8. Explanation and justification of the actions algorithm	
Understanding the content of the text	9. Prediction at the semantic level (the hypotheses about the text content, the disclosure of causality) 10. Highlighting of core elements in the text: main thoughts, facts, and explaining its main idea 11. Finding the primary and secondary information in the text 12. Critical evaluation and interpretation of the facts in the text	
Understanding the language material	13. Defining the meaning of compound words having a familiar part 14. Defining the meaning of unknown words in the context without using a dictionary 15. Defining the meaning of unknown words with the use of references and additional literature	

3.2 Research Tools

The evaluation of skills to act independently in the process of working on a foreign language text was carried out according to the following parameters: psychological attitude to learning activities; self-management of learning activities; social interaction;

understanding the language material; and understanding the content of the text. Each parameter includes 2–4 skills, which were suggestes for evaluation by the following criteria: 2 – the skill is developed, 1 – the skill is poorly developed, the teacher's support is often needed; 0 – the skill is not developed, the learner follows the instructions of the teacher. Quantitative analysis was performed using statistical methods of data processing (the method of variation series, the characteristics of a scatter plot). All of these allowed us determining the level of skills in self-directed reading which can be low, medium and high.

4   Results

Table 2 presents the survey results, according to which it can be concluded that the level of student autonomy remains low, most students are dependent on the teacher and are not quite ready for self-directed learning in general, and reading in particular. Students do not possess rational ways of working with information sources. Moreover, they are not able to independently organize a plan for work and apply it to reading. And, finally, it should be noted that learners lack the self-reflection skills.

Table 2. The levels of self-directed reading skills.

The levels	Number of persons
High	5
Medium	20
Low	31

The ability to think independently is formed in the process of active learning. Therefore, teachers need to create conditions for development of student abilities to apply learning strategies, realize and evaluate their effectiveness and take the initiative of their learning. The teacher role is not reduced, on the contrary, it performs the difficult task of creating conditions in which students may become independent and responsible personalities in the educational environment. The main aim is to develop the student's ability to self-management and training of cognitive activity and the leading subjective qualities of autonomous personality. The teacher's activity in this context lies in encouraging students to solve the following problem:

- To learn how to organize educational activities;
- To develop skills that are necessary for self-directed learning activities;
- To learn to apply them in educational and professional activities;
- To test the effectiveness of the acquired skills in real-world learning and practical situations;
- To learn to exercise self-reflection.

Based on the presented arguments, analysis of psychological and pedagogical literature, the survey results, we formulate the following teaching tasks from the position of self-directed learning activity for reading in a foreign language in the technical university:

- Managing the work of students to obtain new reading strategies;
- Assessment of effectiveness of reading strategies;
- Advising students about opportunities of achieving goals of foreign language reading and results in learning activities;
- Developing the ability of students for goal-setting, self-assessment, self-reflection, self-correction and self-control of personal intrinsic intention to read and cope with all related challenges;
- Developing the ability of students to employ this experience in other types of educational and professional fields;
- Continuous professional education and self-development.

These aspects reflect the main teaching tasks: educational, developing, advising and encouraging that characterize the key role of a teacher *in co-management of the educational pro*cess.

5 Conclusion

To sum up, we came to the following conclusion: the main tasks in teaching reading are, firstly, to develop students' target skills; secondly, to organize their own learning actions; thirdly, to achieve reading goals; fourthly, to learn to apply learning strategies; next, to help students be active and conscious; and, finally, to manage learning activity in foreign language reading.

Enumerated tasks have a great influence on the process of skill development in foreign language reading, based on self-directed and cognitive activity of technical university students.

The paper presents a part of the complete research and the illustrated results show valid and reliable data.

References

1. Anikina, Z.S., Agafonova, L.I.: Rasvitiye uchebnoj avtonomii v obuchenii inostrannym yasykam (In Russian) [The development of learning autonomy in language learning: the history of the issue in foreign pedagogic]. Vestnik Tomskogo pedagogicheskogo universiteta **4**, 23–27 (2009)
2. Bagdziuniene, D., Kazlauskiene, A., Liniauskaite, D., Nasvytiene, D., Sakadolskiene, E., Seckuviene, H.: Evaluating the motivation of students to become teachers during admission into education-related study programmes. Pedagogy **113**(1), 28–44 (2014)
3. Benson, P., Voller, P.: Autonomy and Independence in Language Learning. Longman, London (1997)
4. Boyadzhieva, E.: Learner-centered teaching and learner autonomy. Procedia – Soc. Behav. Sci. **232**, 35–40 (2016)
5. Bimmel, P., Rammpillon, U.: Lernautonomie und Lernstrategien. GIN, München (2000)
6. Cubukcu, F.: The correlation between teacher trainers' and pre-service teachers' perceptions of autonomy. Procedia – Soc. Behav. Sci. **232**, 12–17 (2016)

7. Dickinson, B.: Crafting learner-centred processes using action research and action learning. In: Hase, S., Kenyon, C. (eds.) Self-determined Learning: Pedagogy in Action, pp. 18–32. Bloomsbury Academic, London (2013)

8. Ehlers, U.D.: Open Learning Cultures. A Guide to Quality, Evaluation, and Assessment for Future Learning. Springer, Berlin (2013)

9. Federal state educational standards for higher education. http://fgosvo.ru. Accessed 19 Apr 2017

10. Gros, B., Kinshuk, M.: The Future of Ubiquitous Learning: Learning Designs for Emerging Pedagogies. Springer, Berlin (2016)

11. Kazlauskiene, A., Gaucaite, R., Poceviciene, R.: Preconditions for sustainable changes in didactics applying self-directed learning. J. Teach. Educ. Sustain. **18**(2), 105–118 (2016)

12. Lederman, J.S., Lederman, M.G.: The functions of a teacher. Teach. Educ. **27**, 693–696 (2016)

13. Little, D.: Learner autonomy in action: adult immigrants learning English in Ireland. In: Kjisik, F., Voller, P., Aoki, N., Nakata, Y. (eds.) Mapping the Terrain of Learner Autonomy: Learning Environments, Learning Communities and Identities, pp. 51–85. Tampere University Press, Tampere (2009)

14. McNaughton, S., Billot, J.: Negotiating academic teacher identity shifts during higher education contextual change. Teach. High. Educ. **21**, 644–658 (2016)

15. Nosacheva, E.N.: Osnovy razvitiya professionalnoi avtonomii budushchego prepodavatelia innostrannykh yasykov kak subyekta uchebnoi deyatelnosti (In Russian) [Basis for the development of professional autonomy of the future teacher of foreign languages as a subject of educational activity]. Izvestia rossijskogo gosudarstvennogo pedagogicheskogo univer- siteta im. A. I. Herzena. Psychologicheskie and pedagogogicheskie nauki (psychology, pedagogy, theory and methodology) 100, 37–47 (2009)

16. Oxford, R.: Language Learning Strategies. What Every Teacher Should Know. Heinle & Heinle Publishers, Boston (1990)

17. Pevzner, M., Zaichenko, O.: Pedagogicheskoje konsultirovanije (In Russian) [Educational counseling]. RIS NovSU, IOMKR, Veliky Novgorod (2002)

18. Prokhorets, E.K., Sysa, E.A., Rudneva, E.L.: Teaching of autonomous foreign language reading in technical university: criteria for the selection of textual material. Procedia – Soc. Behav. Sci. **215**, 256–259 (2015)

19. Rampillon, U.: Autonomes Lernen im Fremdsprachenunterricht ein Widerspruch in sich oder eine neue Perspektive? Die Neueren Sprachen **93**(5), 455–466 (1994)

20. Solovova, E.N.: Avtonomiya uchashchikhsya kak osnova rasvitiya sovremennogo nepre- ryvnogo obrasovaniya lichnosti (In Russian) [Autonomy of students as a basis for the development of modern continuous education of a person]. Inostrannye yasyki v shkole **2**, 11–17 (2014)

21. Stronge, J.H.: Effective Teachers = Student Achievement: What the Research Says. Routledge, New York (2010)

22. Tambovkina, I.Y.: To the problem of autonomy of students of foreign language at the pedagogical University. Inostrannye yasyki v shkole **4**, 84–88 (1998)

The Role of Evaluation Tools Kit in Recording of Foreign Language Learning Results

Alexander N. Shamov[ID] and Ludmila V. Guseva[✉][ID]

Minin University, Nizhny Novgorod 603002, Russian Federation
schamow.alexandr@yandex.ru, ludmila_guseva@yahoo.com

Abstract. The article examines evaluation tools kit as means to control the level of foreign language acquisition. Certain principles of such a kit collection are described. With the help of the kit objective evaluation and assessment of foreign language skills in diverse educational environments are possible.

Keywords: Control · Evaluation tools kit · The structure of evaluation tools kit · Principles of evaluation tools kit use · Indicators and descriptors of language skills level

1 Introduction

In didactics and methodology the term "control" is viewed as a process of defining the amount of knowledge and the quality of speech skills a student has developed as the result of completing a series of oral and written assignments in a certain educational institution [1, 2, 4, 5]. On the basis of the results, the assessment of one unit of the course or some academic period is made. Further the assessment will be the basis for a certain grade (mark) which will show the level of foreign language acquisition.

Mastering different kinds of a foreign language speech activity is a gradual and systematic process of speech skills formation. Gradualness and sequence in language skills acquisition is reflected in different levels of language ability. Teacher's controlling activity also has two more functions: (1) to "diagnose" difficulties which occur when students master their language skills; (2) to check the effectiveness of ways and means used in language learning [3, 4, 6].

2 Methodology

In Russian methodology there are two definitions of the term "control" which are given in the "Dictionary of Methodological Terms". The first one views control as a process of defining the level of student's language skills which they've achieved as the result of written and oral exercises, completed during some academic period or the course. The second definition views "control" as a part of a lesson which is devoted to checking the student's knowledge or skills gained during the lesson.

The control over the language level acquired or the quality of language skills mastered can be performed in different forms and ways:

A. Filchenko and Z. Anikina (eds.), *Linguistic and Cultural Studies:*
Traditions and Innovations, Advances in Intelligent Systems and Computing 677,
DOI 10.1007/978-3-319-67843-6_25

- Frontal (in a group) and individual questioning;
- Peer-to-peer questioning;
- Mixed questioning;
- Observation of speech activities during the lesson;
- Interview after a completed project or assignment;
- Testing;
- Traditional quiz;
- Self-evaluation;
- Exam.

To organize a comprehensive procedure of control and to evaluate foreign language learning results adequately, we suggest we use Evaluation Tools Kit.

Evaluation Tools Kit is a complex of controlling and evaluating assignments, tests, quizzes, projects, tasks for creative works such as essays, compositions, etc., aimed at assessing and evaluating the value of professional knowledge and quality of language skills, communication performance and professional competences as a whole. The Kit is created by a teacher for a certain course. The assignments/tools are collected by a teacher during several years of instruction.

Not all the tools will appear in the Kit. To be included in the Kit the tools should have some consistent pattern and follow certain principles. We will define some of them.

1. *Regularity.* The evaluation of student's academic activity should be performed regularly during the whole period of instruction.
2. *Encouraging.* The assignments should encourage students to further development of their academic and cognitive skills.
3. *Fairness.* The assignments should evaluate the whole scope of learner's skills and competences fairly.
4. *Based on competence approach.* The tools should show what a learner knows, what skills possesses, and how they perform.
5. *Representativity.* The evaluation should include and represent the progress results of knowledge acquisition, professional skills development, communicative experience, separate components of language competence development and others.
6. *Control over development of all the components of professional competences.*
7. *Correction assignments.* The Kit should include tools to correct and re-do some tests and exercises to achieve results of higher level.
8. *Combination of traditional and non-traditional tools for evaluation and assessment of student's skills and competences.*
9. *Possibility to automatically (with the help of some gargets) interpret the results.*

The Evaluation Tools Kit can include the following groups of *assignments:*

1. Assignments on certain topics, modules, the whole course.
2. Current control assignments for a module, grammar topic, and vocabulary topic.
3. Final control assignments at the end of semester, term, and year.
4. Assignments for students having individual curriculum (those who have a semester abroad, on a maternity leave, etc.).
5. Assignments for self-evaluation.

6. Placement assignments for working out an individual educational track for a student.
7. Accreditation assignments to be presented to external controlling organizations.
8. International professional tests (optional).
9. Written tests on (1) grammar and lexis; (2) terminology; (3) language activities.
10. Learner's electronic portfolio [9].

Practice-oriented test assignments are often used to control learner's language skills. All the tests are divided into:

- General skill tests
- Progress tests
- Diagnostic tests
- Placement tests
- Competence tests

3 Results

After the tests are completed the results can be evaluated according to the following criteria:

100%–85% of the right answers – "excellent";
84%–70% of the right answers – "good";
69%–55% of the right answers – "satisfactory";
54% and less of the right answers – "poor" [10].

The control and evaluation of foreign language oral and written skills has a very important place in educational process. The control does not only mean counting grammar, lexical and pronunciation mistakes. One cannot check and evaluate oral and written competences with the help of tests. Due to this, a new goal for controlling and evaluation of language skills has come to existence: it is to solve communicative and pragmatic tasks. In the controlling process not only skills of different language activities but successful solving of communicative and pragmatic goals with their help should be evaluated and assessed [9].

The evaluation of speech skills is based on certain indicators. The indicators should be: (a) objective; (b) measurable; (c) usable in everyday foreign language teacher's activities.

To evaluate the practical use of language skills basic and additional criteria are applied. The basic criteria help define the minimum level of a certain language activity. Additional criteria help define a higher and more qualitative level of language skills.

The basic criteria describe qualitative indicators of language activity operation. The indicators are set by the syllabus for each language activity and for each level. The quantitative criteria characterize a language skill in absolute units. There is an example of a speaking skill evaluation in a monologue. The cited below task was completed by the students majoring in economics in Minin University, Nizhny Novgorod, Russia (Table 1).

Table 1. Evaluation criteria for speaking in the form of a monologue and expected results.

The object of control – language activity	Expected results
Speaking Form – a monologue Situation 1 The authorities of the company where you work as a manager assigned you a task to present a new innovative product at an Annual Scientific Innovations Fair. Prepare a presentation and report to the group	KNOWS - Production process of certain goods, new technologies of the production - Topical vocabulary and possible situations - Terms of the given scientific field and production CAN - Present an innovative product in their native as well as in a foreign language - Show distinctive features and advantages of the product - Convince a partner to buy the product and to promote it successfully on the market USES - Lexical and grammatical structures - Terminology - Professional knowledge in the given field

To evaluate the quality of speaking skills in a monologue we can use the following indicators:

1. adequacy of a monologue content to the communication goal given in the situation;
2. adequacy of the monologue to the topic and the situation;
3. coherence and content logic;
4. appropriateness to the goal of communication;
5. diversity of speech patterns;
6. logic in the monologue structure;
7. good timing;
8. presence of creative elements in the presentation;
9. presence of different types of narrative in the presentation;
10. adequacy of grammatical and lexical material used to achieve the communication goal;
11. lexical, grammatical and phonetic accuracy [6, 8].

The given criteria can be divided into (1) basic, and (2) additional. The basic evaluation criteria of monologue skills mastery are 1 to 10. Additional criterion is 11. To control and evaluate language skills in a monologue we can use such ways as:

- to make up a story based on the situation given;
- to describe a picture of a new innovative product;
- to give a visual presentation of the product;
- to create an ad for the product and comment on it.

To evaluate the assignment results we can give grades (marks) taking into consideration how many criteria the monologue has met:

"excellent" – all the basic and additional criteria are met in the monologue.
"good" – one basic criterion is missed.
"satisfactory" – one basic criterion is missed; lack of grammatical or/and lexical or/and phonetic accuracy in the speech;
"poor" – more than one basic criteria are missed; lack of grammatical or/and lexical or/and phonetic accuracy in the speech.

In the process of controlling and evaluating foreign language skills in a dialogue, a role play is considered as a very important means. A role play has many functions in education process. They are learning, educational, motivational, and compensational. In the role play structure we define the following:

– Roles, taken by the players;
– Playing actions as ways to realize a role;
– Substitution of real objects by playing objects;
– Real relationships between the players;
– Content – situation, given reality.

There are several stages in a role play realization:

1. Preparation stage (writing a script, working out a plan of realization, description of a play, and working out characteristics of roles);
2. Explanation stage (distribution of roles, working out the time frame for the play, setting the main goal, defining the problem, choosing a situation, working with additional materials, and psychological preparation for the play);
3. Realization stage (in-class or out of class activity);
4. Reflection stage (analyses of the result, self-analyses, evaluation and self-evaluation, conclusions and recommendations for the future).

The following role play was performed by the first-year students of a non-linguistic major at Minin University, Nizhny Novgorod, Russia (Table 2).

Table 2. Evaluation criteria for speaking in the form of a dialogue and expected results.

The object of control – language activity	Expected results
Speaking Form – a dialogue Situation 1 Your group is getting ready for a Teacher's Day. You get together after classes to discuss how to congratulate your teachers. Each of you has a suggestion	KNOWS - Vocabulary on the topic "Holidays and Traditions" - Speech etiquette for a discussion and exchanging opinions - Facts about the holiday traditions CAN - Describe the holiday traditions - Write a congratulatory card - Prove a point of view and convince everybody of it USES - Vocabulary on the topic - Grammatical structures characteristic of holiday discourse - Speech etiquette formulas in a given situation

Firstly, the students think over the situation, their possible suggestions and replies. Then, they role play the conversation. To evaluate their speaking skills in the form of a dialogue, we use the indicators given below:

(1) Adequacy of the content to the communication goal;
(2) Adequate use of dialogue techniques;
(3) Appropriateness of emotions;
(4) Length of the conversation (according to the syllabus requirements);
(5) Presence of spontaneous creative elements;
(6) Use of speech etiquette formulas and clichés;
(7) Use of strategies of cooperation (formulas to keep the conversation going, to attract attention, to start and to finish the conversation);
(8) Flexibility in developing the topic of the conversation;
(9) Use of different types of dialogues;
(10) Use of non-verbal means of communication;
(11) Adequacy of lexical and grammatical material chosen for achieving the communication goal;
(12) Lexical, grammatical and phonetic accuracy of a dialogue;
(13) Skills to predict and use strategies to avoid cultural differences;
(14) Skills to adapt to foreign environment following the norms of politeness and etiquette, showing respect to traditions and lifestyle of a different socio-cultural community [6].

The basic evaluation criteria of dialogue skills mastery are 1 to 11. Additional criteria are 12 and 13.

To evaluate the assignment results we can give grades (marks) taking into consideration how many criteria the dialogue has met:

"excellent" – all the basic and additional criteria are met in the dialogue.

"good" – one basic criterion is missed.

"satisfactory" – one basic criterion is missed; lack of grammatical or/and lexical or/and phonetic accuracy in the speech;

"poor" – more than one basic criteria are missed; lack of grammatical or/and lexical or/and phonetic accuracy in the speech.

4 Conclusion

Objectiveness is one of the most important criteria of a learner's speech activity evaluation. Judging by the results of each individual, we can make a conclusion about the learning progress in a group as a whole. In its turn, this conclusion can give us grounds to evaluate our teaching results: what techniques and technologies work best, what ways of control, evaluation and assessment show the results more adequately.

Objectiveness in evaluation of learner's results can help the students:

(1) See their progress results in language learning;
(2) See the quality of their speech skills in situations close to real communication;

(3) Cope with a foreign language course at baccalaureate and master's level;
(4) Get ready for a final exam;
(5) Gain experience and skills to participate in international professional exchange programs or continue their education in one of the foreign universities.

References

1. Adey, P.A.: Model for professional development of teachers thinking. Think. Skills Creativity **1**(1), 49–56 (2006)
2. Benson, P., Lor, W.: Concepts of language and language learning. System **27**(4), 459–472 (1999)
3. Cetlin, V.: Kontrol' rechevyh umenij i navykov v obuchenii inostrannym yazykam [Assessment of speech skills in foreign language teaching]. Vysshaya shkola, Moscow (1970). (in Russian)
4. Hofer, B.K., Yu, S.L.: Teaching self-regulated learning through a 'learning to learn' course. Teach. Psychol. **30**(1), 30–33 (2003)
5. Konysheva, A.: Kontrol' rezul'tatov obucheniya inostrannomu yazyku: Materialy dlya specialista obrazovatel'nogo uchrezhdeniya [Result assessment of foreign language acquisition: materials for a specialist of an educational institution]. Karo, Moscow (2004). (in Russian)
6. Klimentenko, A.: Teoreticheskie osnovy metodiki obucheniya inostrannym yazykam v srednej shkole [The theoretical basis of foreign language methodology in a secondaru school]. Pedagogika, Moscow (1981). (in Russian)
7. Mirolubova, A.: Voprosy kontrolya obuchennosti uchatchihsya inostrannomu yazyku [Assessment questions of student skills in a foreign language]. Titul, Obninsk (2017). (in Russian)
8. Perevoshchikova, N.: Modernizaciya obrazovatel'nogo processa: tekhnologiya konstruirovaniya ocenochnyh sredstv dlya ocenki obrazovatel'nyh rezul'tatov. Uchebno-metodichnskoe posobie [Modernization of the educational process: the technology of assessment tools for educational performances]. Minin university Publishers, Nizhnii Novgorod (2016). (in Russian)
9. Shamov, A., Yurlova, N.: Ocenka yazykovoj i rechevoj deyatel'nosti na urokah inostrannogo yazyka [Assessment of language and speech activity at classes in English]. NIRO, Moscow (2014). (in Russian)
10. Shatilov, S.: Metodika obucheniya nemeckomu yazyku v srednej shkole (in Russian) [Methodology of German language teaching in a secondary school]. Prosveshchenie, Moscow (1986)

Linguo-Cultural Competence Formation During the University Courses Learning

Inna A. Cheremisina-Harrer(iD), Veronika M. Rostovtseva(iD),
Nikolay A. Kachalov(⊠)(iD), and Yunona V. Shcherbina(iD)

Tomsk Polytechnic University, Tomsk 634050, Russian Federation
{cheremisina, kachalov}@tpu.ru, vicol@mail.ru,
juna7tokyo@yahoo.com

Abstract. The article considers formation of the linguo-cultural competence in the framework of the developed elective course "US Higher Education". The created methodology for the competence formation is based on a set of principles (personality-oriented, cognitive-cultural orientation, culturally-correlated study of foreign and native languages, reliance on the native culture, productivity, principles of increasing the ratio of students' independence, constructive cooperation of the university courses; principles of monotony and concentricity) and teaching methods (information-receptive, reproductive method of problematic presentation, heuristic and research methods), including a number of functional stages.

Linguo-cultural competence is determined as a structural component of the sociocultural competence within the framework of the communicative competence of students in a language-oriented elective course. The article illustrates the use of the linguo-cultural field of the studied culture as a means of generalization and systematization of knowledge about a foreign culture for specification of the elective course content within the framework of the linguo-cultural approach. The specific content of student's educational activity represents the linguo-cultural interpretation of the text, the task to which requires the students' comprehension of the culturally-labeled lexical units.

Keywords: Linguo-cultural competence · Culturally-marked lexical unit · Teaching principles and methods · Elective course

1 Introduction

The modern higher education determines the language teaching method as a set of forms, methods and techniques for teacher's work, in which they are defined as the instructor's actions and operations, aimed at gaining knowledge, formation and development of skills, and motivation of students to learning activities in order to solve particular tasks of the learning process [1].

The process of development of the methodology for the linguo-cultural competence formation was based on a set of principles that accomplish the goal and the content of foreign language teaching from the position of the linguo-cultural approach for development of an elective course at the department of foreign languages in a non-linguistic

© Springer International Publishing AG 2018
A. Filchenko and Z. Anikina (eds.), *Linguistic and Cultural Studies:*
Traditions and Innovations, Advances in Intelligent Systems and Computing 677,
DOI 10.1007/978-3-319-67843-6_26

institution: personality-oriented, cognitive-cultural orientation, culturally-correlated study of foreign and native languages, reliance on the native culture, productivity, principles of increasing the ratio of students' independence, constructive cooperation of subjects in the educational process; principles of monotony and concentricity.

2   General and Methodological Principles of Teaching

Certain aspects of this topic were covered in the works of the following researchers who studied the problems of Linguo-cultural competence formation [3–9]. The methodology basis should be made up by a set of principles, which are implemented through procedures. The methodology and linguo-didactics determine the principles of teaching as the initial conditions that specify requirements to the learning process. The initial ones in determining the nomenclature of principles of formation of linguo-cultural competence are the personal-activity, cognitive-communicative, linguo-culturological, linguo-cultural, and socio-cultural approaches [1].

In order to identify the system of principles that determine the process of foreign languages teaching from the standpoint of the linguo-cultural approach, we have relied on numerous studies, researching the common didactic and methodological principles of teaching. The study of the problem of the education principles within the framework of culturally-oriented approaches and concepts allowed pinpointing the following principles as the leading principles for the elective course: principles of personality-oriented, cognitive, culturological orientation, culturally-correlated study of foreign and native languages, reliance on the native culture, productivity, principle of increasing the ratio of students' independence, constructive cooperation of of the university courses; principles of monotony and concentricity.

First, let's consider the essence of each of these principles to further elucidate their implementation within the framework of the methodology.

The personality-oriented principle of a foreign language learning [2], which involves change in the status of the student as a real subject of the cognitive activity, responsible for the results of his academic work and formation of personal qualities important for intercultural communication.

The principle of cognitive culturological orientation [6, 9]. In teaching a foreign language from the position of the linguo-cultural approach, this principle is oriented on implementation of a positive transfer of the cultural information, assigned to a language unit and prevention of any possible interference. In accordance with the following principle, cognition and comprehension of the cultural phenomenon occurs due to the "acculturation" of linguistic units [8].

In other words, full comprehension of a cultural fact is ensured by the obligatory observance of a number of conditions:

(A) perception of culturally-labeled lexical units;
(B) research of the cultural meaning of culturally-labeled lexical units through various sources;
(C) awareness of the cultural meaning of culturally-labeled lexical units and through it of the subject, represented by it;

(D) inclusion of a cultural phenomenon in the network of cultural associations of the studied culturally-marked lexical units on the basis of comparison with a similar phenomenon in their native culture and detection of common and different features in two linguo-cultures.

The latter condition is the key to implementation of the principle of cognitive-cultural orientation since compliance with this condition ensures the formation of cognition which is based on comparison of two cultures. The implementation of this principle promotes the development of students' skills of perception and penetration into a deep network of cultural associations of culturally-labeled lexical units, which determine the integral awareness of the linguistic unit, as well as comprehension of the extra-linguistic phenomenon that the culturally labeled lexical unit represents and enables us to understand not only the features of a foreign culture, but also of the native culture.

The principle of culturally correlated study of foreign and native languages [6]. This principle consists in a detailed explanation and constant demonstration of inseparability of the studied language and foreign culture. In accordance with this principle, work with the linguistic material must be accompanied by comparison of the characteristics of two cultures. Such activity is a necessary condition for the formation of consciousness, based on real knowledge, and not on stereotypes, as well as development of the ability of learners to penetrate deeper into their own language and culture. Implementation of this principle in the educational process contributes to the formation of a multicultural worldview of students, their awareness of the fact that there are numerous ways of understanding the world.

The principle of reliance on native culture [5]. This principle complements the previous one and provides the basis for:

(A) prediction and removal of difficulties that are possible when students perceive, understand and interpret authentic texts;
(B) analysis and comparison of cultural phenomena within the framework of two linguistic societies.

The principle of increasing the ratio of students' independence [3] emphasizes the importance of students' self-determination and self-realization, suggests the development of their autonomy and reflection in the process of foreign language learning, development of a conscious attitude to organization of their educational activity up to planning an individual trajectory of learning.

The principle of constructive cooperation between of the university courses [9]. This principle complements the previous principle and implies that the organization of the process of foreign language teaching from the position of the linguo-cultural approach is a process of a joint (teacher and students) solving of creative tasks. In addition, this principle assumes a gradual decrease in the managerial role of the teacher and an increase in the cognitive activity of students with the aim of self-regulation of the educational activities by the latter.

The principle of productivity [9]. This principle aims at improvement of both the material products of education, upbringing and development, and the intangible (in the form of making a semantic decision, increasing knowledge, forming skills and developing skills, and spiritual qualities of an individual).

It is necessary to identify the principles that underlie the linguo-cultural approach and are essential for the formation of the linguo-cultural competence.

The principle of a systematic and holistic representation of cultural phenomena. In the framework of foreign language teaching in higher education from the position of the linguo-cultural approach, this principle is realized through the joint activity of the of the university courses in developing the linguo-cultural field in the process of studying the layer of foreign cultural reality. As a result of this activity, students develop a systematic and holistic view on the essence of the cultural phenomenon being studied at the expense of substantial continuity of linguo-cultural themes within the specified field, as well as the hierarchy of the field structure, which implies a clear distribution of linguo-cultural themes from general to special, and then to particular (individual), from dominant to peripheral.

The principle of monotony. With reference to foreign language teaching in higher education from the position of the linguo-cultural approach, this principle involves immersion in the culture of the country's peoples of the target language through the study of a certain layer of culture, in our case, "Higher Education in the US", which sets the topic of study relevant and interesting for the learners. In accordance with this topic, authentic texts (as well as audio and video materials as an additional means of instruction) of a problem nature are selected, which are full of linguo-cultural themes with thematic affinity (all linguo-cultural themes disclose the essence of higher education in the US as a system). As a result of these texts studying, a linguo-cultural field with a single name is developed, the lexical content of which is united by one common idea, characteristic of each linguo-culture in the field framework, which allows you to separate the thematic language units from foreign units with a different theme. That way the systematicity in the perception and appropriation of cultural facts by students is achieved.

The principle of concentricity. This principle is interrelated with the previous principle and involves the creation of the linguistic and cultural field by the students, which is structured as concentric circles of the main (general) and additional (special). The task of additional (special) circles is to reveal the essence of the given cultural phenomena with additional shades and greater depth. In this way, in each of the following special subfields the cultural information is expanded and deepened. Such organization of the linguo-cultural field facilitates the perception and assimilation of culture-logical material built on the "snowball" principle which involves a sequential expansion of a certain phenomenon cultural background in the reality of a foreign culture represented by subfields of different order.

3   Methodological Basis of the Competence Formation

The choice of the principles for foreign language teaching from the position of the linguo-cultural approach led to selection of methods for linguo-cultural competence formation. In order to define the methods of teaching, a classification of the teaching methods, suggested by Lerner, was used [7]. Under "methods" we understand a system of successive actions of the teacher, organizing and conditioning the cognitive and practical activities of learners in the assimilation of all elements of the education content to achieve the learning goals.

As a criterion for classifying methods, we use their focus on a certain component of the teaching content and the nature of students activity. Based on these positions, the methods are divided into reproductive and productive methods, which are represented by the following nomenclature: information-receptive, reproductive method of problematic presentation, heuristic and research methods.

Information-receptive method. The essence of this method is in the communication of the culturological information (key categories in linguo-culturology; background knowledge about the system of higher education in the USA in general), as well as information related to extra-linguistic features of authentic texts reflecting a different culture, from the teacher and perception by the students.

By this method:

(A) a culturological background is created, which is necessary for further activity of students;
(B) the students are motivated to identify and comment on the characteristics of another culture, as well as to conduct the intercultural comparisons of phenomena in two cultures.

Reproductive method. In the framework of the developed course the reproductive method has different forms of representation: reproduction of cultural material that has been acquired both in practical exercises and in the process of independent research and study of reference literature and internet sources; reproductive conversation, solution of the reproductive communicative cognitive tasks: presentation of the main content of the text and replication of the commentary from the glossary.

Heuristic method. The essence of the heuristic method lies in the fact that a complex educational task is divided by the teacher into a number of more particular ones, and he manages its solution with the help of heuristics (more complicated questions and tasks). The heuristic method is fully implemented in the process of controlled conversation about the text at the stage of its comprehension (Enhancing Comprehension), in the process of analysis of the culturally-labeled vocabulary (Culture Perspective), and creation of a linguo-cultural field.

Method of problem presentation. This method assumes that the teacher through an authentic text presents a problem of intercultural nature to the students. All work on the text is problem-based, starting with a discussion of problematic issues and ending with a cross-cultural discussion of the problems in the text at the Leads-To-Discussion phase. The problem statement motivates the students to perform the intercultural interpretation of the text, conduct comparative analysis of the cultural phenomenon presented in the text in their own and other cultures.

Research method. The essence of this method is that students are invited to conduct a mini-research, collecting a cultural information in the process of independent work with internet resources, development of the linguo-cultural field, compilation of the educational linguo-cultural commentary, and in the process of project assignments preparation to complete the work in the elective course. This method activates the thinking activity of students, increases the positive motivation for learning. The process of active acquiring of knowledge forms a personality that is capable of critical comprehension of one's own and other cultures. The ability of critical thinking, based on knowledge and not on stereotypes, is an indispensable condition for teaching students the research attitude to the culture under study [4].

The research method involves joint activities of the students and the teacher in the process of developing a linguo-cultural field and creating a learning commentary. The facilitator activity provides:

(A) necessary degree of perception and comprehension of the culturological information by students and its objective interpretation;
(B) serves as a necessary prerequisite for the gradual development of the students' ability to conduct research activities independently.

4 Conclusion

Selection and justification of the principles led to a set of teaching methods. Each of the selected methods (information-receptive, reproductive, heuristic, problematic, research) has found its implementation in the reasoned teaching techniques (conversation on the topic of the text, tasks for analysis of the semantic structure of the culturally-labeled lexical units, a conversation on the intercultural problems of the text, collective creation of the linguo-cultural field, etc.).

These teaching principles and methods were used as the basis for developing stages of the formation process of elementary linguo-cultural competence by students:

- introductory conversation (orientation) aimed at formation of the background representations and linguo-cultural knowledge, comprising an indicative basis for the subsequent work with authentic texts in the elective course;
- pre-reading talk aimed at activating students' knowledge about the problem of the text and creating motivation for reading the text at home;
- text-based (while-reading activities) providing organization of perception, reflection and self-control, completion of lexical exercises and collection of information about the culture phenomenon;
- deepening of the text comprehension (enhancing comprehension), suggesting further comprehension of the implicit information in the text;
- linguo-cultural interpretation of vocabulary with a cultural component of meaning (culture perspective) aimed at familiarization of learners with the culturally-labeled lexical units and development of skills necessary for their analysis;
- shortening of the text (summarizing) aimed at developing skills of generalizing and stating the main content of the text;
- discussion of the intercultural problems of the text (discussion) suggesting the development of the skills of comparing phenomena in two cultures;
- development of the linguo-cultural field (cultural web) connected with the organization of generalization and systematization of cultural knowledge by means of a linguo-cultural field, and aimed at development of the skills of language units field research on basis of the linguo-cultural field;
- project work.

References

1. Azimov, E.G., Shchukin, A.N.: Novyi slovar metodicheskikh terminov i ponyatii (teotiya i praktika obucheniya) [New dictionary of methodological terms and concepts (the theory and practice of language teaching)]. Iikar Publisher, Moscow (2009). (in Russian)
2. Bim, I.L.: Llichnostno-oriyentirovannyi podkhod – osnovnaya strategiya obnovleniya shkoly [Personality-oriented approach – the main strategy of school's renewal]. Foreign Lang. Sch. **2**, 11–15 (2002). (in Russian)
3. Bim, I.L.: Kompetentnostnyi podkhod k obrazovaniyu i obucheniyu inostrannym yazykam [Competency-based approach to education and foreign language teaching]: Competence in Education. INEK, Moscow (2007). (in Russian)
4. Cheremisina Harrer, I.A., Kachalov, N.A., Borodin, A.A., Kachalova, O.I.: Sociocultural component in the content of teaching foreign languages to pre-school children. Procedia –Soc. Behav. Sci. **166**, 344–350 (2015)
5. Dikova, E.S.: Metodika formirovaniya medgkulturnoi professionalnoi kompetentsiyi na material reklamnykh tekstov [Methodology for formation of intercultural professional competence of the students on the basis of advertisements texts]. Linguistics University Publisher, Irkutsk (2009). (in Russian)
6. Elizarova, G.V.: Kultura i obucheniye inostrannym yazykam [Culture and Foreign Language Learning]. Caro, St. Petersburg (2005). (in Russian)
7. Lerner, I.Y.: Didakticheskiye osnovy metodov obucheniya [Didactic basics of methods of training]. Pedagogics, Moscow (1981). (in Russian)
8. Vorobyov, V.V.: Lingvokulturologiya [Linguoculturology]. RUDN Publishing House, Moscow (2008). (in Russian)
9. Yazykova, N.V.: Metodicheskiye aspekty obucheniya inostrannym yazykam s pozitsyi medgkulturnogo podkhoda [Methodological aspects of foreign language teaching from the position of the intercultural approach]. BSU Publishing House, Ulan-Ude (2010). (in Russian)

Linguistic Studies

Ten 'Whys' About Intercultural Communication

Nikolay V. Baryshnikov$^{(\boxtimes)}$ and Armen V. Vartanov

Pyatigorsk State University, Pyatigorsk 357500, Russian Federation
Baryshnikov@pgu.ru

Abstract. The author of the article poses ten questions vital for understanding the phenomenon of Intercultural communication. These reveal the most crucial aspects of the concept and the myths around it that are quite firmly established in the popular consciousness. The author believes that it is a matter of the utmost urgency for the world cross-cultural research community to demythologize of the term "Dialogue of Cultures" and to work on scientifically based principles of the parity cross-cultural communication.

Keywords: Intercultural communication · Dialogue of cultures

1 Introduction

The term 'Intercultural communication' experiences a real renaissance today. It was firstly introduced by E. Hole more than 60 years ago in the sphere of highly specialized problems connected with the adaptation of American diplomats to conditions of exotic cultures. Today the term has acquired quite a wide spectrum of knowledge and contexts of research. The phenomenon 'Intercultural communication' is the subject of investigation of various scientific spheres including philosophy, sociology, culture studies, ethnology, pedagogy, language, and culture didactics.

It is well known today that the process of mastering a foreign language is often groundlessly treated as an act of Intercultural interaction or as linguistic preparation to real Intercultural dialogue, what seems to us to be more accurate. We will see ten difficult questions below, and the author's answers to them as a proposal to continue the scientific debate about the multidimensional and contradictory phenomenon 'Intercultural communication'.

2 Literature Review

The response of the higher education to the active development of international networks was the start of specialists training in the field of intercultural communication at the educational institutions. In the period when this field of study was emerging, the scientists used the literature which mainly reflected the point of view of American anthropologists who were at the origin of the development of intercultural communication as a discipline. Since the middle of the last century, the world situation has

© Springer International Publishing AG 2018
A. Filchenko and Z. Anikina (eds.), *Linguistic and Cultural Studies: Traditions and Innovations*, Advances in Intelligent Systems and Computing 677, DOI 10.1007/978-3-319-67843-6_27

undergone significant changes, which entailed the transformation of the very nature of intercultural interaction. Today, the theoretical and methodological bases for writing the article were the studies on the formation of intercultural communication by N.I. Almazova, N.V. Baryshnikov, O.R. Bondarko, G.V. Elizarova, O.G. Oberemko, O.G. Polyakov, V.V. Saphonova, P.V. Sysoev, S.G. Ter-Minasova, V.P. Phurmanova, M. Byram, C. Kramsch, A. Nickols, S.J. Savignon, D. Stevens, etc.

3   Professional Intercultural Communication Implementation

3.1   Why is Verbal and Non-verbal Interaction Between the Representatives of Various Cultures Termed 'Intercultural Communication'? After All, Cultures Cannot Meet Each Other

Answer: As we suppose, there is a rather complete answer in the question itself. In fact, communication takes place between individuals, not between cultures. These are individual representatives of various cultures, who have their special level of communicative abilities and psycho-linguistic characteristics. Thus, it proves that the term 'intercultural communication' is not very accurate, there is some metaphorical effect in it.

Undoubtedly this term should not permit any shades of meaning, it must be neutral and monosemantic. However, historically the term 'interlingual communication' has not been clearly determined by the scientific terminology.

We can presume with certainty that the term 'Intercultural communication' originates from the comprehensive and universal term 'culture'. According to some authors, the term 'culture' has more than 600 definitions. This is the evidence of the fact that the term 'culture', when used in hundreds of contexts, loses its semantics. It is sensible to agree with the fact that one of the greatest minds once said that 'terms shouldn't be argued, they should be agreed upon'. Moreover, there are some very good definitions of 'Intercultural communication' in scientific literature. We think that the best one belongs to the professor Khaleeva. 'Intercultural communication is the combination of special interactions between humans who belong to different cultures and speak different languages. At the same time the participants of the interaction realize that they belong to different cultures, and each of them accepts the foreignness of the other' [8, p. 11].

It's clear from the definition given above that the author associates the term 'Intercultural communication' with the differences in languages and cultures between partners.

Proceeding the research in this field, we paid attention to the fact that specific processes of interaction between people who belong to different cultures are realized by means of common language for both partners. At the same time for one of the partners this language is, as a rule, a mother tongue, for the other - learnt, i.e. a foreign one. Thereby, the communication takes place in a language, but de facto there are two cultures in it. Therefore, we can surely state that in the process of communication partners' position themselves as the representatives of their own cultures. Consequently, it becomes possible to define the interaction between different cultures as 'Intercultural communication'. Consequently, taking all the metaphorical sense of this term into account, it seems not so rational to change one inaccurate term into another.

It's very important to understand the specific character of Intercultural communication, which is often ignored by many Russian and foreign authors. It is about the unequal character of Intercultural communication in cases when one of the participants speaks one native language and vice versa. We do not need to prove the fact that even the professional knowledge of a foreign language (according to the European scale - Level 6) cannot be compared with the knowledge of native speakers. There are many ways to recognize a foreigner who speaks not his/her native language.

Evidently that is why Khaleeva in her definition of Intercultural communication emphasizes the fact that the participants of the interaction realize that they belong to different cultures and each of them accepts the foreignness of the other [8, p. 11].

No doubt there are variants of Intercultural interactions including those in which status-inequality is excluded.

Let's take the Intercultural communication between the representatives of American and Australian cultures as examples. We want to point out that in spite of all the differences between American and Australian variants of English, representatives of these cultures as a rule do not have any difficulties with verbal and non-verbal communication. A similar example is the interaction between the representatives of French and, for instance, Moroccan, Senegalese, Congolese cultures, as far as they speak French as second native language.

Intercultural communication deeply changes its character when it is realized by the representatives of incomparable cultures. The brightest example of such a case is the Intercultural interaction between the residents of West-European countries and the USA with the representatives of Russian cultures (in integrated understanding). Communication is realized in the languages of Russian partners (English, French, Spanish and German), these languages are native for them but not for their Russian counterparts.

Russian system of higher education in its linguistic field is unique. It provides highly qualified linguistic training. Russian students-linguists become proficient in foreign languages. It is a well-known fact. However still we must admit that even the proficiency level of a foreign language is not always sufficient for equal Intercultural communication as far as it presupposes deep knowledge of communicative culture, customs, traditions, and habits of the partner.

Hence, cross-cultural communication in its modern interpretation is not a simple communication between partners who belong to different cultures and speak different languages. It is interaction between partners of the same status where for one of those the communication language is native and for the other it is a foreign one. Therefore, this process is characterized by some specific processes:

- maintenance of parity in Intercultural dialogue on conditions that partners' knowledge of a language is on different levels;
- opposition of interests between partners, which they defend in the process of Intercultural dialogue;
- productivity of Intercultural interaction is provided by the relations between partners on the interpersonal level, but also by the relation between countries which they represent;
- use or misuse of manipulations, deceptions, crooked techniques, and other communicative tricks.

3.2   Why Does Intercultural Communication Acquire Tougher, More Aggressive Character?

Answer: There are a lot of reasons of communicative practices rules tightening, however, the most important of them is the desire of Intercultural communication partners to gain advantages by hook or by crook. A special language of communication, the negotiations, business advice, and other forms of Intercultural interaction are carried out, favors communicative practice tightening.

In the connection to the fact that financial and economic interests are the subject of business communication, each of the participants of business communication strives for making a treaty, contract, agreement by means of communication. Business interests have stipulated the elaboration of manipulations, tricks and deceptions which are widely used in communicative practice.

Bredmayer has worked on the communicative techniques for German-speaking Europe, which acquired the name 'Black Rhetoric' and it is defined by the author as the manipulation by means of all the necessary rhetorical, dialectal, eristic, and rabulistic methods in order to turn the conversation into the necessary course and drive the opponent or the audience to the desired result. [4, p. 12] The author points out directly that those who use Black Rhetoric, breaking the rules of conducting the conversation is a winner.

Black Rhetoric techniques have acquired the wide expansion in Intercultural communication and became one of the most significant tools for waging the information war. Therefore, the usage of manipulations, tricks and deceptions makes the Intercultural communication tough and uncompromising.

In its turn strategies and methods of manipulative techniques' neutralization are being worked out. Still it remains impossible to break a vicious circle: Intercultural communication often represents a communicative duel with the use of communicative attacks and self-defense. Therefore, communicative contacts become inefficient according to its content and aggressive according to its impact on the Intercultural communication partner. We consider further development of 'poisonous communication' as hopeless because it is well-known that a powerful 'communicative antidote' will be opposed to them.

We believe that an optimal variant of Intercultural communication is an open, symmetrical communication realized by eristic strategies, based on humanistic values, such as kindness and beauty. Following academician Likhachev we affirm that 'kindness and beauty are common for all the peoples. They are common in two senses: truth and beauty are eternal helpmates, they are united with each other and equal for all nations. Lie is evil for everybody. Sincerity and honesty, integrity and unselfishness are always virtue [7, p. 10]. Apparently, Likhachev was an idealist and firmly believed in the mankind. However, the civilization developed according to another scenario, consequently Intercultural communication has acquired even more non-symmetrical and tougher character. We should not ingenuously suppose that Intercultural communication is an unsophisticated, smooth and nice interaction between partners, whose aim is to give each other compliments and make concessions.

Alongside we believe that there still exist unused reserves of changing the nature of Intercultural communication.

We mean the unity of all subjects of Intercultural communication under the slogan/appeal 'Intercultural communication is not luxury, but a means of peaceful co-existence of various languages and cultures', and also supervision of our Intercultural partners and our Code of Honor of the participants of Intercultural communication, which is not perfect of course, but can become a uniting basis for the followers of Intercultural communication of the same status.

3.3 Why is Tolerance in Intercultural Communication Realized in Most Cases on the Declarative Level?

Answer: It is a well-known fact that Intercultural communication is based on a number of principles one of which is the principle of tolerance, as the first between the equals. The term 'tolerance', being widely used in scientific literature, is defined as an ability of an individual to perceive without any objections and oppositions the way of life, behavior and peculiarities of other people which differ from his/her own [3]. Such understanding of tolerance is the main hindrance of Intercultural tolerance principle as for every Intercultural partner it is very difficult, practically impossible accept without any objections the opinion of the partner as their points of view often do not coincide and sometimes are absolutely different in its content.

To give the complete and definite answer to this question it is important to get into the deep meaning of tolerant attitude to the position of Intercultural partner. The superficial view of the situation is not enough for understanding the psychological nature of tolerant attitude towards the partner's position. Here the deep analysis of the perception, adoption or aversion of the foreign partner's position is necessary. According to the original statement of Bowt, tolerance is the readiness to equalize your own and others' position on the value-scale. When the positions are equalized, the difference of potentials must be coordinated, and then the search of the neutral decision begins which will satisfy everyone [2]. Thus, at this very point the most important obstacle of tolerant attitude is hidden, as it is quite a complicated matter to find a neutral decision which would satisfy all the partners with their opposite interests. So, the principle of tolerant attitude cannot be realized as it is not so easy to reach the neutral decision. From our point of view tolerance is a cornerstone of Intercultural communication. However, to make tolerance provide a peacekeeping mission, it's necessary to follow two rules of the tolerance principle:

(1) to act tolerantly does not mean you have to accept the position of partner, without any objection, on the contrary, you can and should argue, but in a tolerant form, without fits of anger and annoyance;
(2) the principle of tolerance must be implemented by both sides of Intercultural interaction.

One-sided tolerance leads to the Intercultural dialogue, which doesn't share the same status of the participants.

We believe that 'declarative character' of Intercultural communication is the evidence of the respect between partners, their readiness to reach the consensus in spite of obvious obstacles. We need to be tolerant to listen to an unacceptable position and react to it properly.

The antipode of tolerance is intolerance which means aversion to alien values and positioning your own point of view as an only right one which excels other ones [5].

Thus the realization of tolerance principle is rather difficult due to the considerable difference in values between the representatives of different cultures.

Decades are necessary for the Intercultural partners to be ready to express the readiness to equalize on the hypothetical scale your and other people's position which are, as it was said above, incompatible.

Intolerance in Intercultural communication is a dead-end of Intercultural interaction if it is based on condescending attitude of one partner to another, the violation of equality in the status of Intercultural dialogue, suppression of the culture of one participant by the culture of the other one and as a result Intercultural communication acquires the destructive features.

It is highly probable that tolerance in Intercultural communication will be declared until Intercultural partners learn to adapt an alien position to their own cultural tradition. It would be extremely irresponsible to believe that it will be easy to bring this thing into life.

3.4   Why Should Intercultural Dialogue Be Necessarily Carried Out in One Language?

Answer: Actually, such a requirement is not prescribed by any official document. Most probably a monolingual Intercultural dialogue is a tribute to traditions. Answering this question we should take into account a series of circumstances, which in our opinion have a significant meaning.

First of all, initially the Intercultural dialogue was carried out by 'a tolmach', this was the name for an interpreter in Russia. In the course of development of international contacts, Intercultural interaction became predominantly monolingual. It is still being studied in different kinds of educational institutions: kindergartens, elementary schools, secondary schools, universities, including specialized linguistic universities. Moreover, there is a broad network of linguistic courses, state and private linguistic schools.

As a result of a purposeful linguistic policy, Russian participants of Intercultural communication achieved status of bilingual or polylingual individuals. Therefore, the tradition of implementation of Intercultural communication in native language of a foreign partner has improved, as well as in English as a language of global communication.

At present, it has changed a lot. It is quite proper to suppose that the traditions have been established for years may change. We believe there must be no clampdown on carrying out Intercultural communication solely in one language. Intercultural communication can be quite bilingual. In case of a sudden misunderstanding between the participants of Intercultural communication, they may well turn to another language for communication, including the native language of the other participant or an Interlingua that both participants of the Intercultural dialogue share.

Undoubtedly everything depends on a certain situation emerging in course of Intercultural dialogue. However, in Intercultural communication of the same status it is conceded to turn to other languages, if that improves mutual understanding between the participants of the Intercultural dialogue.

We think that the fundamentally important factor is mutual striving of the participants of Intercultural communication for mutual understanding and for reaching a consensus on questions being discussed without infringement of interests of any of them. This psychological set is a guarantor of efficiency of Intercultural duel, where there is a place neither for a winner, nor a loser.

Studying Intercultural interaction problems on professional level, we formulated some principles of Intercultural interaction, which include the principle of choosing the language for a professional Intercultural communication, where its implementation improves impartial status-equality of Intercultural communication, provides a friendly and democratic style of communication [1, p. 118].

3.5   Why are Cultural Differences of Participants Considered to Be the Main Difficulty of Intercultural Communication?

Answer: We believe that this statement is false. The matter is that cultural differences are not the difficulties, but the basis of Intercultural communication. To substantiate this thesis there is no necessity to carry out research experiments, as long as it is obvious, that if there were no differences between cultures that the participants belong to, their communication could not be classified as Intercultural. Thus, the main difficulty of Intercultural communication is the so-called barriers, including psychological and language barriers.

Experts studying problems of Intercultural communication often claim that linguistic lapses of the participant, for whom the language of communication is not native, are easily forgiven. As for socio-cultural lapses, they are painfully perceived and are hardly forgiven. So, what can we say? On the one hand, everything is correct, but on the other hand it's impossible to agree with it. Actually, it is not about the nature of lapses (linguistic or socio-cultural ones). It is solely about the nature of relations between the partners of Intercultural communication. If there is a mutual desire to understand each other, we have every reason to believe that any lapse will be not only forgiven, but even taken no notice of. And vice versa, if there is no intention to reach a mutually acceptable decision, then even skillful language proficiency of a foreign partner will not work out.

In fact, it is a question of tolerant relations between the partners that we discussed answering one of the previous questions.

3.6   Why Cultural Universals Don't Significantly Influence the Success of Intercultural Communication?

Answer: To give a precise answer to the given question, it is necessary to specify, what 'cultural universals' mean in the context of Intercultural communication.

In scientific literature on linguistics, universals (from Latin *Universalis* - common) – are phenomena that occur in many languages.

Linguistic universals are the basis of multilingual didactics, as they promote forming of transposition, i.e. positive transfer of phonetic, lexical, and grammatical skills from one language into another.

In Intercultural communication universals are interpreted as phenomena, peculiar to cultures which are in contact and compared.

Therefore, cultural universals take place practically in any culture, which is proved by the existence of objective cultural facts in all cultures. Cultural universals are such phenomena as love for mother, paternal house, loss of the dearest and nearest, friendship, love and many others. One can give a lot of examples.

We think that cultural universals don't significantly influence the success of Intercultural communication, because they are not perceived by representatives of different cultures as such, as long as they are ethnically tinged in minds of both of them; and cultural facts of the foreign interlocutor of Intercultural communication are interpreted through the prism of native culture. In this transformed perception of cultural universals one can observe the primary reason for insignificant influence on the success of Intercultural communication. As in case with linguistic interference, the notion 'cultural interference' was introduced as well.

Therefore, it is cultural interference which is the major hindrance to mutual understanding in communication of representatives of different languages and cultures.

3.7   Why is the Question 'Why' Inappropriate for Intercultural Communication?

Answer: We are strongly confirmed, that question 'why' is inappropriate in the process of Intercultural interaction, because the answer does not clear anything up. Such a question is rhetorical deep down, if it concerns not the subject of discussion, but the culture peculiarities of the partner of Intercultural communication. This stipulation is significant. The question 'why' touching upon the cultural principles of one of the partners of Intercultural communication, is an oral display of intolerant attitude to the culture of the other partner. Indeed, what answer can one give to the questions asked by a foreign partner of Intercultural communication, addressing his Russian interlocutor, for example:

Why do children stand up in Russian schools when the teacher enters the classroom?
Why do people address each other using first and patronymic names?
Why do people eat so much bread in Russia?
Why do Russians smile so little?

The given examples of questions and hundreds of similar ones have a conflict nature, as long as they provoke excessive emotionality of the participants of the Intercultural dialogue. Besides, they often come out as an irritating factor.

Answers about something being accepted in Russian culture does not impress some of the foreign interlocutors, because there is no readiness to perceive 'without objections' other cultural facts, differing from standards, rules, and customs accepted in their culture.

Question 'why' in Intercultural communication may indicate an unfriendly attitude towards the partner, and as a rule the Intercultural dialogue comes to a gridlock. However, it does not mean that one should not ask 'why' while discussing business matters on a Intercultural level. This question is quite appropriate, when it is necessary to specify the partner's position and the argumentative basis concerning statements discussed earlier.

It is arguable that Intercultural communication like all other kinds of speech inter-action has an interpersonal character, as long as it is carried out by certain individuals, despite their belonging to different languages and cultures. Thereupon, it makes sense to suppose that in Intercultural communication restrictions are made or not made by the participants of this communication themselves. Much, if not everything, depends on the internal censorship of the partners of Intercultural communication. Thus, special focus is given to a human factor, which we consider to be fundamental, because it lies at the basis of partner's self-non-self-discrimination. Partners of Intercultural communication mark their belonging to different cultures and depending on the goals of the dialogue they look for points of contact to reach mutually acceptable results of Intercultural interaction or vice versa, they demonstrate differences in positions, impossible achievement of a positive result. It is obvious that a fragile communicative balance it extremely difficult to maintain with inappropriate 'why' questions.

3.8 Why Does the Opinion About the Necessity of Isolating from One's Own Culture to Succeed in Intercultural Communication Become Predominant?

Answer: It is better to say unequivocally and unceremoniously, that even if this opinion gains more and more supporters, we consider it to be false and even harmful, that is why we can't accept it. Let us turn to substantiation of our position. Firstly, everything was very correct and attractive, when such an anthropological, cultural, linguistic and linguodidactic phenomenon was discussed as 'a dialogue of cultures' for the first time. A dialogue of cultures is considered either as an interaction form of representatives of different languages and cultures, or as a concept of teaching foreign languages, as a linguodidactic tendency, teaching technology or as a version of Intercultural communication.

On the whole, the dialogue of cultures was interpreted by the specialists from different countries as a teaching of 'understanding the alien', that is as 'Intercultural', based on interaction of two cultures. This understanding of 'dialogue of cultures' in all its hypostases caused no objections. However, with time some authors began to specify the notion 'dialogue of cultures', which seems as a monologue in a dialogue, which is able not only to strengthen but also to ruin understanding. Thus, the dialogue of cultures became more comparable with suppression of one culture by the other. This understanding of 'dialogue of cultures' principle moved to the field of Intercultural communication. Authors advise that while communicating in a foreign language this understanding goes beyond the limits of one own culture, 'without isolating from it', adopting another culture, copying it and rejecting their own one. This understanding of Intercultural communication causes serious objections. First, adopting another culture eliminates Intercultural interaction, as long as communication gains multicultural character. Second, denial of one's own culture turns a person into a cosmopolitan, badly influences his psyche, and may lead to dual personality. Third, the interlocutor rejecting their own culture becomes a boring participant of Intercultural dialogue. Finally, statement of a question about isolating from native culture of a potential participant of Intercultural communication invariably leads to Intercultural interaction of the same status and open suppression of all the cultures by one of them.

3.9   In Certain Sources It is Underlined that in Intercultural Communication It's Necessary to Imitate Strategies of Speech and Behavior of the Foreign Partner. Why?

Answer: We believe that a partner of Intercultural communication imitating strategies of speech and behavior of a foreign partner makes an impression of a bad clown on the circus arena.

We can also answer this question with a question: for what purpose? It's quite difficult to formulate an answer to the last question. In our opinion, the demands about imitating strategies of speech and behavior of a foreign partner and isolating from one's own culture are closely connected. Their aim is to make Intercultural communication of the same status impossible.

Even having a perfect command of foreign language, the partner for whom the language of communication is not native, can't be compared with the native speaker. This is the first reason to claim that a foreigner studying foreign language is permitted to make some linguistic and socio-cultural mistakes and certainly preserve his or her patterns of speech and behavior, that can't affect the success of Intercultural communication, if there is a bilateral (mutual) striving for reaching a mutually acceptable agreement.

The demands about imitating strategies of speech and behavior of a foreign partner, in our view, are dictated by a negative attitude of some anthropologists to Intercultural communication of the same status.

3.10   Why are the Representatives of Russian Culture Considered to Be Uninteresting Partners of Intercultural Communication?

Answer: Most probably one can assume that our answer will differ much from those given by authors and representatives of other languages and cultures.

Following the principle of objectivity, we would like to notice that evaluation of the partner of Intercultural communication – the representative of any languages and cultures, including the representative of Russian culture depends on several factors:

1. Attitude towards the personality of a certain partner of Intercultural communication;
2. Attitude towards the culture which a certain partner represents;
3. Attitude towards the country, in this case to Russia, which a certain partner represents;
4. Language in which Intercultural communication is carried out;
5. Degree of tolerance between the partners of Intercultural communication;
6. Degree of trust between the partners for reaching adequate mutual understanding;
7. Use or nonuse of communicative tricks and manipulations by partners of cross cultural communication;
8. Degree of readiness of the partners of Intercultural communication to status-equal communication.

Factors mentioned above are enough to make sure of the evaluation ambiguity of the partner of Intercultural communication - the representative of Russian culture.

As we can see, the attitude towards any of the given factors can be either objective or subjective. As a rule, objective attitude is positive whereas subjective is negative.

Subjective negative evaluation forms the opinion about the uninteresting character of the partners of Intercultural communication – the representatives of Russian culture.

It is important to point out that in respect of Intercultural communication the opinion about a Russian partner as an uninteresting one emerges depending on the attitude towards the country which he/she represents. The research has been recently held by the BBC: the representatives of 25 countries were asked which country in the world they treated best? The results turned out to be very interesting and conspicuous: only in 8 countries (32%) people are friendly towards Russia. The worst attitude to Russia is shown in France (63%), in Germany (61%), in Great Britain (57%) and in the USA (59%) [6, p. 5]. Taking such an attitude to Russia and its authorized representatives hardly we may expect the positive or even neutral treatment towards Russian Intercultural partners from their foreign colleagues.

Many universally recognized qualities of the Russian Intercultural partners are well-known all over the world. In particular:

(a) the benevolent attitude towards foreign partners;
(b) proficiency in language on a prompt level;
(c) tolerance to the cultural facts of the represented country;
(d) high level of trust in the foreign partner;
(e) high degree of readiness to conduct the communication of the equal status;
(f) exclusion of communicative tricks, manipulations and deceits;

Given above unbiased collective image of Russian Intercultural partner is the evidence to the fact that it is misleading to consider the representatives of the Russian culture being uninteresting. Moreover, it has mostly a subjective nature.

It is quite fair to refer this opinion to the Intercultural communication.

One of the urgent aims of the world Intercultural research community is the demythologization of the term "dialogue of cultures" and working out scientifically based principles of the Intercultural communication parity.

4 Conclusion

The article contains a comprehensive analysis of the conditions and principles for the professional intercultural communication implementation. The author considers the conditions for the intercultural communication implementation without "rose-colored glasses", showing that the socio-cultural background of intercultural dialogue is an information war and propaganda.

The principles of professional intercultural communication in the author's formulation, the principle of cooperation, the principle of protecting the interests of the homeland, the principle of self-esteem, the principle of striving for mutual understanding, etc., are a vivid evidence of the author's concept of professional intercultural communication.

The article presents the author's style of communication to the reader, which does not incline him/her to his/her point of view, but offers the reader to become a worthy interlocutor as the material is read.

References

1. Baryshnikov, N.V.: Osnovy Professionalnoy Medzkulturnoy Kommunikatsiyi (in Russian) [The Fundamentals of Professional Intercultural Communication]. Infra-M, Moscow (2013)
2. Bowt, G.: Ne sprashivaite, i ya ne skazhu. Politicheskaya korrektnost effektivna dadze s nedostatkami (in Russian) [Don't ask and I'll not tell. Political correctness is effective even with drawbacks]. Izvestia (2002). http://web.archive.org/web/20070930171308, http://www. hro.org/editions/press/1102/18/18110261.htm. Accessed 5 June 2017
3. Bondareva, S.K., Kolesov, D.V.: Tollerantnost (Postanovka Problemy) (in Russian) [Tolerance. Introduction to the Problem]. Moscow Psychology and Social University Press, Moscow (2003)
4. Bredemayer, K.: Chernaya Ritorika; Vlast i Magiya Slova (in Russian) [Black Rhetoric: the Power and Magic of a word], 5th edn. Alpina Business Books, Moscow (2007)
5. Leontovich, O.A.: Russkiye i Amerikantsy; Paradoksy Medzkulturnogo Obsheniya (in Russian) [The Russians and the Americans: Paradoxes of Intercultural Communication]. Gnozis, Moscow (2005)
6. Rur, A.: Myagkaya sila Rossiyi (in Russian) [soft power of Russia]. Argumenti Facti **23**, 27–30 (2013)
7. Likhachev, D.S.: Pisma o Dobrom i Prekrasnoim (in Russian) [Letters About kind and Beautiful], 3rd edn. Detskaya Literature Press, Moscow (1989)
8. Khaleeva, I.I.: O gendornykh podkhodakh k teoriyi obucheniya yazykam i kelturam (in Russian) [about gender approaches to the theory of teaching languages and cultures]. Izv. Russ. Educ. Acad. **1**, 11–19 (2000)

Insights into Receptive Processing of Authentic Foreign Discourse by EFL Learners

Olga A. Obdalova(ID), Ludmila Yu. Minakova(ID),
Evgeniya V. Tikhonova(ID), and Aleksandra V. Soboleva(✉)(ID)

Tomsk State University, Tomsk 634050, Russian Federation
alex.art.tom@gmail.com

Abstract. Learning a foreign language nowadays is not just remembering words and grammar rules; it is a process of penetration into the hidden meaning of the message. Context plays a major role in the success of oral comprehension. This investigation looks at the issue of perception of authentic speech from the perspective of discourse approach. The authors study various factors influencing EFL learners' comprehension of authentic English American speech. Our observations and empirical data have led us to believe that there are some major difficulties of comprehension in the level of discourse which include not only unknown words but also lack of cultural awareness and pragmatic competence. Problems arise when the learners cannot infer the right meaning of the utterance because their perception stays in the vocabulary or grammar level without integrating both linguistic and extra linguistic features of the context of communication.

Keywords: Foreign language discourse · Oral comprehension · Processing · EFL learners · Cross-cultural communication · Context

1 Introduction

The 21st century carries a great deal of challenges to the world order. Barriers to effective communication nowadays come from transformation of local and global communities into multilingual and multicultural society. Therefore, ELT requires preparing learners for appropriate intercultural interactions [2, 4, 5, 9, 14, 19, 28, 35, 36].

Understanding what other people as members of different language communities say and write is known as language comprehension. Early theory and research [6, 8, 11, 13, 26, 31–33, 37, 42, 44] revealed some basic knowledge about the nature of listening and comprehension and provided the best practices for teaching this skill in language learners of English. Listening research demonstrates that psychological variables, such as perceptions, learning styles, and level of communicative competence, influence the way and efficacy of the listener's ability to create adequate meaning from the input.

Current research [1, 14–18, 27, 29, 32, 36, 38–41] reveals the importance of the well-coordinated work of cognitive mechanism in processing various aspects of information and communication. Nowadays, effective listening seems to be more complicated in the framework of intercultural communication than it might appear at first. In this regard, comprehending language involves a larger variety of capacities,

© Springer International Publishing AG 2018
A. Filchenko and Z. Anikina (eds.), *Linguistic and Cultural Studies:*
Traditions and Innovations, Advances in Intelligent Systems and Computing 677,
DOI 10.1007/978-3-319-67843-6_28

skills, processes, knowledge, and dispositions that are used to derive meaning from spoken, written, and sign language. In this broad sense, language comprehension includes oral comprehension, which is the basic element of communication in both personal and professional settings. Deriving meaning from spoken language involves much more than just knowing grammar rules, the meaning of words, and understanding what is intended when those words are put together in a certain way. To bear in fully comprehending what another person says it is necessary to take into account various parameters of discourse that influence the processes of both speech producing and listening, and comprehending. This approach focuses on two perspectives – the speaker's and the listener's, and dynamic relationships between them and contexts.

Indeed, a successful comprehension by EFL learners is actually the result of complex cognitive processes. Work in cognitive science unveils added features of the comprehension processing. The auditory input must be decoded, segmented into words and phrases, and integrated into a coherent meaning representation in the listener's mind. At the same time, relevant information from long-term memory needs to be retrieved to help the comprehender choose between alternate meanings and draw inferences. Much of the literature on comprehension suggests that our knowledge is organized around certain patterns (frames, scripts), which humans carry as mental representations of knowledge in the brain [7, 24]. The cognitive process of inferring meaning is based on these representations, which makes it possible to interpret a lot of everyday situations. But if there is lack of relevant mental representations in a person's mind, comprehension may be blocked or impeded.

Many of these processes are assumed to function much less efficiently in inter-cultural communication, in which the "degree of congruence between the cognitions of a listener and the cognitions of a source" [25, p. 70] is inevitably low. Success of listening decreases because information processing involves not only perception of different persons, different cultures, but also engagement of different languages as means of interaction. The reason for this is context that governs the use of language under the influence of the person's identity, social context, and specific context of interaction. This has to do with pragmatics which relies on the rules of the language use [16, 18].

In this respect, discourse plays a very important role in the success of comprehension as it represents a whole process of determining situational interpretation, the entire interplay of linguistic meaning and context, individual and societal factors, prior and actual situational context, linguistic units and situation [15, p. 134].

2 Discourse Factors in Intercultural Communication

2.1 Context Dependency in Reception and Understanding

The way the information is adopted and conveyed by EFL learners depends on the work of their cognitive mechanism. The first stage of processing activity deals with the reception of the input language, and the second – with understanding spoken language. It is essentially an inferential process [33], closely associated with the context. Both stages affect how messages are understood.

In linguistics, context usually refers to any factor – linguistic, epistemic, physical, and social – that affects the actual interpretation of signs and expressions. Context-dependency is one of the most powerful views in current linguistic and philosophical theory [20]. According to the socio-cognitive approach to pragmatics [15, 16], context represents two sides of world knowledge: prior context and actual situational context, which are intertwined and inseparable. Thus, listening perception and listening comprehension significantly depend on success of the mentioned factors. Meaning is assigned based on how adequately the listener takes in authentic foreign speaker's speech, how well the organization and identification of the content and function have been carried out, and how well the construction of propositions forms a coherent image of the message.

2.2 Authentic Discourse Factors in Intercultural Communication

Authentic discourse can be defined as the language enrooted in social practices. Modern linguistics refers to discourse as a broad concept that encompasses various parameters of communicative activity within communication context. According to Gural, "discourse is the realization of language communicative functions in all the diversity of their manifestations" [12, pp. 112–113]. Thus, discourse can be treated as a combination of the process and the result of communicative activity, which possesses linguistic, paralinguistic, extra-linguistic features and is conditioned by the definite situation and the general context.

The concept of 'discourse' is also used to describe "the text as a means of dynamic interaction between communicants" [43, p. 13]. It should be noted that the text as the basic unit of discourse along with the linguistic, cognitive and semantic aspects is characterized with some extralinguistic parameters. This point of view seems proved if we take into account the fact that sociological, psychological and linguistic factors closely interact with each other in the process of communication [3].

According to Geddes and White [10, p. 137], there can be distinguished two types of authentic discourse in educational environment in the context of EFL teaching: (1) unmodified, which represents a genuine act of communication, and (2) simulated, which is produced for pedagogical purposes and exhibits features that have a high probability of occurrence in genuine acts of communication.

The main component of discourse, and therefore the main object of perception is the text, which represents organized integral unit, characterized by purposive nature and pragmatic load. It is some structure in which language signs form a definite system of relations. Thus, EFL learners as recipients of a foreign-language authentic discourse should identify the language tools that ensure the connectivity of the utterance, and, first and foremost, identify the key lexical units that represent both the semantic and the pragmatic meaning of the utterance. Discourse is a linear unit since it develops in time and it is irreversible [21].

Moreover, discourse is the language of communication in socially prefabricated situations [16, 18]. Certainly, these situations are partially prefabricated and partially are developed ad hoc. Research in EFT [23, 35] shows that the more "common ground" the interlocutors have, the more understanding they can achieve. But in the case when one of the participants belongs to a different cultural community, mutual understanding faces serious delimitations. When communication is based on the common knowledge and "the relevant environment of language use" [38, p. 3] the participants demonstrate "socially shared knowledge in talk and thus display their interpretations of the relevant aspects of the situation" [38, p. 85]. Non-native cultural members may experience difficulties in different levels of perception and integration of meaning with respect to various properties of discourse.

Researchers distinguish between such limitations as linguistic knowledge and experience (vocabulary, grammar, idiom level) and discourse level. In the discourse level the following aspects of oral comprehension of authentic foreign language spoken discourse by EFL learners have been identified: (1) inadequate perception of linguistic and extra-linguistic factors of communication; (2) imperfect mechanism of perception, interpretation and comprehension of the foreign language discourse; (3) insufficient volume of culture-loaded receptive and productive vocabulary of students; (4) lack of perceptual experience; (5) lack of discursive-cognitive experience that promotes awareness of the difference of cultures and languages, as well as varied ways of speech activity to infer an adequate meaning of a foreign language utterance [30].

Figure 1 displays the complex interplay of various factors contributing to communication between speakers of different sociocultural communities when the means of interaction is a foreign language. We build strong links between the following major groups of factors: sociocultural background, communication process, foreign language discourse, discourse pragmatics, and communicants' cognitive activity. Each of them contributes to the complexity of listening perception and listening comprehension.

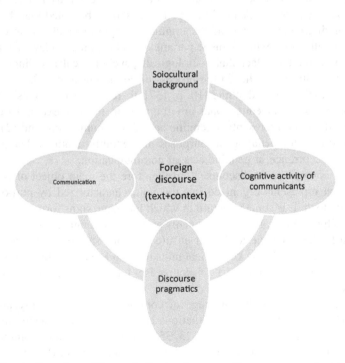

Fig. 1. Interplay of factors in foreign language discourse comprehension.

Basing on the factors, mentioned above, we can conclude that an EFL learner needs to derive the complete information from the situation of communication as a whole. Thus, discourse perception and comprehension is carried out within the situational context which contributes to drawing conclusions, interpretations, and hypotheses within a particular communicative situation [21].

The correlation of objective information with information about the situation of communication becomes one of the most important conditions for an adequate interpretation of a message. Therefore, the object of perception in communication is not only the utterance, but all the discourse components, including verbal, prosodic, paralinguistic, and extra-linguistic. For non-native speakers of English apart from these, it is crucial to form knowledge of prefabricated patterns of communication, typical of the English speaking community, and build situationally relevant understanding of the context.

When considering the situation within intercultural communication EFL learners should take into account, to varying degrees, all the components of the discourse, and construct a collective and integrated understanding about the subject and semantic content of the original message.

3 Findings and Discussion on Authentic Foreign Language Discourse Comprehension Problems

The participants included first and second year non-linguistic students from Tomsk State University, majoring in Science. They were the students of Chemical Faculty (N = 11) and Biological Faculty (N = 11). All the students are taking EFL course within their curriculum which involves 4 academic hours a week. In advance of the research the participants took a language test to prove their language proficiency which had to be not lower than intermediate according to the conditions of the experiment.

In the first stage of the experiment the participants were exposed to the comprehension of samples of dialogues occurring between native speakers of American English [22, 28]. This speech is characterized by predominance of such feature as situationally-bounded utterances. The processing of such discourse demanded form the listeners, taking into account not only linguistic but also the complex of extra linguistic factors.

The exploration focused on the process of comprehension within intercultural communication based on culturally determined authentic unmodified discourse; the factors that affect the success of discourse interpretation by Russian learners of English; and the causes of communicative failures in interaction between representatives of different cultures. The experimental study of the cognitive processes was based on listening comprehension and interpretation of discourse fragments by Russian EFL learners. The procedure was carried out on situation-bound utterances and cognitive-discursive fragments of the contemporary American English language. The research aimed at identifying the impact of context on the students' understanding lexical units; identifying the factors that determine the interpretation process and its success. Empirical data showed the important role of contextual factors and socio-cultural information for the success of an authentic foreign language discourse

comprehension. Situation-bound utterances, given in the situational context, contributed to the optimization of the perception process that promotes the analysis and
synthesis of all the characteristics of a word as a means of language and its environment
(the actual situational context).

The sample of 22 subjects participated in the experiment. Each of them interpreted
10 authentic speech units via interlingual translation. So, we obtained 220 utterances
for discourse analysis. Table 1 demonstrates the difference in the efficacy of comprehension culturally determined discourse by EFL learners.

Table 1. Comprehension of situation-bound utterances depending on the context.

Phrase	Without the situational context		In the situational context	
	Right interpretation	Wrong interpretation	Right interpretation	Wrong interpretation
10				
Total number	95	125	191	29
Ratio to the total number (%)	43	57	87	13

As we can see from Table 1, lack of situational context had a negative impact on
the correctness of culturally determined units of authentic discourse comprehension, as
the number of the right answers was only 95 cases (43%), while that of the wrong
answers was 125 (57%) from the total number of all given variants. Analyzing the data
of Table 1, we can notice that phrases with low situational load, such as *Take a seat;
How are you doing? No problem; What's up?* were interpreted by students correctly
even in the absence of a situational context. On the other hand, the utterances having a
high situational load, such as *You bet, Here you go, Give me a break, Get out of here,
Come again, Be my guest* were miscomprehended in the majority of cases when they
were presented to the students without a situational context. We can suggest that the
reliance on the word-for-word translation and meaning inference according to their
linguistic context, did not contribute to inferring the adequate meaning of a definite
phrase as it has larger dependence on the context and particular situation of
communication.

Only when culturally determined units were presented in the situational context, the
number of the correct answers increased significantly to 87%, and the number of the
wrong answers decreased to 13%. In our opinion, this fact is connected with a great
difference in the meaning of the utterances which are highly dependent on the situation.
Students recognized familiar words which comprised those utterances and considered it
enough for understanding [23, 34].

The major outcome of the experimental set is that situational context which contains culturally-marked language units greatly influence the efficacy of comprehension,
and grammar and lexical knowledge is not enough for the success of information
processing.

The second series of experimental procedures was based on oral direct speech prompts. The sample comprised 22 subjects from the same cohort of the participants. The research aimed at the way information is conveyed by Russian EFL learners as their interpretation of the fragments of spoken discourse. In this part of the research the direct speech utterances, constituting separate talks, served as stimuli. These situations are characterized by the presence of lexical and grammatical structures which represented different types of utterances: statements, questions, and requests or commands. The subjects were given worksheets which contained twelve utterances for interpretation from direct speech into reported. The speech patterns were recorded by native speakers of English and presented in video files. One of the speakers was a male, another one was a female (John and Mary).

The prompts were delivered with the help of videorecording. Each prompt was repeated twice with a 10 s break. The subjects were to watch, listen to each prompt, understand what people on the screen said, and report each utterance in writing as close to the original context as possible. The adequacy of the output was assessed by three native speakers of English.

This part of the experiment was predominantly focused on conveying speech utterances containing specific grammar and lexical devices for rendering direct speech via transformation into reported speech.

When reporting the phrases uttered by native speakers of English, different parameters for estimation of the transformation adequacy were defined. It is important to note that indirect reporting represents not only semantic content, but also its pragmatic load. Thus, the following parameters were applied for the obtained empirical data treatment: (1) content closeness; (2) content words; (3) grammatical, lexical, and spelling aspects. Content closeness refers to the final stage of receptive processing of authentic foreign discourse, namely listening comprehension. Content words and language mistakes reflect the first stage of this receptive process or listening perception.

In spite of the type of mistake (grammar or lexical), the native-speaking assessors marked each interpreted utterance as deviation from context, which corresponds to the worst malfunction of context closeness.

The following Table 2 illustrates how these parameters manifested in the subjects' interpretation of the stimuli speech patterns. Totally 22 students produced 264 sentences when interpreting 12 stimuli.

Table 2. Results of discourse analysis.

	Context closeness				Content words			Language mistakes		
	No change	Minor changes	Paraphrased context	Deviations from context	No significant difference	Minor difference	Major differences	Grammar	Lexis	Spelling
	1	2	3	4	5	6	7	8	9	10
Total number	60	66	22	116	88	110	66	145	84	18
Ratio to the total number (%)	23	25	8	44	33	42	25	59	34	7

75% is made up by columns 5 and 6 and manifest efficacy of perception of the stimuli. This high value speaks for good result of listening perception. It can be seen that at the same stage recipients faced a great number of difficulties, most of all in grammar structures (59% from the total number of mistakes). In the lexical aspect the number of mistakes is much less, only 34%.

The obtained data for this group of subjects show that problems in listening perception affected the efficacy of context delivery. The third and fourth columns taken together give 52% of deviations from original meaning of the utterances.

Thus, we can assume that misperception of lexical and grammatical devices of original speech brings to inadequate interpretation. Moreover, good results in listening perception do not obligatory lead to a successful interpretation of the speaker's meaning. Without sufficient lexical and grammatical knowledge EFL learners are not able to deliver speech adequately as assessed by native speakers. Another pedagogical suggestion deals with the idea that in the current educational situation it is important to integrate various methods of FL teaching for achieving the outcome formulated in the terms of intercultural competence. Communicative framework alone does not allow paying enough attention to the cultivating grammar competence. We suggest using cognitive and discursive approaches to enhance the communicative basis of teaching.

4 Conclusion

The research reviewed, although not exhaustively, current understanding of the nature of listening reception and listening comprehension in psycholinguistics, pragmatics, discourse analysis, and cognitive science. The central issue from linguistic and pedagogical perspective deals with the nature of information processing by EFL learners in the context of authentic foreign language discourse.

Quantitative research has allowed us to clarify the dependence of comprehension and interpretation of the context of authentic foreign language discourse from pragmatic and cultural features of speech in the framework of intercultural communication. Culturally loaded context causes more difficulties for receptive processing by EFL learners, in which case they need not only solid grammar skills but also and foremost intercultural competence. It manifests itself due to the learners' ability to take into consideration definite situational context.

These findings identify the areas for improvement of teaching various skills and capacities in EFL learners. This research can help EFL teachers in organizing the learning environment to promote teaching effective intercultural communication.

Certainly, much more work is needed to get a broad picture on the matter as well as implementation of an integrated approach as we continue to expand our theoretical basis about perception in EFL teaching practice and continue to develop best pedagogical practices.

Acknowledgements. This research was supported by "The Tomsk State University competitiveness improvement programme" under Mendeleev Fund (No. 8.1.13.2017).

References

1. Anderson, J.R.: Cognitive Psychology and its Implications, 7th edn. Worth Publishers, New York (2009)
2. Baker, W.: From cultural awareness to intercultural awareness: culture in ELT. ELT J. 66(1), 62–70 (2012)
3. Bayrak, N.V.: Features of the discourse organization in the invitation situation in British culture (unpublished dissertation), Voronezh (2006) (in Russian)
4. Byram, M.: From Foreign Language Education to Education for Intercultural Citizenship: Essays and Refections. Multilingual Matters, Clevedon (2008)
5. Chapelle, C.A.: If intercultural competence is the goal, what are the materials? In: Proceedings of Intercultural Competence Conference, pp. 27–50. CERCLL, Tucson (2010)
6. Coakley, C., Wolvin, A.: Listening in educational environment. In: Purdu, M., Borisoff, D. (eds.) Listening in Everyday Life: A Personal and Professional Approach, pp. 179–212. University Press of America, Lanham (1996)
7. Edwards, R., McDonald, J.: Schemata theory and listening: a framework for memory research. J. Verbal Learn. Verbal Behav. 11, 671–684 (1993)
8. Field, J.: Key concepts in ELT: "bottom-up" and "top-down". ELT J. 53(4), 338–339 (1999)
9. Furstenberg, G.: Making culture the core of the language class: can it be done? Mod. Lang. J. 94(2), 329–332 (2010)
10. Geddes, M., White, R.: The use of semi-scripted simulated authentic speech in listening comprehension. Audiov. Lang. J. 16(3), 137–145 (1978)
11. Grosjean, F., Gee, J.P.: Prosodic structure and spoken word recognition. Cognition 25, 135–155 (1987). doi:10.1016/0010-0277(87)90007-2
12. Gural, S.K.: Teaching foreign discourse as a sophisticated self-developing system (linguistic university): the author's abstract (unpublished doctoral dissertation) Tomsk (2009) (in Russian)
13. Hasan, A.: Learners' perceptions of listening comprehension problems. Lang. Cult. Curric. 13, 137–153 (2000). doi:10.1080/07908310008666595
14. Kearney, E.: Cultural immersion in the foreign language classroom: some narrative possibilities. Mod. Lang. J. 94(2), 332–336 (2010)
15. Kecskes, I.: Situation-bound utterances as pragmatic acts. J. Pragmat. 42, 2889–2897 (2010)
16. Kecskes, I.: Intercultural Pragmatics. Oxford University Press, Oxford (2013)
17. Kecskes, I.: On my mind: thoughts about salience, context, and figurative language from a second language perspective. Second Lang. Res. 22(2), 219–237 (2006)
18. Kecskes, I.: How can intercultural pragmatics bring some new insight into pragmatic theories? In: Capone, A., Mey, J.L. (eds.) Interdisciplinary Studies in Pragmatics, Culture and Society, Perspectives in Pragmatics, Philosophy & Psychology, pp. 43–71. Springer, Heidelberg (2015)
19. Kecskes, I.: Intracultural communication and intercultural communication: are they different? Int. Rev. Pragmat. 7, 171–194 (2015)
20. Kecskes, I.: Context-dependency and impoliteness in intercultural communication. J. Politeness Res. 13(1), 7–31 (2017)
21. Makarov, M.L.: Osnovy teorii diskursa (in Russian) [Fundamentals of Discourse Theory]. Gnozis, Moscow (2003)
22. Minakova, L.Yu.: Vlijanie situativnogo konteksta na adekvatnost' ponimanija autentichnoj rechi pri obuchenii inojazychnomu diskursu (in Russian) [The Influence of a Situational Context on the Adequateness of Authentic Speech Comprehension in Foreign Language Discourse Teaching Language and Culture] 1(33), 160–170 (2016)

23. Minakova, LYu., Gural, S.K.: The situational context effect in non-language-majoring EFL students' meaning comprehension. Proc. Soc. Behav. Sci. **200**, 62–68 (2015). doi:10.1016/j. sbspro.2015.08.014
24. Minsky, M.L.: A framework for representing knowledge. In: Winston, P.H. (ed.) The Psychology of computer vision, pp. 211–227. McGraw-Hill, New York (1975)
25. Mulanax, A., Powers, W.G.: Listening fidelity development and relationship to receiver apprehension and locus control. Int. J. Listening **15**, 69–78 (2001)
26. Nunan, D.: Listening in language learning. In: Richards, J.C., Renandya, W.A. (eds.) Methodology in Language Teaching: An Anthology of Current Practice, pp. 238–241. Cambridge University Press, New York (2002)
27. Obdalova, O.: Exploring the possibilities of the cognitive approach for non-linguistic EFL students teaching. Proc. Soc. Behav. Sci. **154**, 64–72 (2014). doi:10.1016/j.sbspro.2014.10. 113
28. Obdalova, O., Gulbunskaya, E.: Cross-cultural component in non-linguistics EFL students teaching. Proc. Soc. Behav. Sci. **200**, 53–61 (2015). doi:10.1016/j.sbspro.2015.08.113
29. Obdalova, O., Minakova, L.: Vzaimosvjaz' kognitivnyh i kommunikativnyh aspektov pri obuchenii inojazychnomu diskursu (in Russian) [Interconnection of cognitive and communicative aspects in teaching foreign language discourse]. Philol. Sci. Issues Theory Pract. **7** (25), 148–153 (2013). Gramota, Tambov
30. Obdalova, O., Minakova, L., Soboleva, A.: The study of the role of context in sociocultural discourse interpretation through the discursive-cognitive approach. Vestnik Tomskogo gosudarstvennogo universiteta – Tomsk State. Univ. J. **413**, 38–45 (2016). doi:10.17223/ 15617793/413/6
31. Richards, J.C.: Teaching listening: approach, design and procedure. TESOL Q. **17**(2), 219–240 (1983)
32. Richards, J.C.: Second thoughts on teaching listening. RELC J. **36**, 85–92 (2005). doi:10. 1177/0033688205053484
33. Rost, M.: L2 listening. In: Hinkel, E. (ed.) Handbook of Research on Second Language Learning and Teaching, pp. 503–527. Erlbaum, Mahwah (2005)
34. Soboleva, A.V., Obdalova, O.A.: Strategies in interpretation of culture-specific units by Russian EFL students. Proc. Soc. Behav. Sci. **200**, 69–76 (2015). doi:10.1016/j.sbspro.2015. 08.016
35. Soboleva, A.V., Obdalova, O.A.: Kognitivnaja gotovnost' k mezhkul'turnomu obshheniju kak neobhodimyj komponent mezhkul'turnoj kompetencii (in Russian) [Cognitive readiness for intercultural communication as an essential component of intercultural competence]. Lang. Cult. **1**(29), 146–155 (2015)
36. Soboleva, A.V., Obdalova, O.A.: Organizacija processa formirovanija mezhkul'turnoj kompetencii studentov s uchetom kognitivnyh stilej obuchajushhihsja (in Russian) [Organization of students" intercultural competence formation in view of their cognitive styles] Vestnik Tomskogo gosudarstvennogo universiteta – Tomsk State. Univ. J. **392**, 191–198 (2015)
37. Swain, M.: Three functions of output in second language learning. In: Cook, G., Seidlhofer, B. (eds.) Principle and Practice in Applied Linguistics, pp. 125–144. Oxford University Press, Oxford (1995)
38. Van Dijk, T.A.: Society and Discourse. How Social Contexts influence Text and Talk. Cambridge University Press, Cambridge (2009)
39. Vandergrift, L.: Second language listening: listening ability or language proficiency? Mod. Lang. J. **90**, 6–18 (2006)
40. Vandergrift, L.: Recent developments in second and foreign language listening comprehension research. Lang. Teach. **40**, 191–210 (2007). doi:10.1017/S0261444807004338

41. Vandergrift, L., Tafaghodtari, M.H.: Teaching L2 learners how to listen does make a difference: an empirical study. Lang. Learn. **60**, 470–497 (2010). doi:10.1111/j.1467-9922. 2009.00559.x
42. VanPatten, B.: Attending to form and content in the input: an experiment in consciousness. Stud. Second Lang. Acquis. **12**, 287–301 (1990). doi:10.1017/S0272263100009177
43. Zalevskaya, A.A.: Individual'naja baza znanij i problema perevoda (in Russian) [Individual Knowledge Base and the Problem of Translation]. In: Klyukanov, I.E. (ed.) Translation as a Process and As a Result. Collection of Scientific Articles, pp. 29–36. Kalinin State University, Kalinin (1989)
44. Wilson, M.: Discovery listening-improving perceptual processing. ELT J. **57**(4), 335–343 (2003)

Cognitive and Pragmatic Interpretation of Terminological Fragments in the Professional Discourse

Galina G. Galich[1] and Anna M. Klyoster[2(✉)]

[1] Omsk State Dostoevsky University, Omsk 644077, Russian Federation
gggalich.2014@mail.ru
[2] Omsk State Technical University, Omsk 644050, Russian Federation
annaklyoster@mail.ru

Abstract. The research of terminological fragments – the fragments of the text or discourse containing the terms, which designate special objects of a certain knowledge area is considered in this article. Using a method of terminological fragmentation authors define a fragment functional intensity in respect of cognitive and pragmatic contents. Cognitive interpretation is carried out concerning fund of knowledge of the author and the recipient. The definitions of scientific concepts are considered by the author of the text in the field of cognitive interpretation with the indication of their hypo-hyperonymic relations; correlation to the frame structure of knowledge of the corresponding field is emphasized. Pragmatic interpretation concerns the intentions of speaker, intensions for informing or assessment, statement impacts on the addressee, influence of a speech situation on subject and communication forms. Set of rules of cognitive and pragmatic interpretation allows specialists to establish interrelations between concepts, to recreate cognitive model and to present professional communication as the integrated multidimensional language phenomenon.

Keywords: Professional discourse · Terminological fragment · Term functioning · Semantics · Pragmatics

1 Introduction

Modern linguistics is characterized by the tendency to the fullest and versatile representation of the language phenomena combined with multiple aspects and complexity of the scientific description. Cognitive linguistics and linguistic pragmatics are possibly its most relevant directions. These two aspects of the person- and activity-oriented functional language theory hold central position because they are recognized as the main and most important ones for language communication with its cognitive and communicative functions.

One can argue which of the two functions is primary, highlighting this or that aspect of the linguistic description. Each existing point of view can be sufficiently substantiated and has the right to exist, however, to develop a common general theory of language, it is important to consider its real properties, namely, the fact that

© Springer International Publishing AG 2018
A. Filchenko and Z. Anikina (eds.), *Linguistic and Cultural Studies:*
Traditions and Innovations, Advances in Intelligent Systems and Computing 677,
DOI 10.1007/978-3-319-67843-6_29

linguistic units are complex in nature and accordingly their description should be complex and multi-aspect.

The primary differentiation of the aspects and their specialized description cannot be the ultimate goal and constitute only a preliminary stage of the research aimed at further generalization at a higher level and leading to more complete and adequate description of the language as a functioning system. Naturally, this approach presupposes the solution of global problems requiring the efforts of many scientists. The position outlined in the article continues the scientific traditions developed in the works of Shcherba [14], Katznelson [7] and other representatives of the St. Petersburg school. Actually, the method of cognitive and pragmatic interpretation goes back to the works of Susov [16, 17].

According to the concept of Susov, pragmatic and semantic meanings of discursive structures "can be combined within a single meaningful (contensive) stratum" [17]. The relationship of coded human activity structures in these two forms with the phonological form of the utterance is realized through the lexical-grammatical mechanisms of the language fixed in its system-structural paradigm. The mechanisms can be represented as "the rules of interpreting formal language structures relative to the structures of consciousness and structures of activity. The totality of the language formal and semantic rules and pragmatic interpretation rules allows us to see the language in its integrity, i.e. as a pragmatically interpreted (or socially interactive) system" [17]. The functional mechanisms unity of the utterance cognitive and pragmatic interpretation correlates with the unity of the same forms of the speaker's activity in the perception, awareness and discursive embodiment of the reflected real - and, in particular - professional situation.

Modern terminology is characterized by an increasingly active penetration of functional aspects of the study. The study of terminological units in the aspect of their role in discourse, cognitive and pragmatic functions, and the setting of communicative tasks is typical for it. Various pragmalinguistic and cognitive models of certain types of institutional discourse are in the focus of attention for linguists, for example, speech therapy discourse [3], military discourse [11], engineering psychology [8], automobile [13] and economic [15] discursive practices, and others.

Within the article, one of the options for the solution of new complex problems can be shown in the field of terminological units functioning in professional discourse in the perspectives universal for different discursive practices.

The main tool of the term complex cognitive-pragmatic study in its discursive functioning in this article is the terminological fragment - a fragment of a text, a discourse containing the name or interpretation of a special object related to a certain field of scientific knowledge [9]. The semantic-cognitive and communicative properties of the term can be described more fully when they are considered in the context of a sufficient volume, i.e. it can be called a terminological fragment. The fragment volume can vary depending on its content and functions: from the synonym explaining the term to the text devoted to the expanded interpretation of the term.

Functional load of a terminological fragment in the aspect of cognitive and pragmatic content is fully revealed in the text/discourse. Its components can be the following: introduction of a new term to the scientific use that carries a certain quantum of special knowledge in its meaning, informing the scientific community about obtaining new results of professional activity, scientific polemics on some issue, report for students – undergraduates or postgraduates – the elements of the knowledge system in a specific subject areas, popularization of knowledge among a wide range of interested people, etc. Outside the special text, the term definition in the glossary or encyclopedic dictionary is the cognitive function as a source of reference information about some professional sphere of human activity.

2   Cognitive Interpretation of the Terminological Fragment

The cognitive content of the term included in the special text is involved in production of deep semantic-syntactical structures of the text, primary unfolding of mental information at its formation and segmentation of knowledge elements. These structures integrate a certain set of discrete elements and text generating mechanisms: concepts, representations and notions, and also structures of intentional valency. Dynamic character of formation mechanisms for information and semantic structure of the text and its fragments determines their expression with verbal knowledge terms and their substantive correlates: *erkennen – Erkenntnis, erfahren – Erfahrung, einsehen – Einsicht, beobachten – Beobachtung, betrachten – Betrachtung, wahrnehmen – Wahrnehmung, bedeuten – Bedeutung, feststellen – Feststellung, bezeichnen – Bezeichnung, untersuchen – Untersuchung, forschen – Forschung, verallgemeinern – Verallgemeinerung* and others.

The titles of the text sections themselves participate in their semantic fragmentation as well, such as *Einführung, Prozessbeschreibung, Resultat/Resultate der Forschung, Vorteile und Nachteile der Methodik, Zusammenfassung*, etc.

Cognitive terms can be given a definition of the special text, as well as any other terms (see below), but the function of the text/fragment and its semantic structuring construction and organization is more specific to them: *von der Erkenntnis ausgehen, ich erwarte eine tiefere Erkenntnis, dahinter steht die Erkenntnis, dieser Hinweis scheint allein nicht auszureichen für die Erkenntnis, dass.., ein gemeinsames Verständnis haben wir noch nicht erreicht, der Verfasser äußerte Verständnis, dass.., diese Einsicht ist weit verbreitet, dies bringt die Einsicht, das ist von grundsätzlicher Bedeutung, die durchgeführte Untersuchung hat gezeigt... der neue Stoff hat die Bezeichnung "Teflon" erhalten*, etc. Such fragments are sometimes called reference points or metatext operators – the indicators of the text informational and semantic structure introducing particular cognitive information into the text and defining its place in the author reasoning sequence [6].

Cognitive interpretation of a terminological fragment is performed concerning the knowledge base of the author and the recipient. In the sphere of cognitive interpretation text definitions are considered as the most explicit nominations – scientific concepts definitions given by the text author with their hypo-hyperonymic and other relations being indicated: *Denjenigen Bereich der Technik, der sich mit der Umwandlung von Arbeit in Wärme befasst, pflegt man als Energietechnik zu bezeichnen* [4]. Defined term *Energietechnik* acts as a more particular concept, a hyponym to more general (hyperonym) of *Bereich der Technik*.

The borrowed term semantization by means of synonymous unit in the receiving language is widespread [10]: *«Usability»? – Damit ist Gebrauchstauglichkeit gemeint* [4].

The short explication of the term contents in a terminological fragment with the minimum volume is possible. Therefore in a subordinate part of one statement, three of such fragments can be selected at once, where brackets are used for allocating the definition of the attributive term through a synonym:...*wenn diese Untersuchungen valide (wahre, gültige), representative (verallgemeinerungswürdige) und relevante (nützliche) Ergebnisse und Erkenntnisse zeitigen sollen* [4].

The correlation of the knowledge frame structure to the corresponding speciality is often emphasized in a definition: *Je nachdem, ob die betreffenden Prozesse oberhalb oder unterhalb der Umgebungstemperatur ablaufen, spricht man von Wärme- oder Kältetechnik* [4].

3   Pragmatic Interpretation of the Terminological Fragment

Pragmatic interpretation of the terminological fragment is based on considering pragmatics as the science about "the behavior of language signs in real communication processes" [12], where there are the subjects of speech production and the subjects of its reception, each having their own inner world, knowledge, experience, social roles, motives, intentions etc [1, 2]. Thus, the object of pragmatic interpretation is the language that is the verbalized professional cognitive and communicative activity [5, 18]. Language forms act in this environment as a means of exchanging information in the processes of people joint cognitive activity, a way of verbalizing human experience and its awareness.

In accordance with the modern understanding of language functioning, we accept one of the main principles of pragmatics about the fact that the reference of the word, including the term, the name of the object by the word/term is made by the speaker, not the linguistic expression. The author of the scientific text is regarded as a linguistic personality - an active participant in public, labor, and cognitive activity, a subject of communication - including a linguistic one [16].

The pragmatic interpretation of the terminological fragment is performed with regard to the communicants activity structures: the speech functions of the terms are interpreted in the context, in combination with the attributes which modify and clarify their meaning, with predicates and other components of syntax and super syntax structure of an utterance, a super-phrasal unity and a text as a whole.

The factors of not only linguistic, but also subject-and-practical activity, for example, the history of development of the corresponding science, the field of knowledge are also taken into account. The elements inherent to the subject-and-practical activity of an author of speech/text and a recipient can be found in language structures and be revealed in the process of interpretation in two kinds of fragment content. They can form a cognitive-communicative complex typical for the language: on the one hand they can be directly nominated in the utterance as a part of its reference (what the author of the speech informs about), and on the other hand, receiving an explication, or remaining implicit they can act as an incentive for communication (why the author says something).

For example, in the following interview fragment, the reason for presenting the information about the domain of the engineering psychology is the inquiry about such information by an uninformed person, presenting a group of people with the lack of knowledge in this sphere and the corresponding interest to it: *Ingenieurpsychologie ist ein eher unbekanntes Betätigungsfeld. Was genau tun Ingenieurpsychologen? – Sie untersuchen das Verhältnis von Mensch und Technik* [4]. The specialist being interviewed has to choose the strategy of the science popularization.

In pursuant to the activity structure in the process of pragmatic interpretation of the terminological fragment, the aims, motives, strategies, reference characteristics of the speech\text author, knowledge volume and recipient characteristics, communicant roles, components of the situation and other parameters can be defined.

In general, the aims and the motives for professional (scientific) communication are usually defined by the specificity of this speech genre and include primarily the intention to inform the readers or listeners about new results of special studies, the inclusion of some knowledge into a system, the message of the scientific community of new knowledge, motivated by the account of the knowledge fund of the recipient or recipients: *Eine technisch befriedigende Lösung des Problems ist erst Vor kurzem gefunden worden. Die Möglichkeit, einen solchen Prozess durchzuführen, wurde zuerst von C. v. Linde erkannt. Dieser Prozess trägt daher seinen Namen. Die Verwirklichung dieses Prozesses stellt zugleich den Beginn ... dar.* The information reported in the statement can be localized in time (*erst vor kurzem, zuerst, den Beginn von etwas darstellen,* etc.), its place in the subject-practical activity is attributed to some subject (*C. v. Linde*) or is determined with respect to the history of the science development: to be the beginning of something.

The examples of the subject speech activity - the definition of a concept called a term in which a new cognitive content is comprehensively presented in combination with the author's intention to inform about the introduction of a new term or refinement of its existing definition, usually placed in the introductory part of the text, - are given above in the section on cognitive interpretation of the terminological fragment.

The intention to inform can be realized by the author as the subject of speech both applied to the results he/she personally received and to his scientific heritage: *Hier können aber bereits eigene produktive Ideen und Schlussfolgerungen theoretischer Art formuliert werden. Die beiden Autoren erzielten folgende Ergebnisse ...* [4].

In the latter case, the author usually establishes the role and place of this information in the process of the corresponding science development, can subject the prospects for further research to the motivated evaluation: *... soll diese Problematik nochmals einer differenzierten Betrachtung unterzogen werden, weil die besagten Autoren insbesondere keine Ausführungen zu den methodischen Modalitäten der Versuchsführung tätigten* [4].

Information can be combined with detailed criticism of some authors personal positions, that in its turn implicitly indicates the components of the pragmatic image of a speaker or writer, his/her professional beliefs and attitude to the situation reflected: *Viele prominente Vertreter dieser Gilde schreiben von anderen Kollegen einfach ab, weil sie selbst keine eigenen produktiven Ideen haben... Sie "schmoren" so gewissermassen im eigenen "Saft"* [4].

More often the critical purpose of the speaker/writer is expressed in a more correct form and is directed not so much against the authors, but against the positions put forward by them. In this case, the disagreement of the discourse subject can be explicitly motivated: *In jüngster Zeit ist zunehmend die generelle Gültigkeit dieser Bestimmungen in Frage gestellt worden, weil* ... The motivation for changing attitudes to the views presented follows below.

The intention of the speaker to inform the recipient can be realized in the form of limiting the amount of information reported for some reasons, especially often associated with a fixed amount of time for the report or place in the article: *Auf weitere Details kann hier leider nicht eingegangen werden, es wird auf die Literatur verwiesen.* The second half of the given fragment, in accordance with the principle of courtesy, is addressed to the recipient, who, according to the subject's prediction, expects the information to be continued and may be disappointed by the interruption of the scientific communication flow. This part of the author's cooperative behavior is unfortunately missing.

The above mentioned attribution to the pragmatic interpretation of both linguistic and subject-and-practical activity can be illustrated by the following example, where in the same activity both the linguistic side (*formulieren*), and practical one are represented (*gemacht werden*): *Fragestellungen sind entscheidbar zu formulieren, wobei die allgemeine Fragestellung kaum entscheidbar gemacht werden kann.* The above example is also interesting due to the fact that it clearly serves as a context for understanding the term, particularly as an attribute of a term that can significantly affect its cognitive content. The role of the attribute is similar in the terminological text, preceding to the given one: *Es wird zwischen allgemeiner und spezieller Fragestellung unterschieden.*

The pragmatical material interpreted in this article includes not so many special scientific terms as general meta-textual terminological apparatus capable of servicing the discursive functioning of any terminology system that sequentially organizes terminological fragments within the framework of the author-selected strategy of constructing the informational-semantic structure of professional discourse or fixing this text discourse. Pragmatically, such type of terms is the most "saturated" and important for the author's information presentation in the discourse and facilitating its perception by the recipient. This apparatus includes, first of all, verbs of speech and thought (*behandeln, besprechen, darlegen, darstellen, beachten, diskutieren, formulieren, hervorheben, resümieren* etc.), and special clichés, explicating the discourse structure (*Äußerst wesentlich ist, dass... im Mittelpunkt der Untersuchung stehen... sind von fundamentaler Bedeutung, besondere Rücksicht auf etwas nehmen* etc.).

In accordance with the role structure of the professional discourse, not only the intentions of its author are subject to interpretation, but also the specificity of the addressee. The account of the addressee's factor can be considered a kind of good style for the subject that generates the discourse. The addressee's factor includes the recipient's knowledge fund, his/her expectations in the process of scientific communication, the ability to make sense from the direct meaning of terminological fragments taking into account the context, readiness to expand the recipient's awareness to acquire the content of new terms, etc. A scientific statement can have different objectives depending on the orientation of the discourse: if it is intended for a scientific community, it is dominated by the communication of new knowledge for this scientific field, new facts, new research results, new generalizations of discussion facts, their consideration, comparison and possible bringing into a new system. The intentionality of fragments targeted at theory and practice-driven specialists is different: *Die angezielten Probleme besitzen eindeutig ingenieurpsychologischen bzw. ergonomischen Charakter, wobei mit differenter Praxisrelevanz. Hier beginnt eigentlich bereits eine Aufwand- und Nutzenbilanzierung.*

The addressee's personality (especially coinciding with the consumer-practitioner) as a full participant of the speech and subject-and-practical activity of communicants can be explicitly nominated in the terminological fragment as a part of free or cliched constructions (structures): *Welches Vorwissen bringen die Nutzer mit? ...im ständigen Informationsaustausch mit dem Anwender; vom Anwender gewünschte Forderungen and so on.*

If the discourse is addressed to students, graduate students, etc., its main intention may be the desire to transfer the system of concepts in the relevant field, to include new knowledge in the frame structure of a particular science (for examples, see the section on cognitive interpretation). The elements of this discourse can have the following fragments-recommendations: *Dabei ist der Erfahrung ein relativ hoher Wert einzuräumen; Bei stark praktisch orientierten Untersuchungen... sollte das Literaturstudium nicht über Gebühr strapaziert werden* and so on.

4   Conclusion

The described possibilities of complex multidimensional (multi-aspect) consideration of the term as a part of a discursive fragment allow us to clarify the features of knowledge and intentions of subjects expression in the discourse. The given alternatives of the cognitive and pragmatic interpretation of terminological fragments in professional discourse are included in the extensive open system of its parameters and can be further supplemented by a study of the relationship between the cognizing subject and the object of cognition, the expression of knowledge and opinion, description and evaluation, detailing the needs of the cognitive situation and its correlation with systems of knowledge, values and interaction rules of communicants.

References

1. Arutyunova, N.D.: Pragmatika (in Russian) [Pragmatics]. In: Jarceva, V.N. (ed.) Lingvisticheskij Jenciklopedicheskij Slovar, pp. 390–391. Bol'shaja Rossijskaja jenciklopedija, Moscow (1990)
2. Arutyunova, N.D., Paducheva, E.V.: Istoki, problemy i kategorii pragmatiki (in Russian) [The origins, problems and categories of pragmatics]. Novoe v Zarubezhnoj Lingvistike **16**, 3–42 (1985)

3. Beilinson, L.S.: Pragmalingvisticheskie modeli opisanija logopedicheskogo diskursa (in Russian) [Pragmalinguistic models of description of speech therapist discourse]. Vestnik Voronezhskogo Gosudarstvennogo Universiteta. Serija: Filologija. Zhurnalistika **1**, 8–11 (2009)
4. Charwat, H.J.: Lexikon der Mensch-Maschine-Kommunikation. Oldenbourg, München (1994)
5. Fedorova, M.A., Vinnikova, T.A.: Teorija Jazyka Smi: Sozdanie Obraza Cheloveka i ego Vosprijatie (In Russian) [Theory of Mass Media Language: Forming a Person Image and its Perception]. Omsk STU Publishers, Omsk (2015)
6. Galich, G.G.: Kul'turologicheski obuslovlennye znanija i rechevoe povedenie lichnosti (in Russian) [Culturally conditioned knowledge and speech behavior of personality]. In: Gicheva, N.G. (ed.) Voprosy Issledovanija i Prepodavanija Inostrannyh Jazykov [Issues of Research and Teaching of Foreign Languages], pp. 39–46. Omskij gosudarstvennyj universitet im. F.M. Dostoevskogo, Omsk (2001)
7. Katsnelson, S.D.: Tipologija Jazyka I Rechevoe Myshlenie (In Russian) [Typology of Language and Speech Thinking]. Nauka Publishers, Leningrad (1972)
8. Klyoster, A.M.: Koncept «sostojanie» v nemeckojazychnom professional'nom diskurse inzhenernoj psihologii (in Russian) [Concept "condition" in German professional discourse of engineering psychology]. Gumanitarnyj Vector **11**(5), 80–83 (2016)
9. Klyoster, A.M., Galich, G.G.: The fragmentation of professional discourse. Proc. Soc. Behav. Sci. **206**, 56–61 (2015)
10. Kobenko, YuV, Ptashkin, A.S.: Equivalence and appropriateness: divergence characteristics of categories under translation. Mediterr. J. Soc. Sci. **5**(27), 1694–1697 (2014)
11. Latu, M.N.: Voennaja terminologija v sovremennom politicheskom diskurse (in Russian) [Military terminology in modern political discourse]. Politicheskaja Lingvistika **3**, 98–104 (2011)
12. Naer, V.L.: Pragmatika nauchnyh tekstov: (verbal'nyj i neverbal'nyj aspekty). Funkcional'nye stili: lingvometodicheskie aspekty (in Russian) [Pragmatics of scientific texts (verbal and non-verbal aspects)]. In: Cvilling, M.J. (ed.) Functional Styles: Lingvo-Methodical Aspects, pp. 13–19. Nauka Publishers, Moscow (1985)
13. Revina, J.V.: Professional'naja leksika v avtomobil'nom diskurse (in Russian) [Professional vocabulary in the automobile discourse]. Sovremennye Issledovanija Social'nyh Problem **1** (13), 97–100 (2013)
14. Shcherba, L.V.: Jazykovaja sistema i rechevaja dejatel'nost' (in Russian) [Language system and speech activity]. Nauka Publishers, Leningrad (1974)
15. Shpetnyi, K.I.: Pragmaticheskie parametry professional'nogo diskursa (na materiale BBC News/news front page) (in Russian) [Pragmatic parameters of professional discourse (based on information materials of BBC News/news front page)]. Vestnik Moskovskogo Gosudarstvennogo Lingvisticheskogo Universiteta. Serija: Gumanitarnye Nauki **596**, 214–234 (2010)
16. Susov, I.P.: Lichnost' kak subjekt jazykovogo obshhenija (in Russian) [Personality as a subject of language communication]. In: Susov, I.P. (ed.) Lichnostnye Aspekty Jazykovogo Obshhenija [Personal Aspects of Language Interaction], pp. 9–16. Kalininskij gosudarstvennyj universitet, Kalinin (1989)
17. Susov, I.P.: Dejatel'nost', soznanie, diskurs i jazykovaja sistema (in Russian) [Activity, consciousness, discourse and language system]. In: Susov, I.P. (ed.) Jazykovoe Obshhenie. Processy i Edinicy, pp. 7–13. Kalininskij gosudarstvennyj universitet, Kalinin (1988)
18. Zheltukhina, M.R., Vikulova, L.G., Serebrennikova, E.F., Gerasomova, S.A., Borbotko, L. A.: Identity as an element of human and language universes: axiological aspect. Int. J. Environ. Sci. Educ. **11**(17), 10413–10422 (2016)

The Functional-Semantic Field of the Mental Deviation Subcategory as a Euphemistic Phenomenon

Nina S. Zhukova[1]([⊠]) [iD] and Ludmila A. Petrochenko[2] [iD]

[1] Tomsk Polytechnic University, Tomsk 634050, Russian Federation
shukovans@mail.ru
[2] Tomsk State Pedagogical University, Tomsk 634061, Russian Federation
lapetrochenko@tspu.edu.ru

Abstract. Deviation is a multidimensional category including technical, legal and juridical, moral and ethical, biological and some other subcategorial components. The analysis of the human aspect of mental deviation subcategory is carried out and its three constituent parts are determined. These are medical terms; words of everyday language, expressing various mental disorders and aberrations; words of everyday language, expressing the underdevelopment of mental capacities. The human aspect of mental deviation subcategory is rich in linguistic means which represent not only the essence of mental disorder or underdevelopment but also one's attitude to it. Every constituent part has its own core and a periphery zone composed of euphemisms, dysphemisms and slang words. Dysphemisms and most slang words include name-calling and any sort of derogatory comment directed towards others in order to insult them or to express one's distaste or contempt. Abnormal behavior and deviations from normal mental activity may be also presented nonverbally. Most of the words and expressions of the periphery zones are lingual units of dualistic nature; their nature is ambivalent from the point of view of situations, participants and their tolerant or hostile attitude to the situation and each other. In English, alongside the true dysphemisms and insulting slang words one can find a lot of trite words and expressions of dysphemistic and slang stock which lost their novelty, becoming stereotypes. Nevertheless they can provide patterns for the formation of new expressions representing stronger, more emotional negative attitudes.

Keywords: Mental deviation · Functional-Semantic field · Periphery · Euphemisms · Trite dysphemisms · Stereotypes

1 Introduction

In English, the deviation category and its subcategories can be expressed by many linguistic means, among which metaphors and euphemisms are a common way of dealing with 'complex and mysterious phenomena' such as illness or mental disorder.

Deviation as a noticeable difference from what is expected, especially from accepted standards of behavior, is a category of multidimensional nature that includes technical, legal and juridical, moral and ethical, biological and some other components,

© Springer International Publishing AG 2018
A. Filchenko and Z. Anikina (eds.), *Linguistic and Cultural Studies:*
Traditions and Innovations, Advances in Intelligent Systems and Computing 677,
DOI 10.1007/978-3-319-67843-6_30

forming the corresponding subcategories. The biological deviation subcategory, in its turn, can be divided into non-human and human parts which include such interconnected and often interdependent phenomena as physical (bodily), behavioral and mental deviations.

On the whole, the deviation category consists of several levels of hierarchy. The highest level is represented by the word *deviation* as the main categorial name; the first lower subcategorial level is formed by technical, legal and juridical, moral and ethical, biological and some others kinds of deviation; the second lower subcategorial level, relevant for this research, includes the biological aspects of deviation - physical (bodily), behavioral and mental deviations. The human part of the mental deviation subcategory is much richer in linguistic means which are required to express not only the essence of mental disorder but also one's attitude to it.

2 Theoretical Background

The mental disorder subcategory is represented in English by a functional-semantic field as other categories and subcategories. The field consists of at least three constituent parts: (1) medical terms; (2) words of everyday language, expressing various mental disorders and aberrations; (3) words of everyday language, expressing the underdevelopment of mental capacities.

The first constituent part, as it was mentioned above, includes numerous medical terms.

The words of the second constituent part point at the symptoms of abnormal behavior, some limited or severe pathological deviations from normal mental activity.

The words of the third part are used to denote the real or imaginary cases when people are regarded as mentally underdeveloped.

"While English is an exceptionally rich language, most behaviors and things have only a few names. Bodily diseases usually have one or two medical names, as well as a few others in everyday language and perhaps slang. For mental diseases, however we have thousands of words, some ostensibly scientific, others in ordinary language, still others in slang" [29].

The functional-semantic field of mental deviation subcategory is a polycentric structure. Every constituent part of the field has its own core and a periphery zone composed of euphemisms, dysphemisms and slang words.

The terms 'mental illness', 'mental disease', 'mental deviation' (used here as synonyms) cover an enormous assortment of conditions, ranging from mildly eccentric or neurotic behavior, to severe psychotic disorders where a patient might lose total contact with reality. Traditionally mental illness is viewed not so much as a disease, but more as a moral failure, some sort of weakness of character, etc. Because of the vagueness of the term 'mental illness', even very minor disorders tend to carry the same negative stigma as the severe psychotic cases. The result of all this is that mental illness still remains an area of abundant euphemism [2].

All the definitions of 'euphemism' have one component of the meaning in common: avoiding offensive words and unpleasant topics.

Euphemism is an example of the use of a pleasanter, less direct name for something thought to be unpleasant [26].

Euphemism refers to the use of deliberately indirect, conventionally imprecise, or socially "comfortable" ways of referring to taboo, embarrassing, or unpleasant topics [13].

Dysphemism is the opposite of euphemism. Like euphemism, it is sometimes motivated by fear and distaste, but also by hatred and contempt. Dysphemistic expressions include curses, name-calling, and any sort of derogatory comment directed towards others in order to insult or to wound them. Dysphemism is also a way to let off steam; for example, when exclamatory swear words alleviate frustration or anger [3]. Whereas euphemisms seek to soften the impact of some horrific event or taboo subject by indirect language and calming metaphors, *dysphemisms* are starkly direct, macabrely metaphorical, or gruesomely physical. An obvious element of black humor is also apparent, since the bizarre metaphors strip away any notion of human dignity. Instead of the classical lexis generally prevalent in euphemisms, the core vocabulary is highly apparent, often in idiomatic phrases [13].

"*Slang*, very informal language that includes new and sometimes not polite words and meanings, is often used among particular groups of people, and is usually not used in serious speech or writing" [26].

"Much of the general slang vocabulary is viewed as fun to hear and fun to use. Many slang expressions are synonyms of, or nicknames for, widely known, standard words and expressions. < ... > Other slang expressions are called dysphemisms. A neutral or good term is replaced by one with some degree of negativity" [25].

"Euphemisms are in a constant state of flux. New ones are created almost daily" [21]. And so are dysphemisms.

The boundaries separating the euphemisms and dysphemisms of everyday speech from scientific language or slang are loose and constantly shifting. "Evidently, it's better to think of euphemisms as a variety of processes rather than a collection of expressions; and, given the variety of human judgments, it's more than likely that one man's euphemism will be another man's dysphemism" [1].

The analysis of special research in this field shows that many slang words and expressions are used as dysphemisms, while the nature of euphemisms may also be regarded as ambivalent. The use of euphemisms and dysphemisms as lingual units which differ in connotation is a topic of special interest to the goals of this study.

3   Research and Discussion

The investigation is concerned with the peripheral zones of the constituent components in the functional-semantic field of mental deviation subcategory, represented only by its human part. In connection with this, the materials for the study were taken from the British National Corpus (BYU-BNC) [6], the Corpus of Contemporary American English (BYU-COCA) [7], English and American dictionaries and contemporary fiction texts, which can help to estimate the frequency of recurrence of the investigated phenomena in everyday speech practices [19, 20]. The choice results from the authors' attempt to show that the peripheral zones under study are made up of euphemisms, dysphemisms, slang words which are typical of both British and American usage.

For this purpose *the contextual method of analysis* and *the semantic interpretation method* were used. Combination of the revealed words and phrases into corresponding groups was conducted on the basis of their functions as core words or peripheral ones. For the identification of their functions along with the above-mentioned methods *the descriptive method* and *the functional analysis* were applied.

The major focus of the investigation is the analysis of the following constituent parts:

(a) medical terms and euphemisms; (b) words and expressions representing in everyday speech the symptoms of abnormal behavior and severe deviations from normal mental activity: mad, crazy, insane, etc.; (c) words and expressions used to denote the real or imaginary cases when people are regarded as mentally underdeveloped, i.e. dull, retarded, etc.

The first constituent part of mental deviation subcategory is represented by medical words. It has the core including a great number of professional terms [14, 30] and a comparatively smaller periphery zone consisting of euphemisms and neutral words.

Using substitute words when discussing health matters is a long standing practice [15]. Attaching some sort of diagnostic label to someone's suffering is not always a positive thing. In some cases, patients may find themselves branded with a label, even after treatment is complete. Mental illness carries such a stigma [2]. That's why doctors and officials are usually very cautious when speaking about somebody's mental illness. They tend to use euphemisms and neutral terms, such as *mental health, mental health problem, mental state, mentally challenged, mentally disordered, mentally disturbed, mentally unstable, mental (health) hospital, psychological distress, etc.*

E.g.: (1) It is specifically for carers of people who have a *mental health problem* (BYU-BNC: K52 1).

(2) We were going to put down things like er, *mentally challenged*, and things like that, but it didn't go down too well (BYU-BNC: K74 1).

(3) ... further reduction of *mental hospital* beds will be to the detriment of the chronic *mentally disturbed* person (BYU-BNC: CS7 20).

(4) She seemed *mentally disturbed* and should be handled with care (BYU-COCA: FIC 2012 21).

(5) He asked practically every girl in school, even a couple of the *mentally challenged* girls from the Special Education program (BYU-COCA: FIC 2005 19).

The tradition of using euphemisms when discussing health problems can become a basis for jokes, as in the following dialogue:

(6) "You shouldn't say she's blind," Connie chided. "You should say '*visually impaired*'." - "Correct. And you shouldn't say I smoke too much. You should say I'm *nicotinically challenged*." - "And I should say you're *mentally deprived* instead of calling you *a nut*." We both giggled [23].

The second constituent part formed by words and phrases expressing abnormal behavior and severe deviations from normal mental activity may be presented as an enormous list of words, the number of which still remains incomplete. On the basis of several English and American dictionaries the following functional-semantic field was made up. In the field the formal synonymic words standing closer to the word 'mad', chosen as the hypernym, may be recognized as components of the core. The lower position of words and phrases in the list shows that they are more likely to belong to the periphery zone, i.e. the zone of euphemisms, dysphemisms and slang words.

MAD: insane, deranged, demented, disturbed, unbalanced, unhinged, crazy, lunatic, maniacal; (*informal*) daft, bonkers, barmy, crackers, mental; of unsound mind, mentally ill, certifiable, out of one's mind (or head), not in one's right mind, sick in the head, mad as a hatter, mad as a March hare, away with the fairies; off one's head, off one's nut, nuts, nutty (as a fruitcake), nutcase, off one's rocker, not right in the head, round the bend, stark raving mad, batty, bats, bats in the belfry, dotty, cuckoo, cracked, loopy, loony, doolally, bananas, loco, dippy, queer, screwy, schizoid, touched (in the head), gaga, up the pole, off-the-wall, not all there, not right upstairs, having a screw loose; *Brit. informal* barking (mad), whacky, round the twist, nobody home, off one's trolley, not the full shilling; *N. Amer. informal* nutso, screwball, go off the deep end, on the thin edge of reason, out of one's tree, meshuga, wacko, gonzo, etc. [9, 16, 22, 24, 32].

E.g.: (7) "The people in the town may say 'Oh, he's *not all there*, you know,' but that's just their little joke" (BYU-BNC: HWC 5).

(8) "But, Doctor," Ace interrupted, "this Pool thing's completely *hatstand*, isn't it?" - "I beg your pardon? *Hatstand?*" The Doctor clutched his own hat protectively. - "*Bonkers. Barmy. Out of its tree. Round the bend, out to lunch.* You know." - "Well," the Doctor's brow furrowed, "let's just say that Pool is a little *out of touch...*" (BYU-BNC: F9X 30).

(9) If you mean he is *off his rocker*, then we're all *off our rockers*. He's as sane as you (BYU-BNC: CKC 2).

(10) If Wilko is thinking of this deal he must be going totally *off his trolley* (BYU-BNC: J1F 50).

(11) The lady with the upright hair who Gloria said had *a screw loose*, clattered into the bedroom holding out a steaming jug on a tray (BYU-BNC: AC5 3).

(12) His wife is going *round the bend*, thinking he's under suspicion (BYU-BNC: HNJ 46).

(13) Dimly, she grasped that fatigue and emotional stress had sent her right *off the deep end*, but she had no strength left to haul herself back up into a sane reaction (BYU-COCA: FIC 2013 14).

(14) Mrs. Westmore is not the easiest client to handle, and those children of hers are *off the wall* [23].

(15) "Okay, the woman was *certifiable*. How was she going to orchestrate this?" [27].

(16) After all the shoplifting and the talk about her *not being quite right in the head*, they even suspect her of murder [28].

(17) Herman said, "Woman's *sick in the head...*" [ibid.].

(18) He went a little *loopy* toward the end [11].

(19) "From what I've heard she's *a nutcase*" [27].

(20) The only thing I'm certain about is that he's *nutty as a fruitcake* and should be taken off the road (BYU-COCA: FIC 2007 4).

(21) Frank had her infiltrate a cult headed by some *nutso* prophet [8].
Abnormal behavior and deviations from normal mental activity may be also presented nonverbally, e.g.:

(22) I could see a marshal on the start/finish line flapping his yellow flag at me and *screwing a finger at his temple*. His meaning was unmistakable. "Slow down, you maniac" (BYU-BNC: FBL 4).

(23) "Badly injured in a car accident in his final year." Then he *tapped his temple* significantly. "Never right afterwards, I'm afraid" (BYU-BNC: H8T 1).

(24) The bridge officers thought I was *nuts...* One of the officers *twirled his finger at his temple*. "What do you think you're doing?" he shouted (BYU-COCA: FIC 2013 2).

(25) Bill had turned back to the shelves, and Edmund *twirled his index finger at his temple* and pointed to him [18].

(26) "Is there something wrong with her... well... up here?" I *tapped my temple* [28].

Most of the words and expressions of the periphery zone are lingual units of dualistic nature. Their nature is ambivalent, euphemistic or dysphemistic, from the point of view of situations, participants and their tolerant or hostile attitude to the situation and each other.

The third constituent part of mental deviation subcategory is represented by words and phrases expressing the underdevelopment of mental capacities. The set of these lingual units remains incomplete, because only in the domain of slang words "there are nearly one thousand terms for 'fool' (and another 650-plus for 'stupid')" [10]. On the basis of English and American dictionaries the following functional-semantic field was formed, the word 'dull' chosen as its hypernym.

DULL: dull-witted, obtuse, stolid, stupid, slow, slow-witted, thick, dense, witless, unintelligent; backward, retarded, odd, dolt, doltish, crass, dumb, simple, foolish, feeble-minded; *informal* dim, dim-witted, dingbat, dink, dippy, dooley, dumdum, nitwit, weak in the head; *Brit. informal* birdbrain, brickwit, chernie, cluffy, derk (durk), nerk, wally, etc.; *N. Amer. informal* dimwit, dullard, dim bulb, blonde, scatter-brained, a little off, clambrain, cupcake, dingleberry, ditsy (ditzy), nerd, pothead, etc. [9, 16, 22, 24, 32].

E.g.: (27) He certainly did not have the intellectual capabilities of his brothers and sisters. He was *dull*, untidy in his books, arrogant and morose (BYU-BNC: ANC 76).

(28) She turned to me and spoke to me like a teacher to a beloved but *dull-witted* child (BYU-BNC: CDX 9).

(29) However, it was my misfortune that McNeile selected C.O. Bevan as my classical tutor. A *stolid*, red-faced clergyman without wit or humour... (BYU-BNC: H0A 59).

(30) I wondered if he was too *obtuse* to pick up what I was driving at (BYU-BNC: H9N 12).

(31) My son is *mentally retarded*. I know I am not alone in having a close relative who is *mentally impaired* (BYU-BNC: K52 24).

(32) The greengrocer spoke with exaggerated patience as though explaining something to a *backward* child (BYU-BNC: EVH 5).

(33) … lions seem to have a particular fondness for drunks. But then they also catch the *simple and delusional* outcasts: *schizophrenics* and the *slow-witted* (BYU-COCA: ACAD 2009 15).

(34) He knew I didn't like him talking like that to the *dimwit* girl (BYU-COCA: FIC 2007 16).

(35) Sometimes I feel *a little off*, and I might not remember things that I usually would have remembered (BYU-COCA: MAG 2015 14).

(36) The girl is so *dense* she'd have trouble pouring water out of a boot [4].

(37) He was the local *pothead*, a grade or two ahead of her. Each grade had taken him a couple of years to complete [12].

The above-mentioned second and third peripheral zones also contain several specific phrases which express *'lacking (missing) some part (or parts) of the whole'*, i.e. *one brick short of a load, not a full load; not playing with a full deck, two/three cards short of a full deck; one sandwich short of a picnic; two bob short of a quid; his elevator doesn't go to the top floor, etc.* They are usually treated as euphemisms [2].

This list can be expanded by other expressions of the same kind because this pattern gives full scope to one's imagination: *a few bricks short of a full load, a few bales short of a full load, a couple buckets short of a full load; a couple of buns short of a burger; one fry short of a Happy Meal, a few fries short of a Happy Meal; two cans short of a six-pack, some beers short of a six-pack, etc.* [31].

E.g.: (38) "She is *crazy*," one woman said. "His mama must be *two cans short of a six-pack*" (BYU-COCA: NEWS 2011 1).
In this example the phrase *'two cans short of a six-pack'* is unlikely to be thought of as a pleasanter name for the word *'crazy'*.

(39) Oddly, Blake saw no inconsistency in hinting Kyle was *a few beers short of a six-pack* for accompanying him (BYU-COCA: FIC 2003 5).

(40) There were those who thought the old woman was *a few bales short of a full load* (BYU-COCA: FIC 2001 1).

(41) This is a human being you're talking about (although, judging by your question the child might be *a couple buckets short of a full load*) (BYU-COCA: MAG 2000 2).

(42) This is a guy who offered today to take truth serum to refresh his memory. He is simply *one fry short of a Happy Meal*, this guy (BYU-COCA: SPOK 2000 3).
A "Happy Meal" is a meal package for kids, sold at the fast-food chain McDonald's.

(43) Having read that, the rest of you probably now think that the author is *a sandwich short of a picnic*. Trust me - *my hamper is full* (BYU-BNC: A5Y 1).

The author uses this expression for comic effect: 'hamper' is a basket for carrying food or drink (a picnic hamper).

(44) This sweet old lady was obviously *a few slices short of a loaf...* [17].

The context of this example indicates the use of the phrase '*a few slices short of a loaf*' as a trite dysphemism rather than a mere euphemism.

(45) "I always thought Plug was *missing a gallon from his water tank.* Now I know it" [5]. The characteristic is given to the chief of the fire department known by his nickname, Plug.

(46) "To put it in a way you can appreciate, she's *a few glazed donuts short of a dozen.*" - "Why don't you just fire her?" I asked [4].

In this case the owner of a restaurant complains of her waitress to the owner of a donut shop who usually delivers donuts to her customers in boxes by the dozen.

In examples 38–46 one can see seven expressions of conventional usage (*two cans short of a six-pack, a few beers short of a six-pack, a sandwich short of a picnic, a few slices short of a loaf, a couple buckets short of a full load, etc.*) and two new ones, coined by the speakers in accord with circumstances (*missing a gallon from his water tank (a gallon short of his water tank), a few glazed donuts short of a dozen*).

(47)

(a) "Tell me, do you know Marjorie Bilson?"
"*Batty* Bilson?" he asked.
"*Batty*?" she repeated.
"Yeah, I went to school with her," he said. "She has *issues*" [18].

(b) Lindsey had no doubt that Batty was hunting her down for another little chat. As it was, she had no intention of engaging in another go-round with *a nut burger* [ibid.].

(c) "Yeah, well, we're *not* dealing with *the most rational person in town* now, are we?" [ibid.].

(d) "Well, she's not exactly operating *at full capacity*," Lindsey said [ibid.].

(e) Everyone knew the woman was *a few slices short of a loaf*; how could they listen to her? [ibid.].

(f) "Does Majorie have a history with firearms?" - "I don't know," he said. "I mean everyone knows *her elevator doesn't reach the top floor*, but it's not like..." [ibid.].

(g) She looked like a scared little kid, and Lindsey felt a pang of sympathy for her. Majorie couldn't help it if she was *a few cards short of a full deck* [ibid.].

(h) It was apparent that everyone thought Marjorie was *a few chapters short of a book*, but was she dangerous? The call certainly gave Lindsey the heebie-jeebies... [ibid.].

The descriptions refer to one person, Majorie Bilson, who is known for her provocative words and aggressive behavior. The words and expressions '*issues (problems), not at full capacity, not the most rational person in town*' can be regarded as euphemisms. '*Batty, nut burger*' are insulting slang words [16]. Besides, we have every reason to believe that such phases as '*a few cards short of a full deck, a few slices short of a loaf, the elevator doesn't reach the top floor, etc.'* belong to the periphery zones as dysphemisms. The gamut of emotions shows that the attitude to this *mentally disturbed* person is uncertain, ranging from anger to pity.

The above-mentioned dysphemisms represent traditional usage, often found in everyday language. As trite expressions of dysphemistic stock they became stereotypes, their pejorative expressive force lost to a considerable degree. At the same time they can provide patterns for the formation of new expressions representing stronger negative attitudes.

The phrase *'a few chapters short of a book'* belongs to a librarian; *'missing a gallon from his water tank'* to the deputy fire chief; *'a few glazed donuts short of a dozen'* to the owner of a restaurant.

The new expressions of this kind are either explicit dysphemisms, or borderline cases of the euphemistic-dysphemistic dichotomy.

4 Conclusion

The category of deviation has a complex paradigm including several subcategories, i.e. biological one represented by physical, behavioral and mental deviations. The human aspect of mental deviation (1–47) includes at least three constituent parts. The functional-semantic field of mental deviation subcategory is a polycentric structure. Every constituent part of the field has its own core and a periphery zone. The euphemisms, dysphemisms and slang words form the corresponding periphery zones of the functional-semantic fields, the full specification of which requires further analysis. Of special interest are the patterns of trite and commonplace dysphemisms that can be used for producing more expressive phrases showing a whole range of possible emotions (ridicule, mockery, scorn, indignation, anger, etc.) with regard to concrete situations and their participants.

Summing up the results of the given research and at the same time specifying the future study prospects, we should say that the description of the means representing the functional-semantic field of mental deviation subcategory in English must include:

– the definition of all the constituent segments and zones the field consists of;
– the study of the lingual units of periphery zones, including neologisms;
– the study of the peculiarities of the relations between euphemisms, dysphemisms and slang words.

References

1. Adams, R.M.: Soft soap and the nitty-gritty. In: Enright, D.J. (ed.) Fair of Speech: The Uses of Euphemism, pp. 44–55. Oxford University Press, Oxford (1985)
2. Allan, K., Burridge, K.: Euphemism & Dysphemism: Language Used as Shield and Weapon. Oxford University Press, Oxford (1991)
3. Allan, K., Burridge, K.: Forbidden Words: Taboo and the Censoring of Language. Cambridge University Press, Cambridge (2006)

4. Beck, J.: Tragic Toppings. St. Martin's Paperbacks, New York (2011)
5. Bolin, J.: Threaded for Trouble. Berkley Prime Crime Books, New York (2012)
6. Brigham Young University: The British National Corpus (BYU-BNC). http://corpus.byu.edu/bnc/. Accessed 15 June 2017
7. Brigham Young University: The Corpus of Contemporary American English (BYU-COCA). http://corpus.byu.edu/coca/. Accessed 15 June 2017
8. Cates, B.: Magic and Macaroons. Obsidian, New York (2015)
9. Fergusson, R.: The Penguin Dictionary of English Synonyms and Antonyms. Penguin Books, London (1992)
10. Green, J.: The Big Book of Being Rude: 7000 Slang Insults. Cassell, London (2002)
11. Hamilton, V.: A Deadly Grind. Berkley Prime Crime Books, New York (2012)
12. Hamilton, V.: Bowled Over. Berkley Prime Crime Books, New York (2013)
13. Hughes, G.: An Encyclopedia of Swearing: The Social History of Oaths, Profanity, Foul Language, and Ethnic Slurs in the English-Speaking World. M.E. Sharpe, New York (2006)
14. Jacobson, J.W., Mulick, J.A., Rojahn, J.: Handbook of Intellectual and Developmental Disabilities. Springer Science-Business Media, New York (2007)
15. Keyes, R.: Euphemania: Our Love Affair with Euphemisms. Little, Brown and Co., New York (2010)
16. Kipfer, B.A., Chapman, R.L.: Dictionary of American Slang, 4th edn. HarperCollins Publishers, London (2007)
17. McKinlay, J.: Books Can Be Deceiving. Berkley Prime Crime Books, New York (2011)
18. McKinlay, J.: Due or Die. Berkley Prime Crime Books, New York (2012)
19. Petrochenko, L.A., Zhukova, N.S.: Kateroriya «mera»: osobennosti reprezentatsii v sovremennom angliyskom yazyke (in Russian) [The Category of Measure: Peculiarities of Representation in Modern English]. Vestnik TGPU 4(157), 26–30 (2015)
20. Petrochenko, L., Zhukova, N.: Constructions expressing inaccurate quantity: functions and status in modern English. J. Lang. Educ. 2(1), 48–55 (2016). National Research University Higher School of Economics, Moscow
21. Rawson, H.: A Dictionary of Euphemisms & Other Doubletalk: Being a Compilation of Linguistic Fig Leaves and Verbal Flourishes for Artful Users of the English Language. Crown Publishers, New York (1981)
22. Rawson, H.: Wicked Words: A Treasury of Curses, Insults, Put-Downs, and Other Formerly Unprintable Terms from Anglo-Saxon Times to the Present. Crown Trade Paperback, New York (1989)
23. Sanders, L.: McNally's Gamble. Berkley Books, New York (1998)
24. Soanes, C., Waite, M., Hawker, S.: The Oxford Dictionary, Thesaurus, and Wordpower Guide. Oxford University Press, Oxford (2001)
25. Spears, R.A.: McGraw-Hill's Dictionary of American Slang and Colloquial Expressions. McGraw-Hill Companies, New York (2006)
26. Summers, D.: Longman Dictionary of English Language and Culture. Longman Group UK, Harlow (1992)
27. Sweeney, L.: The Cat, the Quilt and the Corpse. Obsidian, New York (2009)
28. Sweeney, L.: The Cat, the Lady and the Liar. Obsidian, New York (2011)
29. Szasz, T.: A Lexicon of Lunacy: Metaphoric Malady, Moral responsibility, and Psychiatry. Transaction Publishers, London (1993)
30. Thackery, E., Harris, M.: The Gale Encyclopedia of Mental Disorders, vol. 1–2. The Gale Group, Detroit-New York (2003)
31. The online slang dictionary. http://onlineslangdictionary.com/. Accessed 13 June 2017
32. Thorne, T.: Dictionary of Contemporary Slang, 3rd edn. A & C Black Publishers, London (2007)

The Event Concept Categorial Network

Anna M. Klyoster[1]([✉]) [iD] and Natalia Ju. Shnyakina[2] [iD]

[1] Omsk State Technical University, 644050 Omsk, Russian Federation
annaklyoster@mail.ru
[2] Omsk State Pedagogical University, 644099 Omsk, Russian Federation
zeral@list.ru

Abstract. The mental representations study is one of the main problems in cognitive linguistics. The language in which the person's special view of life is recorded can be the basis for the human cognition studying; in this context the various knowledge formats structuring attract interest. The objects of reality and inner world of a human differently exist in his brain. For mental representations of the propositional objects (event, situation and process) the following terms can be used: concept, scheme, scenario, scene, script and frame). The frame as a typical structure of such a type of knowledge contains some semantic units: obligatory and optional slots. Obligatory slots form a stereotype situation; they become optional slots by describing the actual one. The content of a set of information nodes (semantic roles) corresponding to the main categories of human life such as the "subject", "object", "action", "result", "instrument", "purpose", etc. can be considered as a form of knowledge arrangement. The complex of such nodes can be studied as the categorial network by means of which a person structures the knowledge about the event.

Keywords: Event concept · Information node · Categorial network · Category · Mental representations

1 Introduction

The knowledge about the nature of mental representations, their types and structure can be complemented by the data obtained from modern linguistic papers. In the cognitive linguistics which has broadened the sphere of its interests into the human consciousness field, the problem of language and thinking interaction is fundamental. In addition, expansionism, anthropocentrism, functionalism and explanatoriness [18] recognizing the language studying as a cognitive and communicative phenomenon are the main principles of the modern linguistic paradigm. A person's thought existing in the consciousness being objectified in the language form can be considered as a multidimensional construction capable of new data on the person's cognitive apparatus structure and functioning.

The idea that the language structure shows the human brain work is represented in a number of Russian cognitive scientists papers [3, 4, 10, 13, 15, 16, 19, 25] and foreign ones [7, 11, 21, 22, 27, 34]. The analysis of the text fragments describing events (processes, situations, actions, circumstances) to a certain extent makes it possible to understand how these manifestations of reality are comprehended by the person.

© Springer International Publishing AG 2018
A. Filchenko and Z. Anikina (eds.), *Linguistic and Cultural Studies:*
Traditions and Innovations, Advances in Intelligent Systems and Computing 677,
DOI 10.1007/978-3-319-67843-6_31

Mental representations are the information structure of our consciousness, the organized knowledge fixing the gained experience. The papers of the linguists studying the person's conceptual system through the language are devoted to the mental representations issue [17, 20, 33]. The propositional knowledge representing a complex structure including at least two elements interconnected by predication, causation etc. is considered as a special type of knowledge in the above mentioned papers.

To define the propositional knowledge, the researchers use the terms "event concept" [24], "scheme" [28], "frame" [12, 23], "scenario" [30], "scene" [7], "script" [29].

The operational units of the conceptual system are the concepts in which the person's knowledge about the environment fragments including subject's emotions, his \her national specific characteristics, associations, etc. is fixed. The term "concept" became well-known in the research base of cognitive linguistics and is used for the person's mental sphere units studying. Thus, the concept represents the generalized form of mental representation in which its features are not differentiated. The concept can be studied by the modeling approach (the conceptual analysis) that consists of the semantic units identified on the basis of linguistic meanings and the conceptual characteristics revealed in the process of the corresponding words lexical meanings interpretation [31]. Mental representations are the structures of knowledge, and their various objectifications by the language system resources allow us to speak about different formats of this conceptual content.

One of the propositional content existence forms is the *frame* representing the data structure for the stereotype situation presentation [23]. The way, how the frame is structured, can be schematized differently. Typically the frame is presented in the form of the structure consisting of slots (nodes): obligatory slots containing both the typical (stereotype) situation about an event, and optional slots which can be complemented with the new information characterizing the actual, perceived or new, unfamiliar situation [26]. The familiar (stereotype) frame structure represents a peculiar support which is able to perceive a new situation and comprehend it adequately.

Consequently, it is necessary to formulate some points important for the event concept as the frame structure consideration: firstly, the frame represents the structure in consciousness frame of a typical situation; secondly, the frame is necessary to consider as the formal principle of mental representation and not the representation itself (its content); thirdly, the frame structure arranges the concept semantic "fabric" in the form of slots complex.

Moreover, other formats of knowledge including scripts, scenarios, and scenes are successfully applied for the event mental representation studies. According to Dem'jankov, a script is necessary to understand as "... one of the consciousness structures types, a form of the frame performing some special task in natural language processing: the familiar situations are described by the script as stereotype changes of events" [5]. Subsequently, the theory of scripts was expanded: they began to be considered as representations based on the stereotype structures support, but all the same, in relation to the specific situation taking into account the circumstances and intentions of participants [29]. In Schank and Abelson's view, the script consists of the interchangeable scenes: *scenes* represent script bricks [30]. The scenario is larger information format applied to texts.

The opinions described above on the specificity of possible knowledge arrangement forms about the event allow to understand the correlation between the concepts: concept, frame, script, scene, scenario in terms of the event concept studying by linguistic methods.

2 The Event Concept Frame Structure

First of all, it is important to understand that the concept "mental representation" can be applied both to the current event image and to the one kept in memory including the prototype image not accepted perceptively. Therefore, it is necessary to distinguish between the typical idea of the situation kept in long-term memory as the stereotype that is the familiar knowledge allowing structuring the new one, and the new knowledge understood as actual, perceived at present moment or simply describing the situation. In both cases the term "concept" is appropriate to use because the idea of an event is considered as an information structure consisting in the human conceptual system. The structure of this concept can be described as the frame containing information slots. However in the first case the issue is about the uppermost (obligatory) slots constituting the familiar stereotype information and in case of the current event conceptualization it is about the terminal slots filled with new, original information. Slots correspond to the global ontological categories reflecting basic cognitive spheres such as the subject, object, instrument, space, time, purpose, etc. Categorial nodes are connected with the global sections of the human experience.

A set of categories is defined by the situation and speech intentions of communicants. In the process of verbalization of an event-concept in person's consciousness there is a set of these categories presented in slots and a frame-stereotype in which some of them are reflected. On the assumption of the real situation, the frame-stereotype "acquires" details, some slots can disappear, and other ones emerge. The verbalization process assumes the movement from the general knowledge (category) to the private one (its members); it is demonstrated in the categorial nodes specification through the use of linguistic means. The categorial network expressed by the lexical parameters represents the complex of the global ideas reflecting human life highlights. The connections between these ideas are intercategorial ones generating the relations of predicativity, causation, etc.

Regarding the way of the frame structure existence, the propositional knowledge is inclined rather to terms "scene" and the "script", proceeding from their interrelation shown in the form of scenes change and understanding of this sequence as a script.

3 The Categorial Nodes of the Event Concept

The sentence is the minimum material expression of the event concept, the integration of linguistic and categorial world segmentation [32]. Taking into account the structuring form of the event concept described above in the form of the frame, knowledge objectified in the sentence can be distributed into the separate slots representing the categorial nodes of the considered mental representation. The identification of these nodes on the basis of corresponding words lexical meanings makes allows to model the event concept structure.

The attempts of the fundamental utterance meanings identification were repeatedly made in linguistics [6, 8, 9, 14]. The most important semantic categories of the event mental representation are the following: "subject", "object", "subject attribute", "objects attribute", "cognitive action", "instrument", "result", "space" and "time". These semantic coordinates form the obligatory (uppermost) slots of the frame-stereotype and represent the ideal structure of situational knowledge, which is specified depending on the current state of affairs. During the utterance formation the obligatory slots acquire the status of optional ones through conceptual signs.

Let us consider the following example: *Jetzt riecht es ganz schlimm nach verbranntem Gummi oder nach Plastik* [1]. Here, the frame element "time" is represented by the word "jetzt". Concept signs of the node "result", which have the form of the qualitative assessment representation, on the one hand, and the existing form of the attribute of the object, on the other, are instantiated by the expression *"ganz schlimm"* and by the verb *"riechen"*, correspondingly. Structural components of the sentence *"nach verbranntem Gummi oder nach Plastik"* correspond to the obligatory slot «object».

Therefore the cognition process verbalization, modified during the interpretation, includes the following components: "object", "result", and "time".

As the second example we will analyze the following fragment: *Die Nachbarn würden noch lange von dem Geruch reden, der ihnen entgegenschlug, als sie die Wohnungstür öffneten* [2]. In this sentence, we should note the following categorical nodes: "subject", "feature of the object", "result", "space", and "time". These information nodes are generalizations of more specific concepts. The noun *"die Nachbarn"*, becomes the objectification of the categorical node "subject"; the noun *"der Geruch"* is regarded as the verbalization of the object feature. The knowledge about the cognition result is expressed by the verb *"entgegenschlug"*, reflecting the dynamic characteristics of the feature. The adverbial sentence *"als sie die Wohnungstür öffneten"* objectifies the space-time framework of the situation.

4   Conclusion

Modern linguistic paradigm presupposes the consideration of linguistic phenomena from the position of the subject which is generating them. A person is the carrier of conceptual structures objectified in language. A way of knowledge structuring should be considered as a way of the world-modeling.

Propositional knowledge, reflecting the specific nature of a person comprehension of the situation, process, event, state of affairs, etc., is a multidimensional structure consisting of a set of coordinates, which nodes form a categorial network of propositional knowledge. A nodes-filled frame is a scene, while a complex of scenes can be viewed as a script. In the case of filled obligatory (uppermost) slots it is a stereotyped event, while in the case of moderately filled terminal slots a current (new) event is considered.

The rules of personal impressions fixation by means of language are mainly determined by the categorizing activity of human consciousness: any perceived phenomenon has a categorization phase - referring it to one or another experience rubrics (categories). The identification of these categories by cognitive interpretation of the context is an important step towards the study of the event concept.

References

1. Berliner Zeitung (a). http://www.berliner-zeitung.de/geschichte-einer-groesseren-und-vieler-kleiner-katastrophen-der-wasserschaden-15924440. Accessed 11 May 2017
2. Berliner Zeitung (b). http://www.berliner-zeitung.de/newsarchive/archiv/2005-11-12. Accessed 11 May 2017
3. Boldyrev, N.N.: Principy i metody kognitivnyh issledovanij jazyka [Principles and methods of cognitive language research]. In: Boldyrev, N.N. (ed.) Principy i metody kognitivnyh issledovanij jazyka, pp. 11–29. Tambovskij gosudarstvennyj university im. G.R. Derzhavina, Tambov (2008). (in Russian)
4. Burenkova, S.V.: K voprosu o dinamike konceptov (na primere nemeckih konceptov ORDNUNG i SICHERHEIT) [To the problem of the dynamics of the concepts ORDNUNG and SICHERHEIT in German]. Voprosy kognitivnoj lingvistiki 4(45), 30–38 (2015). (in Russian)
5. Dem'jankov, V.Z.: Skript [Script]. In: Kubrjakova, E.S., Dem'jankov, V.Z., Pankrac, JuG, Luzina, L.G. (eds.) Kratkij slovar' kognitivnyh terminov, pp. 172–174. Filologicheskij fakul'tet MGU im. M.V. Lomonosova, Moscow (1997). (in Russian)
6. Fillmore, Ch.J.: The case for case. In: Bach, E., Harms, R.T. (eds) Universals in Linguistic Theory, pp. 1–25. Holt, Rinehart and Winston, London (1968)
7. Fillmore, Ch.J.: Frame semantics. In: Han'guk Ŏ.H. (eds) Linguistics in the morning calm: Selected papers from the SICOL-1981, pp. 111–137. Hanship, Seoul (1988)
8. Furs, L.A.: Sintaksicheski reprezentiruemye koncepty: avtoref. dis. …dokt. filol. nauk [Syntactically representable concepts]. Unpublished these abstract (2004). (in Russian)
9. Galich, G.G.: O sistematizacii kategorij jazykovogo soderzhanija s pozicij antropocentrizma [On the systematization of language content categories from the view of antropocentrism]. Vestnik Omskogo universiteta 2(64), 372–375 (2012). (in Russian)
10. Galich, G.G., Shnyakina, N.Y.: Categories of immaterial objects in mind and language. Procedia Soc. Behav. Sci. 206, 30–35 (2015)
11. Jackendoff, R.S.: Semantics and Cognition. MIT Press, Cambridge (1983)
12. Kintsch, W.: The Representation of Meaning in Memory. Halsted Press, New York (1974)
13. Klyoster, A.M.: Kognitivnyj podhod k izucheniju mezhotraslevoj terminosistemy [Cognitive approach in studying of interdisciplinary terminosystem]. Omskij nauchnyj vestnik 4(111), 171–174 (2012). (in Russian)
14. Klyoster, A., Galich, G.: The fragmentation of professional discourse. Procedia – Soc. Behav. Sci. 206, 56–61 (2015)
15. Kobenko, Yu.V., Ptashkin, A.S.: Equivalence and appropriateness: divergence characteristics of categories under translation. Mediterr. J. Soc. Sci. 5(27), 1694–1697 (2014)
16. Kobenko,Yu.V., Zyablova, N.N., Gorbachevskaya, S.I.: Towards translation naturalism at translation of english film production into modern German. Mediterr. J. Soc. Sci. 6(3), 494–496 (2015)

17. Kravchenko, A.V.: Javljaetsja li jazyk reprezentacionnoj sistemoj? [Is the language a representative system?]. Studia Linguistica Cognitiva. Jazyk i poznanie: Metodologicheskie problemy i perspektivy **1**, 135–156 (2006). (in Russian)
18. Kubrjakova, E.S.: Nachal'nye jetapy stanovlenija kognitivizma: lingvistika – psihologija – kognitivnaja nauka [The initial stages of the development of cognitivism: linguistics - psychology - cognitive science]. Voprosy jazykoznanija **4**, 34–47 (1994). (in Russian)
19. Kubrjakova, E.S.: Ob ustanovkah kognitivnoj nauki i aktual'nyh problemah kognitivnoj lingvistiki [On cognitive science and actual problems of cognitive linguistics]. In: Novodranova, V.F. (ed.) V poiskah sushhnosti jazyka: kognitivnye issledovanija, pp. 13–35. Znak, Moscow (2012). (in Russian)
20. Kubrjakova, E.S., Dem'jankov, V.Z.: K probleme mental'nyh reprezentacij [On the problem of mental representations]. Voprosy kognitivnoj lingvistiki **4**, 8–16 (2007). (in Russian)
21. Lakoff, G.: Women, Fire and Dangerous Things: What Categories Reveal About the Mind. University of Chicago Press, Chicago (1987)
22. Langacker, R.W.: Cognitive Grammar: A Basic Introduction. Oxford University Press, New York (2008)
23. Minsky, M.L.: A framework for representing knowledge. In: Winston, P.H. (ed.) The Psychology of Computer Vision, pp. 211–277. McGraw-Hill, New York (1977)
24. Pechetova, N.J.: Interpretacija sobytijnogo koncepta v mediatekste [The interpretation of the event concept in a media text]. Vestnik Severo-Vostochnogo federal'nogo universiteta im. M.K. Amosova **7**(2), 144–149 (2010). (in Russian)
25. Popova, Z.D., Sternin, I.A.: Kognitivnaja lingvistika [Cognitive Linguistics]. AST: Vostok – Zapad, Moscow (2007). (in Russian)
26. Romashina, O.Ju.: Frejmovyj analiz semantiki frazeologicheskih edinic (in Russian) [Frame analysis of the semantics of phraseological segments]. ACTA LINGUISTICA **2**(2), 33–38 (2008)
27. Rosch, E.: Principles of categorization. In: Rosch, E., Lloyd, B. (eds.) Cognition and Categorization, pp. 27–48. Lawrence Erlbaum Associates, Hillsdale (1978)
28. Rumelhart, D.E.: Schemata: the building blocks of cognition. In: Spiro, R.J., et al. (eds.) Theoretical Issues in Reading Comprehension, pp. 33–58. Lawrence Erlbaum Associates, Hillsdale (1980)
29. Schank, R.C.: Dynamic Memory: A Theory of Reminding and Learning in Computers and People. Cambridge University Press, New York (1983)
30. Schank, R.C., Abelson, R.P.: Scripts, Plans, Goals and Understanding: An Inquiry into Human Knowledge Structures. Lawrence Erlbaum Associates, Hillsdale (1977)
31. Shnjakina, N.J.: Jetapy modelirovanija sobytijnogo koncepta (na primere verbalizovannoj situacii poznanija zapaha) [Modeling stages of event-concept (case study of verbalized smell cognition situation)]. Sovremennye problemy nauki i obrazovanija **2**(434) (2015). (in Russian). http://www.science-education.ru/122-21340. Accessed 11 May 2017
32. Shnjakina, N.J.: O verbalizacii sobytijnyh konceptov [Verbalization of event-concepts]. Istoricheskaja i social'no-obrazovatel'naja mysl' **7**(5–2), 283–288 (2015). (in Russian)
33. Tameryan, T.Y.: Ponjatie mental'noj reprezentacii v kognitivnoj lingvistike [Mental representation in cognitive linguistics]. Bjulleten' VIU. Vladikavkaz **17**, 37–41 (2006). (in Russian)
34. Vezhbickaja, A.: Jazyk. Kul'tura. Poznanie [Semantics. Culture. Cognition]. Russkie slovari, Moscow (1996). (in Russian)

Neologisms in American Electronic Mass Media

Tatiana A. Martseva[(⊠)] (iD), Anastasia Yu. Snisar (iD),
Yuriy V. Kobenko (iD), and Ksenia A. Girfanova (iD)

Tomsk Polytechnic University, Tomsk 634050, Russian Federation
t.martseva@mail.ru

Abstract. The paper analyses media neologisms in modern American English considering their pragmatic, functional, stylistic, structural and semantic features. The analysis is based on the onomasiological approach to studying neologisms with the focus on their representative function in mass media. The authors establish selection criteria for neologisms as well as their classification. The research follows the paradigm of neological approach to describing vocabulary surplus, which draws attention to the following aspects of neologisms: their standardization, graphical form, semantics, stylistic function and productivity.

Keywords: Neologism · Neology · Media discourse · Electronic environment · Internet · Classification of neologisms · Author's neologisms · Occasional words · Non-usual word-formations

1 Introduction

Linguists are currently highlighting the fact that modern English is neologizing rapidly which reflects global trends. Such process is defined as a neological boom [5]. This phenomenon is connected with an intensified pace of life, scientific progress and distribution of advanced technologies. Lexicographers do not manage to capture new vocabulary appearing due to rapid development of American English which has encouraged the researchers to look into the issues of neology in English language. The problems connected with lexical neologisms become clearer; they include issues of codification, typology and translation of neologisms into other standard languages. It is important to emphasize that a substantial number of new lexical units has appeared due to computer technologies which allows us to admit that the thematic group connected with Internet communications is the most susceptible to neologisation. By Internet communications we mean communicative environment where people create and share their philosophy, knowledge and values. General public access of to such information technologies as Internet forums, social networks and chats allows a person to discuss and comment on the events worldwide, which has also caused a burst of neologisation. Many neologisms have got a connotative meaning and are only used within the media environment [13, p. 445].

© Springer International Publishing AG 2018
A. Filchenko and Z. Anikina (eds.), *Linguistic and Cultural Studies:*
Traditions and Innovations, Advances in Intelligent Systems and Computing 677,
DOI 10.1007/978-3-319-67843-6_32

The analysis of new vocabulary developments, research of the ways and means of their coming into the language is a priority task of describing lexical changes in modern American English. Modern American media neologisms are studied because Internet sphere rapidly reflects new phenomena and thus creates positive environment for new words and thematic groups. Nowadays Internet resources have become the major and sometimes the only source of news. The only chance to find out information about some events is to search online. As long as English is a language of international communication and often the official language of some events it neologises more rapidly than the other ones and sets the word-building patterns borrowed into the other languages. At the same time, language and style of the texts taken from various American Internet sites have got their distinguishing features which allow us to set them apart from the other language spheres. The material of the research is represented in a corpus of 2000 lexical units, collected by a continuous sampling method from 500 American Internet texts over a period of 2011–2016. The samples were selected from the following tabloids: USA Today (http://www.usatoday.com/), The Wall Street Journal (http://www.wsj.com/europe), The New York Times (http://www.nytimes. com/), Daily News (http://www.nydailynews.com/), New York Post (http://nypost. com/), Newsday (http://www.newsday.com/), etc.

2 To the Essence and Typology of Neologisms

While developing English language constantly acquires new words, which appear to designate new objects and concepts. Continuous changes in American English vocabulary as well as acquiring new lexical units that we can observe nowadays are creating a precondition for a thorough linguistic research of a new lexical layer.

Initially newly coined words are used by a narrow group of people, further if they are expressive enough and correspond to the American English phonetics they are acquired by wider professional and social groups of users. Getting adapted in the language neologisms accumulate new meanings and aim for monosemy, i.e. separation of a content plane [8, pp. 19–20]. They can either become common words and after a while penetrate fundamental lexis or fall out of use. After Fischer, a neologism is a word which has lost its status of an occasional word but is still the one which is considered new by the majority of a speech community [3, 6].

Neologisms are divided into lexical ones (newly coined by derivative means of a language), words and phrases borrowed from other languages and semantic neologisms (new meanings of the existing words). Alongside with general neologisms there are individual and stylistic or occasional ones.

Neologisms mainly perform a nominative function, although they can also be used for stylistic purposes. Dobrosklonskaya states that a creative linguistic personality produces texts pragmatically aimed at creative communication, using occasional words that symbolize late XX – early XXI centuries [3, 4].

Neologisms do not only reflect changes in different areas of modern life, they also assist in interpreting the reality and understanding the peculiarities of its perception in a definite linguocultural community. The peculiar features of neologisms as well as existing words are based on national identity as their usage reflects national and cultural attitude to the world [4, p. 48].

Neologisms are mainly created in the worldwide web or the Internet. Many linguists focus their scientific attention on such phenomena. Modern researchers mainly use onomasiological approach to studying neologisms. According to D. Crystal, modern scientists who deal with Internet linguistics face a problem of great volume of texts kept on the Internet. This volume exceeds the collections of texts in all libraries worldwide [2]. The amount of Internet-texts is growing rapidly. They also become more informative with the increase in number of countries, subcultures and therefore individual users joining the Internet. The second problem is diversity of linguistic material kept online: web-pages, texts of emails, instant messages from chats, blogs, social networks, Twitter posts, etc.

New vocabulary appears online regularly mostly due to the international status of English in the IT sphere. English has become a mediator between the Internet users from all over the world; they switch to this language in case of any communication problems with the representatives of other nations [9, p. 494]. New collocations, that first appeared online, very often become an important part of vocabulary used for regular communication. For instance, a lexical unit *blog* (a website where people write regularly about recent events or topics that interest them, usually with photos and links to other websites that they find interesting) has formed a word *blogosphere* – a community of bloggers and their environment. For the past six months hundreds or even thousands of new words have been created online. We have attempted to find out the sources and reasons of such trend.

New Internet resources, technologies and devices have caused the burst of neologisation of English. For example, a *gadget* – a new lexical unit used to designate any innovative technical device. These days this term is applied to any small devices. The Urban Dictionary defines this unit in the following way: «*an electronic device that fits in the palm of your hand*». Other new words stem from this unit: *gadget-holic* – a person who cannot function without an electronic gadget, as a smart phone, netbook, iPad or some other eDevice. The latter word means an electronic device that can be connected to the Internet. Another example is the word g*adgetry* – the need for gadgets to assist in everyday life and situations.

– I want the new I-phone so I can have access to Pandora at all times!
– No, you don't need it. That's just gadgetry.

Many neologisms result from Internet-users' creativity. For instance, *demotivators* (framed pictures with a comment or slogan) were initially used to mock motivational posters. American demotivators have rapidly spread worldwide. There are millions of them nowadays. Another example is a *meme* (an image, a video, a piece of text, etc. that is passed very quickly from one Internet user to another, often with slight changes that make it humorous). Memes frequently contain unusual collocations and occasional words.

Due to the popularity of social networks, we can observe a vast number of specific symbols and new lexical units to designate them, e.g.: hashtag *#* – a symbol used in Instagram and Twitter to mark the content. When used after a hashtag, the words blend. Urban Dictionary offers the following example: «*Take this hashtag for example:*

#worstjobeverhad. This Hash tag would compel many others to share the worst jobs they've ever had, thus contributing to a fun conversation. It can be used for specific searches or individual twitters that begin them for their followers». Blends are used in informal communication to create humorous context.

Many new lexical units appear in social networks to designate the interaction between the users: *to text* – to send somebody a message; *to tweet* – to create a message in Tweeter; *Tweetheart* – a popular user of Tweeter (similar to sweetheart); *Tweet tooth* – someone having a craving to post a new tweet on Twitter; *a tweetup* - an organized or impromptu gathering of people that use Twitter. Other examples of Internet neologisms are: *to sofalize* (sofa + socialize) – to communicate with people while staying at home; *a screenager* (screen + teenager) – a young person who spends all free time in front of a computer; *netiquette* (net + etiquette) etiquette of Internet communication; *Weblish* (web + English) – the language of Internet-users. Weblish contains a lot of acronyms, lacks punctuation marks and capital letters.

As it was mentioned before, the Internet plays a crucial role in informing its users of the latest news. For instance, looking at the sphere of *Paralympics*, we can see that the majority of events for disabled athletes are broadcast online. Many electronic mass media have a section *"Disability Sport"* where detailed information about para-competitions and para-athletes can be found. Commenting on Paralympic events journalists have to find a compromise between politically correct nominations of athletes with impairments, a necessity to describe the peculiarities of various para-contests and an urge to impress the readers with the para-sportsmen's achievements [12, p. 161]. Therefore they come up with new lexical units to solve these problems.

Professional discourse has also created a large number of new terms to designate changes in modern life [7]: *a webinar* (web + seminar); *entreprenerd* (entrepreneur + nerd) – a person having an IT start-up; *to transition* – to relocate; *to calendirise* – to schedule some events; *to Ctrl-Alt-Delete* – to reconsider something or to make a fresh start.

Some well-known words get new meanings or turn into other parts of speech: *to friend; to photoshop, to google.* Major part of many people's life is connected with the Internet. They communicate, work, and get many services online therefore a lot of neologisms have been created: *to telework, to telecommute* – to work remotely; *teleshopping, telebanking, teleconferencing*, etc.

Rapid neologisation is only one example of language changes caused by the Internet. New technologies are erasing the difference between conversational and written speech. Written speech is expanding, forming a specific system of graphics, looking for the new ways to deliver the pragmatic context of direct communication. Merging of conversational and written speech has led to new graphical and orthographic derivatives which can distort lexical units or affixes. Such examples of anti-orthography are widely represented in informal electronic communication and can be viewed as the markers of Internet discourse.

New words appear due to pragmatic needs: the addresser analyses the lexical material and selects the means helping him/her express the ideas and attitudes in the best possible way. If such means do not exist in an addresser's vocabulary, existing words are adapted or new words are created. New lexical units appear in speech to express a definite communicative intention, they are spontaneous and not targeted at the expansion of vocabulary. The fundamental nominative function of the language, responsible for documenting socially-important experience, is realized via constantly developing system of various tools of lexical nomination. If a new lexical unit has been adapted in the language, it can produce derivatives. Derivative lexical units are those created by word-building means (affixation, blending, conversion, reduction, etc.). Every derivative word belongs to a definite group and usually follows the pattern which makes it easy to identify its meaning. Every language system has got neologisms targeted at expressing some ideas.

We can distinguish lexical (new words) and semantic (new meanings of existing words) neologisms. New words are created in different ways: 1. extension of the meaning: a word which is used in different contexts receives new shade of meaning and in some cases a completely new meaning; 2. affixation; 3. conversion; 4. univerbation; 5. word reduction; 6. analogy with existing words; 7. addition of a participle or adjective to a noun; 8. loanwords.

Initially all new words are considered occasional or potential words. Their studying has resulted in a complex classification where their sphere of usage is primarily ana-lyzed. Neologisms are used in various genres of speech. For example, there are many individual neologisms or author's neologisms in fiction; they are quite common in science fiction as a writer needs to name non-existing phenomena. They are created by word-building regulations of a language, although the combination of the morphemes can be quite unusual. Such new lexical units create a communicative system used by the characters of anti-utopias. Such system is named "Newspeak", the term offered by an English writer George Orwell, the author of an anti-utopia "1984". The writer came up with the analysis of this new language in the appendix to the novel "The Principles of Newspeak". According to Orwell, Newspeak is characterized by a continually diminishing vocabulary.

Individual neologisms are the words which do not exist in the language, they are created by a definite author for specific purposes (e.g. to express meanings and con-notations). Such neologisms have been named differently in linguistic literature: author's neologisms, individual neologisms, occasional words, verboids (the term offered by Shansky N.M.), etc. Such variety of terms reveals different properties of neologisms emphasizing their origin, creativity and occasional appearance in the texts. We consider neologisms to be similar to regular lexical units as they have a phonetic and grammar form, their semantic slot, but they also have some properties which distinguish neologisms from the regular words.

However we would like to differentiate author's neologisms from occasional words as occasional lexical units are either formed with the help of non-productive word-building patterns, or with significant deviations from word-building regulations. Author's neologisms should also be differentiated from other types of neologisms. According to Babenko, an author's neologism is a speech phenomenon, while a neologism is a new word in a language [1, 4]. Although some author's neologisms still get into a lexical system of a language such examples can be considered the exceptions. On the whole, neologisms and author's neologisms have different functions. The former is coined for nominative purposes; the latter fulfills an esthetic purpose. Eventually a neologism can become a commonly used word; it can get dated or even come out of use. Author's neologisms are usually not dated.

Analyzing various examples of author's neologisms in the US media, we have attempted to identify the motivation of their authors to start a word-creating process. Firstly, we can speak of an urge to innovate the language, to create something that was not created before. It results in emergence of different eccentric lexical units [10, p. 45]. Such trend can be observed in various types of texts, in particular – op-ed articles, which are replete with author's neologisms. Secondly, we can observe the urge to attract a reader's attention to the article by means of one neologism. Newly coined lexical units are frequently placed in the title of an article, where they have minimal or non-existent context. Finally, the creators of new lexical units tend to be concise; therefore they come up with neologisms that can substitute long descriptive passages. E.g. "*Meet the Blade Babe*" (an interview with a Paralympic runner on blade pros-theses); "*Russia's Para Bloopers*" (a report about a doping scandal connected with Russian Paralympic team); "*Cybathlon: Battle of the bionic athletes*" (a report about a new kind of competition for disabled athletes who use state-of-the-art assisting devi-ces). All above mentioned titles contain new lexical units created by the journalists to attract readers' attention and express a wide range of ideas in one meaningful phrase.

Author's neologisms have a few distinguishing features: 1. they can produce derivatives; 2. they belong to a conversational style; 3. they are a part of an author's ideoglossary; 4. they are single-use; 5. they are not reproduced; 6. they are non-normative; 7. they are expressive; 8. they have nominative optionality; 9. they have synchronic and diachronic diffuseness [10, 11].

Individual and author's neologisms do not necessarily have all the above listed features; nevertheless it helps us to identify them as non-standard speech units. Derivation is one of the optional features. An author's neologism is made by com-bining at least two morphemes. For instance, *Netizen* (net + citizen) is a person who spends time online. Moreover citation of author's neologisms in media does not equal its repeatability. This is a very important remark as long as nowadays IT technologies make it easy to calculate the frequency of lexical units' usage. It is doubtful that the number of hyperlinks to an author's neologism can demonstrate its repeatability. Such repeatability would probably demonstrate how often the authors use their own neologisms.

As for normativity, it is important to distinguish normativity vs. non-normativity of the author's neologisms and normativity vs. non-normativity of their derivation patterns.

Lexical aspect of neologisms is more interesting if we want to look into their style in a media text. However if we are more interested in derivation patterns which reflect a neologism's descriptiveness we should focus on a grammar aspect of a non-usual word. Shansky believes that a non-usual word is always a neologism, coined as a result of derivative processes [14], therefore derivative properties are crucial for understanding a new lexical unit. In many cases the idea of a neologism's normativity is quite relative: lexical and derivative non-normativity are not the same. Nevertheless neologisms are commonly perceived as something unfamiliar, unconventional and stand out from a language usage. Generally, neologisms that break derivation patterns stay apart from the new lexical units which look similar to the existing ones. Such new words only partially reproduce the patterns. There are also newly coined lexical units that follow the pattern invented by an author, they are also non-normative. Any author's neologism is a product of a creative process. People certainly use the resources of a language system to make new words, but they still create something new.

Synchronic and diachronic diffuseness is one of the major attributes of author's neologisms. According to Babenko it demonstrates their synchroneity [1]. Lykov offers a more extended description of this attribute; he mentions a peculiar position of neologisms at the synchronic and diachronic axes of a language system [11]. On the one hand, an author's neologism is synchronic as it is linked semantically and grammatically with the other lexical units surrounding it within a sentence or a text. Readers usually understand the meaning of a new word even if they come across it for the first time. On the other hand, author's neologisms are diachronic as they are created due to the linear processes on the time axis. It is quite easy to identify their origin and derivative patterns. For example, *abdorable* (abhor + adorable) – dumb, fat, lazy, arrogant, and totally, completely, irresistibly cute (usu. about pets). Synchronic and diachronic aspects coexist in author's neologisms, the fact of creating them aligns with the fact of their existence, and it makes author's neologisms quite outstanding. They are bound to be constantly perceived as the new lexical units.

In conclusion we would like to present a classification of author's neologisms by their derivative patterns. Looking into these patterns we can distinguish at least three types of non-usual lexical units.

1. Not-usual word created with productive derivatives (e.g. *adimpression*);
2. Not-usual word created with non-productive derivatives (e.g. *emoji*);
3. Not-usual word created with a definite derivative pattern (e.g. *Obamatopia*).

Newly-coined lexical units sound unfamiliar which leads to de-automation of a reader's perception but the majority of author's neologisms do not contradict the derivation principles. Their novelty lies in their lexical meaning. This ambiguity of author's neologisms makes them a valuable source of American media vocabulary which requires further studying.

3   Conclusion

Neologisms are created in American media texts due to pragmatic needs of the speakers, development of new technologies and peculiarities of media interaction. Newly coined words appear in media discourse quite rapidly: any change in a human life is reflected in a language. Mass communications help people structure and rationalize their experience and beliefs. Besides, mass media do not only provide the information, but influence everyday life and expertise of communicators.

Onomasiological analysis of neologisms offered in this article has demonstrated that blending and affixation are the major means of creating them. Such word-building patterns are typical for other languages too. On the other hand, a lot of factors (a definite communicative or creative aim, specific descriptors that can be presented in a new word, need to express an emotion or a specific connotation in one word) can influence the ways of media neologisation. As a result the above mentioned means of neologisation are complemented by means of a language game (paronomasia, allusion, rhyming, etc.) which adds to pragmatic, functional and stylistic profilisation of a lexical unit. Thus newly coined units in American Internet discourse are determined by the peculiarities of virtual interaction and further the formation of Weblish (a new type of English).

Word-creation as a means of expanding English vocabulary and a type of a linguistic game is spreading due to Internet communication. Availability and variety of ways to join transboundary virtual reality, including social networks and various collaboration projects aimed at developing new words database as well as word-creation contests (Urban Dictionary, The Unword Dictionary, Verbotomy: The create-a-word-game, etc.) help to realize the potential of a media-individual and activate the projective function of the Internet community, which is shaping the language of future.

References

1. Babenko, N.G.: Okkazional'noe v hudozhestvennom tekste [Occasianal Words in a Literary Text]. Strukturno-semanticheskij analiz. Kaliningrad State University Publishers, Kaliningrad (1997). (in Russian)
2. Crystal, D.A.: Dictionary of Linguistics and Phonetics. Wiley, Oxford (2008)
3. Dobrosklonskaya, T.G.: Medialingvistika: sistemnyj podhod k izucheniju jazyka SMI [Medialinguistics: System Approach to Studying Mass Media Language]. Media-Press, Moscow (2008). (in Russian)
4. Dubenets, E.M.: Lingvisticheskie izmenenija v sovremennom anglijskom jazyke [Linguistic Changes in Modern English]. Glossa-Press, Moscow (2003). (in Russian)
5. Elina, E.N., Tibbenham, P.J.: Changes & tendencies in contemporary native English. Vestnik RUDN, serija: Russkij i inostrannye jazyki i metodika ih prepodavanija **3**, 79–87 (2009)
6. Fischer, R.: Lexical Change in Present-Day English: A Corpus Based Study of the Motivation, Institutionalization, and Productivity of Creative Neologisms. Narr, Tübingen (1998)
7. Klyoster, A., Galich, G.: The fragmentation of professional discourse. Procedia – Soc. Behav. Stud. **206**, 56–61 (2015)
8. Kobenko, Y.V.: Jazyk i sreda. Opyt sistematizacii dannyh mezhdisciplinarnyh issledovanij [Language and Environment. An Attempt to Systemize the Data of Interdisciplinary Studies]. Tomsk Polytechnic University Publishers, Tomsk (2017). (in Russian)
9. Kobenko, Y.V., Zyablova, N.N., Gorbachevskaya, S.I.: Towards translation naturalism at translation of English film production into modern German. Mediter. J. Soc. Sci. **6**(3S2), 494–496 (2015)
10. Kostomarov, V.G.: Vkus jazykovoj jepohi: Iz nabljudenij nad rechevoj praktikoj mass-media [Taste of Linguistic Epoch: Observing Mass Media Speech Practice]. Zlatoust Publishers, St. Petersburg (1999). (in Russian)
11. Lykov, A.G.: Sovremennaja russkaja leksikologija (russkoe okkazional'noe slovo) [Modern Russian Lexicology (Russian Occasional Word)]. Vysshaja Shkola, Moscow (1976). (in Russian)
12. Martseva, T.A.: Nekotorye osobennosti ispol'zovanija paralimpijskoj leksiki v britanskih SMI [Some peculiarities of using Paralympic vocabulary in British mass media]. In: Proceedings of XVI International Conference "Linguistic and Cultural Studies: Traditions and Innovations", pp. 158–163. TPU Publishers, Tomsk (2016). (in Russian)

13. Muravyev, N., Panchenko, A., Obiedkov, S.: Neologisms on Facebook. In: Komp'yuternaya Lingvistika i Intellektual'nye Tekhnologii Proceedings, pp. 440–454. Russian State Humanities University Publishers, Moscow (2014)
14. Shansky, N.M.: Leksikologija sovremennogo russkogo jazyka [Lexicology of Modern Russian Language], pp. 225–236. LKI Publishers, Moscow (2007), (in Russian)

Standard German Hybridization in the Context of Invasive Borrowing

Yuriy V. Kobenko$^{(\boxtimes)}$ ⓘ, Petr I. Kostomarov ⓘ,
Tatiana I. Meremkulova ⓘ, and Daria S. Poendaeva ⓘ

Tomsk Polytechnic University, Tomsk 634050, Russian Federation
serpentis@tpu.ru, {petrkost,poendaevads}@yandex.ru,
sonata_tanya@rambler.ru

Abstract. The article focuses on the structure changes of standard German accumulatively forming a tendency for its hybridization in the process of invasive borrowing of English words with relevant tendencies for internationalization, univerbation and aposteriorisation of the lexical means. The analysis of terminological hybridism in the borrowed fund of fixed Anglo-Americanisms is carried out and the degree of hybridization in the nucleus, adaptive and peripheral spheres of lexical-semantic system of standard German is determined. Hybridization of the standard German system as a component of exoglossic language situation in contemporary Germany is symptomatic and occurs mainly due to a multilevel asymmetry of contiguous idioms and functional aspect of English-German bilingualism. Close attention is paid to the theoretical foundations of the hybrid phenomenon in general and in German literary language in particular. Hybridization of the lexical structure in the German language is recognized as the maximum in the peripheral layers of the system and is a parallel phenomenon of its (terminological) neologization.

Keywords: Hybridization · Invasive borrowing · Standard German

1 Introduction

Hybridism is an immanent characteristic of the standard languages system, which functioning is conditioned by the importance of the adaptation of borrowed material related to prestigious donor languages. There are several types of hybridism: pragmatic, stylistic (genre), derivative, etc. All of them show a constant complexity of the language and its phenomena. This article is dedicated to the phenomenon of genetic (resource) hybridism of standard German as a result of a number of exoglossic language situations in its genesis.

The term "exoglossic language situation" means a system of heterogeneous and discriminatingly functioning components of a certain state of language, the most important one tends to be the imported idiom (metalect) with the index of the communication capacity not less than 0,5. According to a survey carried out by Allensbach Opinion Research Center in 2008, in the language situation in the Federal Republic of Germany the most powerful idiom is the English language with the index of 0.58. Longtime dominance of the imported metalect within a certain language situation

A. Filchenko and Z. Anikina (eds.), *Linguistic and Cultural Studies:*
Traditions and Innovations, Advances in Intelligent Systems and Computing 677,
DOI 10.1007/978-3-319-67843-6_33

determines invasive borrowing whereby the number of borrowed (prior) or the units formed on the basis of borrowed material (posterior) dominate over the new lexis created due to the resources of the target language. In the first case it is a unit in the system of the latest effective means, while the other is the system of effective resources that lead to the hybridization of standard German vocabulary, i.e. resource heterogeneity [11].

The term "invasion" is used in contact linguistics with reference to foreign influences, the intensity of which determines the peculiarity of borrowing of linguistic means, in which the citation of the donor language in speech practice exceeds the use of the means of the target language (more than 0.5). The invasion is a natural consequence and symptom of the dominance of foreign metalect in particular language situation. The invasions are not rare in the history of the German language. The replacement of language situations is the basis for the development of literary languages accompanied by the change of prestigious metalects and, hence, donor languages, it can be stated with confidence that the invasions are an impulse to change the system of the German literary language through the hybridization. Thus, the history of the German literary language can be represented as successive change of the periods of Latinization (with the epochs of Golden and Silver Latinism) [2], Gallization (with three waves of borrowing of Gallicisms) [15], Anglicization/Americanization with all its invasions [28], Slavization/Russification (especially the language of the GDR) [5, 19, 21], namely the change of imported metalelects.

Hybridization, known as "language blend", features a combination of the target language material and the donor language citation. Autochthonous means of the German language in this case perform as a functional undifferentiated unit, while interfering Anglo-Americanisms – as units of the vertical medium, success and prosperity markers [14]. Within the framework of language politicization, hybridization of standard German can be initiated by the permissive actions of the subjects (organs) of language policy or institutions possessing an exponent function of standard setting, compare, for instance, the hybrids in the modern German advertising: 'Farbdisplay', 'Organizer-Fach', 'United-Sondermodell', 'ICT-Kapazität', 'Food-Lüge', 'Outdoor-Uhr', 'Euro-Jackpot'.

The hybridization of literary languages seems a quite natural process. The language rhizome (the term of Deleuze and Guattari [4]) branches asymmetrically, centrically, nonlinearly, its development occurs, contrary to the common belief of cognitivists, not in the framework of certain concepts of language development, but spontaneously and distinctively. It means, in particular, that the saturation of the system by foreign-language citations will be the more active, the more the rhizome is diversified. Thus, the hybridization of the literary language system (a standard language) directly depends on the diversification of the language capabilities, which are combined with the search for new, more convenient ways of expressing, including the borrowing.

The hybridization is expressed in different ways: through affixation, formation of composites, and calque. Special case of hybridity is the mixing of the letter and numbered parts of lexical units, for example, as a result of the universality (see below). This kind of the phenomenon can be characterized as an iconic or linear hybridity, which eventually can become stable expressive possibility of a language system.

Shishigin points out that the development of languages can be characterized by fusional character of the hybridization, i.e. the enlargement of productive opportunities in various structural subsystems of literary languages [25]. Ovsyannikova stresses that the hybridization is natural result of constant rapprochement of languages. Borrowing some units from each other, languages adapt to each other in order to reach efficient exchange of information. The history of language development is the history of how it alternately acts as a beneficiary (using the wealth of other languages), then the beneficiant (donor). This phenomenon can be compared with children from mixed marriages with a "pleasantly dark" appearance, as Sholokhov's Nagulnov notes. Some languages are considered more susceptible to the hybridization than others [18]. Donec considers hybridism as an attribute to a foreign language that influences one's native language during a contact or when studying in and thereby brings it closer to the sociolinguistic notion of "metalelect" [6]. Solntsev describes among universal characteristics of a language system such its quality as heterogeneity, completely identical in content to hybridization. Discreteness and heterogeneity are the fundamental and most common system-forming characteristics of language elements that persist throughout the lifetime of a language system and are necessary for its functioning [26].

In the vocabulary of literary languages foreign elements that facilitate the hybridization are quite dispersed. There is a lexical core in the center of the vocabulary, in which basic stock is concentrated – the words that cannot be dispensed with in daily communication. Of course, there are borrowings from other languages here, but their appearance is hardly distinguishable from the appearance of autochthonous units. Above the «core» a thick "outgrowth" was formed – additional layer of the vocabulary reflecting the evolution of the lexical-semantic level as the communicative environment becomes more complex (aggregation). Initially, the vocabulary is served only by the interaction of speakers within the ethnic environment (nuclear layer), then with the appearance of interethnic contacts new symbols come into the language, which adapt perfectly to the peculiarities of the language system at the early stage of a language development. Therefore, additional layer of the vocabulary, which forms concentric ring over the nuclear structure, is also appropriate to designate the adapted one. Peripheral spheres (Fachsprachen) with high content of low-assimilated borrowings of a terminological nature with the tendency to neomorphism appear in the language system during rapid development of technology and internationalization in many areas of knowledge. Their "petals" do not form separate spheres of the vocabulary, but are superimposed both on the two previous layers, i.e. even in closed professional communities people do not use only terms (like the termosphere of renewable energy sources in terminology of energy). If the changes in the center occur very slowly, then the processes of updating and obsolescence (dying out) of lexical units are accelerated as much as possible towards the peripheral layers. It should be noted that in the German lexicological tradition all three spheres are considered exclusively in the aspect of borrowing, which point the role of invasive borrowing in the development of lexical system, cf. German-speaking trichotomy Erbwortschatz (autochthonous), Lehnwortschatz (adapted) and Fremdwortschatz (non-assimilated vocabulary) [16].

Field structure of lexical system and its components is a result of the cumulation of the language experience in the direction from the center to the periphery (from the general to the particular). The criteria for the transition from the core, in which the maximum set of field-forming features is concentrated, to the diffuse periphery with partial set of such features were developed by the Prague Linguistic Circle. The linguists use the antinomy "center vs. periphery" to describe the principle of the language rhizome diversification, where it is not easy to single out autonomous microsystems. The theory of the field was proposed by the Germanist J. Trier (Münster) in 1931, although it does not claim to be exhaustive interpretation of the structure of lexical and semantic system, but in its basic formulations is entirely in tune with centrifugal principle of describing the diversification of the language system and the mixing of autochthonous and borrowed registers.

The theory of language mixing originated on the basis of the ideas of H. Schuhardt, who did not accept the concept of the genealogy of A. Schleicher, believing that there is no single language free from crossings and alien elements [23]. A similar position was held by Baudouin de Courtenay, who emphasized that there is not and cannot be single pure, non-mixed language whole, and stressed that not only the exchange of new elements exists during the language mixing between contacting languages (words, syntactic phrases, forms, pronunciation), but also the weakening of the degree of discernibility. In order to prove his theories Baudouin de Courtenay referred to cases of language contacts of Slavic tribes with native speakers of Germanic and Romance languages [3]. Along with Baudouin de Courtenay, language mixing was recognized by Polivanov, who considered the hybridization as the processes that caused the change of the substratum language [1]. Shcherba proposes to call "language mixing" mutual influence of languages, because the term "mixing" suggests that contacting languages are involved in the formation of new language [24].

In modern linguistics language mixing, crossing or hybridization is treated as process in which the donor language continues in the target language, while own resources of the target language reduce [29]. The hybridization of languages is as inevitable as spontaneous: language contacts which are characteristic of the sociology of any language leave "imprints" of foreign influences in the system of language that form its resource heterogeneity.

2   Tendencies for Standard German Hybridization

Schmidt singles out the following complementary tendencies for hybridization of standard German: the influence of functional lexis on the language, the tendency for specialization, generalization, rationalization and economy, integration, internationalization and differentiation of linguistic means [22].

The first tendency is consistent with "popularization" mentioned by Wolff, meaning the infiltration of lexical units from various fields of human life and branches of knowledge into standard German. It is necessary to point out that there is usually a shift of meaning along with the extension of the people's mental lexicon [30]. Among other factors, this tendency is exoglossic, considering the fact that among units chosen from the lexical peripheral there exist English borrowings, e.g.: "Timing", "Jogging", "Tee-Break" (from the field of sport and leisure); "Trouble", "Speed", "Action" (youth dialect), etc. The tendency that is described above is accompanied by the second one – "specialization", explained by Schmidt as a tendency for a formation of polycomponent substantive composites that fall into the vocabulary [22]. A significant observation is made by Fedortsova, who points out scale-up linearity of substantive composites as the most appropriate for specialization and reduction of the number of international components within a compound word [8]. Specialization is added up by generalization as there is a need to express a lot of particular units with the help of a general one that leads to polycomponent blends, e.g.: Fahrschein-Entwerter-System. Blends are typical for the following tendencies by Schmidt: rationalism and economy (speech effort), that mean a quantitative reduction of the utterances by morphemic compression, e.g.: sie fahren nach Berlin → die Fahrt nach Berlin → die Berlinfahrt. Integrative tendency coincides with Wolff's popularization of lexical units from the peripheral spheres of standard German vocabulary. In the context of the discussion about internationalization, Schmidt presents details of his typology by an observation of international morphemes of Latin origin that serve as the platform for the process of neologization, e.g.: Interface, Terminal, Basic [22]. The last part of this typology is a tendency for differentiation of German dialects by communication conditions, for instance, youth dialect is significantly different from the lawyers language.

Alongside with the tendency for internationalization that is mentioned by most of the researchers the most important tendency at the lexical-semantic level is aposteriorization of the borrowed material of the English and American origin as a special case of hybridization of borrowed and autochthonous units. Aposteriorization or a posteriori derivation is a formation of new units or meanings based on the borrowed donor language material and the derivatives are unclear, unpopular and unknown in this meaning in the donor language, e.g.: Twen, Ego-Shooter, Dressman, Handy, Beamer, Talkmaster (units); toppen, killen (meanings). Aposteriorization is a symptom of losing the "reciprocity principal" according to Kabakchi [10]; it means that it is impossible to genetically identify the formed unit in the target language with the system of the donor language that must be used as the source of borrowing.

Along with the artistry the aposteriorization is the most important quality of literary languages. Artificial languages, created with the purpose of structural improvement of natural languages, are traditionally considered to be posterior [7]. However, the aposteriorization has long gone beyond laboratory language projection. Today in acute competition for the right to be called "world" only that language wins which users themselves select all the most necessary from other languages. The principle of structural enrichment by means of the hybridization formed the basis for the rigid selection of competitors for the status of common language and significantly extended the limits of the aposteriorization. Now the aposteriorization extends to borrowing (including pseudo borrowing), if the goal is to copy the advantage of the donor language for its productive realization (mainly through the derivation) in the system of the target language. There is an opinion that such posterior language has become, at least in Europe, international English, which has absorbed the characteristics of number of language systems and helps the speakers of many other European languages to discover something of their own [17]. Schäfer points out that English only became international language because its composition has about 60% of borrowings which are accessible and understandable to the speakers of other (European) languages. However, own words of the English language are in the minority [20]. If the artificiality allows the literary language to adapt to the needs of any communicative environment, the aposteriorization is called upon to make this adjustment as fast and streamlined as possible. Moreover, thanks to the development of the literary language the borders of the communicative environment can be expanded even in the presence of historically unshakable national borders. This is the basis for any language expansion implying the expansion of communicative radius of the language outside the territory of the ethnos-medium and euphemistically is considered to be "prestigious". Prestigious foreign languages (English, German, French) are learnt in school and university, but are not the most wide spread, and the only beneficiaries of the expansion of the communicative environment are its inhabitants (Anglo-Saxons, Germans, French).

In a point of fact, the tendency for internationalization is mainly dependent on aposteriorization, the evidence of that is found in the works of Fedortsova, who point out that one of the specific features of word-building on international basis in comparison with German word-building is the presence of an extensive and open list of bound bases in the function of word-building elements, determining a special character of international word-building [8].

Within this article, "a posteriori" denotes the units taken from lexical-morphemic exoglossic material with a disrupted reciprocity principal. Among the a posteriori units in standard German that were formed on the basis of borrowed language material of Anglo-American origin there should be mentioned hybrid terminological units. The latter tends to be complicated compound units (composites) that are created for special purposes and consist of heterogeneous stems of English and German origin. According to their structure and semantics, they are divided into copulative (coordinative) and determinative (subordinate) composites. Copulative terminological hybrids feature low-cycle combination of equal components of English and German origin (1,66% of the whole number of Americanisms), e.g.: "Tuner-Recorder-Empfänger". Among the components of determinative terminological hybrids main and secondary ones are distinguished, e.g. "Color-Scheibe" (tinted window), "Break-Even-Analyse" (break even analysis). The components "Scheibe" and "Analyse" which are performing the function of hyperonyms (genus proximum) turn out to be the main components while the determinant components "Color" and "Break-Even" which are performing the function of hyponyms (differentia specifica) become secondary.

The components of hybrid composites are outlined with Latin capital letters "E" for English words and Anglo-Americanisms and "D" for native and assimilated units as well as for assimilated internationalisms of non-German origin (Kreatinin – protein metabolism end-product).

According to the element of hybrid units, the most common are the composites of the following types: 1) E+D (Hot-Labor, 36.17%), 2) E+E+D (Time-Sharing-Betrieb, 20.22%), 3) D+E (Werbespot, 24.72%), 4) D+E+D (Freizeitjob, 16.45%). The least frequent ones are the hybrid units of the type D+E+D (Rumpel-Leed-Test) or E+D+E (H2-Rezeptorenblocker) (total 2.44%).

The number of determinative terminological hybrids with a German component in the function of the main element in standard German comprises 57.61% while the ones with an English component in the same function comprise 42.39% respectively. Anglo-Americanisms prevail in equal proportions as the secondary element of determinative hybrid composites, among them there are elements of the second order (E2) that comprise 21.44%.

According to the type of composition Anglo-Americanisms are divided into simple, compound units, acronyms, hybrid and monolingual clusters (contaminations), linear reductions (contractions). Monolingual contaminations include prior units (borrowings) of international pattern, e.g.: "Speating" from "speaking" and "eating" (in English it means a play form of foreign language teaching (mainly English) while eating), "Infotainment" from "information" and "entertainment" (in English it means a mode of information transfer with elements of entertainment).

The tendency for univerbation, that is observed in many modern standard languages and implies the formation of polycomponent composites of word groups, increases the formation of mixed composites of borrowed and autochthonous units, e.g.: "Livesendung", "Reiseboom", "Powerfrau", "Werbespot". A special feature of such hybrids is their reciprocity, i.e. their ability to transform into composites and back into collocations: Livesendung ↔ live senden, Powerfrau ↔ eine Frau mit Power. In this context, it is important to highlight that the need for nominalization of certain units is dictated by fixed language mechanisms of content compression as well as by an intention to condense speech semantic intensity per unit time. In order to reach this aim the English language, e.g., has conversion and gerund while the German language forms composites.

The products of univerbation also include hybrid composites that do not have any stable spelling rules, e.g. "Bohr-Boom", "Mega-Party", "Umweltdesaster", "Stoppuhr", "Overhead-Projektor". Univerbation units feature composites of determinative type with ulterior dependence relative to the genetics of the main and determined components. In hybrid determinative composites, denoting a young animal, there observed a tendency for a shift of the main and determined words. This tendency is caused by the need to end a compound word by a motivated component that tends to be the main one in a German composite. Motivation absence of the final component of a compound word represented by a borrowing "Baby" leads to an implicit incompleteness of a compound unit that is compensated by the shift of its components, e.g.: "Ich sah gestern eine Babymaus durch den Gang huschen" by analogy with "I saw yesterday a baby mouse darting through the corridor", instead of "Ich sah gestern ein Mausbaby/Mausjunges/Mäuschen durch den Gang huschen". "Habt keine Angst, Männer, es ist nur ein Babykrokodil!" by analogy with "Don't be afraid, men! That's simply a baby crocodile", instead of "Habt keine Angst, Männer, es ist nur ein Krokodilbaby/Krokodiljunges!". The choice of exoglossic variant (Mausbaby и Krokodilbaby) is explained by the tendency for imitation which results in translation automatism [12].

The spelling of the hybrid composites is determined by general rules that means that a hyphen can divide heterogeneous components or all the elements of the composites together with the option of joined-up writing, e.g.: "Computerfachabteilung" (as well as: "Computer-Fachabteilung"), "Cornedbeefbüchse" (also: "Corned-Beef-Büchse", "Cornedbeef-Büchse"). A univerbation hybrid word-building model with an elliptic initial component is gaining in popularity, e.g. "E-Mail" (Electronic-Mail), "T-Girl" (Transsexual-Girl). It is vital to highlight that the graphic pattern is not stable; that is why, elliptic components have a tendency for spelling with lower case letters and/or joined-up writing, e.g.: "Willst du meine eFrau werden?". It is worth noticing that such a word-building model coexists with a similar one where the first element has iconic origin or is shortened with the purpose of euphemization, e.g. "T-Test" (statistic hypothesis text, from the base formula t = Z/s).

In order to estimate the hybridization density of standard German basic (nuclear), adaptive (additional) and peripheral spheres of its lexical-semantic system were analyzed. In the basic sphere of lexical-semantic system represented by a lexical-semantic group of the grape growers language and winemaking with the whole number of 12 520 units there are 24 Anglo-American borrowings, 12 of which are determinative hybrid composites, 11 of them are of type E+D (Foxgeruch), 1 of them is of type D+E (Lederfrack) and 2 of the determinative composites are non-motivated lexical units that still have a clear morphemic boundary allowing to establish the determinativity in the donor language (Bulldog, Bulldozer).

In the adaptive sphere of lexical-semantic system of standard German represented by medical vocabulary with the total word stock of 184 974 lexical units there are 4 619 Anglo-Americanism that have in their structure 346 units that tend to be hybrid determinative composites of the following types:

(1) E+D with the initial component of acronym (HDL-Cholesterin, HIV-Antikörper), or appellative (Dumping-Syndrom, Computertomographie);
(2) D+E (Kreatinin-Clearance) or D+D+E (Rechts-Links-Shunt);
(3) E+E+D (Pacemaker-Implantation) or the units that have both initial elements of the English origin represented by foreignisms with characteristic spelling in accordance with graphics and the spelling rules of the donor language (burning feet-Syndrom, Burn(-)out-Syndrom, Bypass-Operation);
(4) E+D+E (H2-Rezeptorenblocker);
(5) E+E (Pacemaker, Bypass).

128 determinative substantional composites of adaptive sphere of standard German have proper names in its structure. There are composites of the E + D type with one proper name (Sheehan-Syndrom, Murphy-Zeichen) or prefixes (McBurney-Punkt, Conn-Syndrom, Coombs-Test); with several proper names, e.g. with two last names D+E+D (Rumpel-Leede-Test) and also E+E+D (Sheehan-Simmonds-Syndrom, Paul-Bunnell-Reaktion, Epstein-Barr-Virus), with a proper name and a last name E+E+D (Adam-Stokes-Anfälle) or E+E+E+D (Paul-Bunnell-Davidsohn-Reaktion); one of the composites has a toponym as the first initial element and a test character as the second initial element (Coxsackie-B-Virus). There are 5 combinations with partial acronymy on the pattern of "genus proximum + differentia specifica" (gender indicator + aspect indicator), that can perform in the role of determinative composites with the English proper name as the initial component, e.g. M. Addison (Morbus Addison, also Addison-Morbus). There are 17 foreignisms with different variations in spelling of the borrowed elements in the structure of the determinative composites: sexually transmitted diseases (with the graphic of the donor language), (hybrid foreignisms) Superficial Spreading Melanoma (with the graphics of the target language). For easier usage many foreignisms are turned into acronyms (STD, BSE).

The peripheral part of standard German vocabulary is formed by a lexical-semantic group "Internet vocabulary" and its main tendency is "direct borrowing of Anglo-Americanisms with an incomplete degree of assimilation (the absence of a clear part-of-speech tagging, low grammar development and graphic adaptation)" [27]. There are the following stable characteristics of lexical units referring to this lexical-semantic group: the extension of the sign system by adding another register of numerals and figures as secondary semiotic means, compare: 0–9, !, ?, &, $, <...>, etc., intersigned, interlingual hybridism of the units, acronym domination, emotivity and as a result, capitalizing (ACK – Acknowledgment), communicative character and distinctive mimitivity, etc. There are 2 718 acronyms that can repeat the word in a new meaning forming neomorphisms, e.g.: Italy – I trust and love you. Among the acronyms there are 1 547 hybrid units, comprising sign-lettered contaminations of two types: 1) full lexemes (awhfy? – are we having fun, yet?) and 2) parts of lexemes with homophonic substance with the help of numerals (b4 – before). Hybrid units also include proper names (trademarks, names of the products and their producers, games, etc.): 100BaseTx, ZX 80, Yellow Book, Acer, Yahoo, Ubisoft, Quake (game). All the hybrid units are a priori here, i.e. they are borrowed units [13].

Lexical sphere of information security lying on the periphery of the vocabulary of the German literary language also reveals interesting development dynamics. The total number of English-language borrowings in this terminology system comprises 52%; i.e. more than 0.5. Among the recorded seven hybrid models in the corpus of 104 terminological hybrids, the largest share belongs to the E+D formation (Cyber-Raum) and comprises 70%. The terms of E+E+D model (IT-Governance-Beauftragte) and D+E model (Schadprogramm) are almost equally represented and have 11% and 10% of the total number of units correspondingly. The shares of 4% and 3% are occupied by E+D+D hybrid model (IT-Grundschutz-Katalog) and E+D+E model (Bluetooth-Protokoll-Stack) respectively. The smallest number of terms created by D+E+D model (Empfänger-E-Mail-Adresse) and E+E+E+D model (Internet Control Message Protokoll) comprises 1%. Obtained data indicate that the most susceptible hybridization sphere is the peripheral structure of the vocabulary of the German literary language inclined to terminological neologization and the formation of multicomponent composites. The fluctuation of the spelling contour and consequent weak normative development exacerbate the suspicion of the incomplete character of the branching of the language rhizome of the German literary language and ongoing process of its hybridization in conditions of invasive borrowing of the prestigious Anglo-American lexicon.

3 Conclusion

Within the functional bilingualism that exists in the modern language situation in Germany, the blend of contact languages is quite predictable and natural. Donor language citation in target language in any case leads to a tendency for hybridization. It is impossible to fully evade borrowing and blending of contact languages due to earlier established exoglossic resources in standard German. It is possible only to prevent further blending by dividing the registers. According to Haarmann, the crucial part in that can be made by language planning measures of research and administrative character, opposed to less effective group and individual measures [9]. However, it is absolutely impossible to exclude hybridization even by law because the historical connection of the two languages is always characterized by their nonequivalent blend due to multilevel asymmetry of the language material. Therefore, hybridization is laid in the standard language nature as one of the word-building mechanisms. As nowadays

Germany has become the world leader of energy supply technology and use of alternative (renewable) energy sources, many terminological units of endoglossic pattern (not borrowings) from the field of energy supply appeared in standard German, e.g.: "Endlager", "Heizalternative", "Verbraucherberatung", "Abwärmeberechnung", "Gezeitenkraftwerk", "schneller Brüter", "erneuerbar", "schadstoffarm", "auslegungsüberschreitend". Among them, of course, there are some Anglo-Americanisms, but they are less numerous and more often tend to be hybrid terminological units, e.g.: "Strommix", "Best-Practice-Beispiele".

Thus, the hybridity is the peculiarity of evolving language systems because it increases the chances of individual language idiom to survive by borrowing, "tamped" by history into heterogeneous layer in the thickness of the language rhizome.

References

1. Alpatov, V.M.: Istoriya lingvisticheskikh ucheniy (in Russian) [History of linguistic studies], 4th edn. Yazyki slavyanskoy kultury, Moscow (2005)
2. Ax, W.: Text and style: studies on ancient literature and their reception. Steiner, Munich (2006)
3. Baudouin de Courtenay, I.A.: Izbrannyye trudy po obshchemu yazykoznaniyu (in Russian) [Selected works on general linguistics]. Vol.1. Academy of Science of the USSR, Moscow (1963)
4. Deleuze, G., Guattari, F.: Rhizome. Merve, Berlin (1977)
5. Domashnev, A.I.: Sovremennyy nemezkiy yazyk v ego natsional'nykh variantakh (in Russian) [Modern German in its national variants]. Nauka, Leningrad (1983)
6. Donec, P.N.: Foreign language as a hybrid language. In: Heterogeneity and hybridization as an object of German studies. German studies in Russia: Yearbook of the Russian Germanists' Association, pp. 251–256. Yazyki slavyanskoy kultury, Moscow (2013)
7. Dulichenko, A.D.: Mezhdunarodnyye vspomogatel'nyye yazyki (in Russian) [International auxiliary languages]. Valgus, Tallinn (1990)
8. Fedortsova, V.N.: International'nyye slovoobrazovatel'nyye modeli v nemezkom yazyke (in Russian) [International models of word formation in German]. Russian Academy of Science, St. Petersburg (1994)
9. Haarmann, H.: Instruction: languages and language policies. In: Sociolinguistics. An International Manual on the Science of Language and Society, pp. 1660–1678. Walter de Gruyter, Berlin (1988)
10. Kabakchi, V.V.: Osnovy angloyazychnoy mezhkulturnoy kommunikatsii (in Russian) [Foundations of intercultural communication in English]. Studia Linguistica 7, 32–55 (1998)
11. Kobenko, Yu.V., Sharapova, I.V.: Polyglossia through the prism of exoglossic nature of the German literary language development. Mediterr. J. Soc. Sci. 6(1S1), 500–505 (2015)
12. Kobenko, Yu.V., Zyablova, N.N., Gorbachevskaya, S.I.: Towards translation naturalism at translation of english film production into modern German. Mediterr. J. Soc. Sci. 6(3S2), 494–496 (2015)
13. Kobenko, Yu.V.: Yazykovaya situatsiya v FRG: amerikanizatsiya I ekzoglossnyye tendentsii [Language situation in Germany: Americanization and exoglossic tendencies]. Tomsk Polytechnic University Publishers, Tomsk (2014). (in Russian)
14. Kostomarov, P.I., Ptashkin, A.S.: Explication of peculiarities of the hypergenre in the text production of Russian German of Siberia. Mediterr. J. Soc. Sci. 6(1S1), 407–411 (2015)

15. Meremkulova, T.I., Kobenko, Yu.V.: The French waves in history of the German literary language (the XI–XVIII centuries). Phil. Sci. Issues Theory Pract. **3**(3), 131–133 (2017)
16. Munske, H.H.: Is German a mixed language? The position of foreign words in the German language system. In: German vocabulary. Lexicological studies, pp. 46–74. Walter de Gruyter, Berlin (1988)
17. Ohnheiser, I.: Languages in Europe. Language and situation and language policy in European countries. In: Ohnheiser, I., Kienpointer, M., Kalb, H. (eds.) Innsbruck Contributions to Cultural Studies, p. 300. Institute of Linguistics, Innsbruck (1999)
18. Ovsyannikova, E.V.: Gibridizatsiya angliyskogo yazyka kak problema perevoda (in Russian) [Hybridization of English as a problem of translation]. Sumy State University Bulletin: Philological series **4**(63), 93–97 (2004)
19. Pirojkov, A.: Russicisms in the Contemporary German: Corpus, Condition and Developmental Tendencies. Weißensee, Berlin (2002)
20. Schäfer, W.: About Handys and Erbex. Towards the discussion of Anglicisms in today's German. DAF 2, 75–81 (2002)
21. Schippan, T.: Lexicology of the Contemporary German Language: A Study Book. Niemeyer, Tübingen (1992)
22. Schmidt, W.: History of the German Language. Hirzel, Stuttgart (2004)
23. Schuchard, H.: Creolic Studies. C. Gerold's Sohn, Graz (1882)
24. Shcherba, L.V.: O ponyatii smesheniya yazykov (in Russian) [On the concept of mixing languages]. In: Language System and Speech Activity, pp. 60–74. Nauka, Leningrad (1974)
25. Shishigin, K.A.: Hybridization of a language system (exampled by Yiddish verbs with the prefix tsu-). Kemerovo State Univ. Cult. Bull. **29**, 111–116 (2014)
26. Solntsev, V.M.: Yazyk kak sistemno-strukturnoye obrazovaniye (in Russian) [Language as a systemic and structural formation], 2nd edn. Nauka, Moscow (1977)
27. Stenschke, O.: Internet terminology and common language stock. Lang. Commun. New Media **7**, 52–70 (2006)
28. Viereck, W.: The German in the language contact: British English and American English – German. In: Language History, pp. 938–948. Walter de Gruyter, Berlin (1984)
29. Vinogradov, V.V.: Osnovnyye etapy istorii russkogo yazyka (in Russian) [The main stages of the Russian language history]. In: Selected Works. The Russian Literary Language History, pp. 10–64. Nauka, Moscow (1978)
30. Wolff, G.: History of the German Language. Francke, Tübingen (1999)

A Relationship of Chulym Turkic to the Peripheral and Uralian Kipchak Languages According to the Leipzig–Jakarta List

Innokentiy N. Novgorodov[1]([✉]) [iD], Nikolay N. Efremov[2] [iD],
Spiridon A. Ivanov[2] [iD], and Valeriya M. Lemskaya[3,4,5] [iD]

[1] North-Eastern Federal University, Yakutsk 677000, Russian Federation
i.n.novgorodov@mail.ru
[2] The Institute for Humanities Research and Indigenous Studies of the North,
Yakutsk 677007, Russian Federation
{nik.efrem50, spiridon_ivanov}@mail.ru
[3] Tomsk Polytechnic University, Tomsk 634050, Russian Federation
lemskaya@gmail.com
[4] Tomsk State Pedagogical University, Tomsk 634041, Russian Federation
[5] Tomsk State University, Tomsk 634050, Russian Federation

Abstract. Background. This article is about a relationship of the Chulym Turkic language to the peripheral and Uralian Kipchak Turkic languages. The Chulym Turks are the people of the South-East of the West-Siberian Plain. The number of the Chulym Turks is around 365 people in Russia.

Materials and Methods. Research materials are words of the Leipzig–Jakarta list of the Turkic languages. These most resistant words were written out from dictionaries and publications. The Leipzig–Jakarta list is a 100 word list to test the degree of a relationship of the languages. In this survey the comparative method is used as the main method. A quantitative method is applied to count the similarities and discrepancies in the Leipzig-Jakarta list of the Turkic languages. Also an interdisciplinary approach and different data to study the classification status of the Teleut (as peripheral) and Siberian Tatar (as Uralian) idioms are employed.

Discussions. Previously we reached a conclusion that Chulym Turkic is of the Kipchak Turkic language origin according to the Leipzig-Jakarta list. This article sets the questions to which modern Kipchak languages Chulym Turkic is related or whether it is a separate Kipchak language group. Also, the classification status of Teleut and Siberian Tatar is discussed.

Conclusions. Authors come to the conclusion that the Chulym Turkic language is more similar to the peripheral Kipchak languages. The Teleut and Siberian Tatar idioms are the separate Kipchak Turkic languages where the first one is peripheral and the second one is Uralian.

Keywords: Languages · Leipzig-Jakarta list · Turkic · Kipchak · Peripheral · Uralian · Chulym Turkic

© Springer International Publishing AG 2018
A. Filchenko and Z. Anikina (eds.), *Linguistic and Cultural Studies:*
Traditions and Innovations, Advances in Intelligent Systems and Computing 677,
DOI 10.1007/978-3-319-67843-6_34

1 Introduction

Altaic Studies are of the current interest among modern researches on the Euro-Asian peoples which are held by the leading countries worldwide according to their geopolitics.

To study processes of the Altaic language community convergence [6, 16, 17] it is reasonable to explore the Leipzig–Jakarta list [5, 10] and after that to conduct a comprehensive study of the Turkic, Mongolian, Tungus-Manchu, Korean and Japonic peoples, taking into consideration the achievements of linguistics, history and genetics.

The Leipzig-Jakarta list on several Turkic languages has already been published [9, 10].

It should be mentioned that we have previously come to the conclusion that the Turkic languages are divided into two main groups according to the Leipzig–Jakarta list [9]. The first one is the Yakut and Kipchak languages, and the second one – the Chuvash and Oghuz languages.

As for the Chulym language we have reached a conclusion that Chulym Turkic is of the Kipchak Turkic language origin according to the Leipzig–Jakarta list (Novgorodov et al.) [9]. Herein we raise a question which of the modern Kipchak languages Chulym Turkic belongs to or whether it is a separate Kipchak language group.

As it is known, the modern Kipchak languages are classified into different groups: peripheral (Kyrgyz and Altai), Cuman or Ponto-Caspian (Karachay-Balkar, Kumyk, Karaim), Tatar-Bashir or Uralian (Tatar, Bashkir) and Kangly or Aralo-Caspian (Kazakh, Karakalpak, Nogai) [13, p. 35].

Since Chulym Turkic is Siberian it is more reasonable to begin the discussion from those Kipchak languages which are present in Siberia nowadays or spread from Siberia not long ago. The peripheral group and Uralian Kipchak languages meet this criterion.

Before discussing the relationship of Chulym Turkic to the peripheral and Uralian Kipchak languages, a few lines about their speakers should be written.

The Chulym Turks are the people of the South-East of the West-Siberian Plain that inhabit the lower and middle flow of the River Chulym. The major part of the Chulym Turks are present settlers of the Russian Federation's Teguldet Region of Tomsk Oblast and Tjuxtet Region of Krasnoyarsky Krai, mainly in Pasechnoye and Chindat villages. The number of Chulym Turks is around 365 people according to official statistics. The Chulym Turkic language consists of the Middle and Lower Chulym dialects, the Middle Chulym sub-dialects are Tutal and Melet. The given differentiation goes back to the historical existence of indigenous provinces, i.e. "volosts". At present the Lower Chulym Dialect is considered to be totally extinct.

The peripheral Kipchak languages include Kyrgyz and Altai. Kyrgyz is mainly present in Kyrgyzstan of Central Asia and spoken by nearly 4.5 million people. Altai is spread in the Republic of Altai, Russia and has around 60 thousand speakers.

The peripheral group includes the idioms of the Telengits, Tubalars, Kumandins and Chelkans whose varieties have close connections with the Altai language [2].

A few words about the Teleut idiom should be written because different ideas are presented in linguistics about its classification status.

Teleut as the Altai language dialect was previously considered to belong to the Kipchak languages according to the Baskakov conception [2] and Siberian group of the Turkic languages that also included the Khakas, Shor, Saryg-Yughur, Yakut, Tuvinian, Altai and other languages according to Mudrak's opinion [7]. Therefore, we shall express our ideas concerning this question under the subtitle 'Discussions' in the article.

The Teleuts are the Turkic people inhabiting southwestern Siberia. The majority of the Teleuts live in Kemerovo Oblast, Russia. According to 2010 All-Russian census, there were 2.643 Teleuts. Around 2.600 people speak the Teleut idiom.

Thus the peripheral group of the Kipchak languages is present in Central Asia (Kyrgyz) and Siberia (Altai).

As for the Uralian group of the Kipchak languages, the Bashkir and Tatar ones must be listed among the varieties under present study.

The Bashkir language is one of the Uralian Kipchak Turkic languages spoken by the Bashkirs (1.584.554 people according to the latest All-Russian census of 2010) the majority of whom live in the South Urals. The Bashkir language is the official language of the Republic of Bashkortostan, Russia.

The Tatar language is one of the Uralian Kipchak Turkic languages spoken by the Tatars (5.310.649 people according to the latest All-Russian census of 2010) the majority of whom settled in the middle and lower reaches of the Volga River. The Tatar language is the official language of the Republic of Tatarstan, Russia. It must also be mentioned that the Tatar language is a language of interethnic communion; for example, cases of the Bashkir-Tatar bilingualism in North-West districts of Bashkortostan at the border with Tatarstan are revealed, as well as instances of the Chuvash-Tatar bilingualism in places where the Chuvash live along with the Tatars.

According to Zakiyev, the Tatar language has the Middle, Western and Eastern dialects. The Eastern dialect consists of the Tatar idioms of Siberia and includes the territorial groups of Tobol-Irtysh, Baraba and Tomsk all being within West Siberia [18].

Siberian Tatar as a separate language was considered to belong to the Kipchak languages according to Mudrak's opinion [7].

Besides, the different points of view concerning the Siberian Tatar classification status as the Turkic idiom, there is no exact data on the number of Siberian Tatars: from 6.779 people who identify themselves as the Siberian Tatars [4] to 323.052 people considering themselves Tatars of West Siberia [1].

As Turkology holds a discussion on the classification status for the language of the Siberian Tatars, we also make an attempt to identify the classification status of the Siberian Tatars and their idiom based on the Leipzig-Jakarta list and interdisciplinary complex data under the subtitle 'Discussions' of this publication.

The Siberian Tatar idiom indicates that the area of the Uralian Kipchak Turkic languages spreads beyond European Russia to Siberia.

There are also languages of various Kipchak diasporas in Siberia as well, for instance, Kyrgyz, Kumyk and Kazakh as the language of the indigenous people of Siberia. The present research will not analyze the languages of the Kipchak diasporas in Siberia. A relationship of Chulym Turkic to the Cuman or Ponto-Caspian

(Karachay-Balkar, Kumyk, and Karaim) and Kangly or Aralo-Caspian (Kazakh, Karakalpak, and Nogai) according to the Leipzig-Jakarta list will be presented in the following publication.

Based on these data in order to establish the relationship of the Chulym Turkic language among the peripheral and Uralian Kipchak languages, we shall take into consideration the Tatar and Bashkir languages (of the Uralian Kipchak group) and the Kyrgyz and Altai languages (which belong to the peripheral Kipchak group).

2 Materials and Methods

Published materials are used in the present study of the Chulym Turkic language [9].

The materials of the Tatar, Bashkir languages and the Siberian Tatar idiom (which belong to the Uralian Kipchak group) and the Kyrgyz, Altai languages and Teleut idiom (which belong to the peripheral Kipchak group) are taken from different publications.

The Leipzig-Jakarta list is a 100 word list to test the degree of languages relationship by comparing words that are resistant to borrowing [9, 14].

The indicated 100 most resistant words are used here to establish a relationship of the Chulym Turkic language to the peripheral and Uralian Kipchak Turkic languages.

Here the Leipzig-Jakarta list is data for a quantitative method. It is used to count the similarities and discrepancies in the Leipzig-Jakarta list of the Turkic languages in order to reveal a degree of homogeneity of comparable objects which is important in studying the relationship of different idioms.

For the convenience of analysis the Leipzig-Jakarta list was taken from open electronic sources rather than [14] due to material arrangement: e.g., the alphabetical order of vocabulary and a single lexeme for identifying each vocabulary sample.

Before presenting materials of the Leipzig-Jakarta list, it should be noted that '1' is a number of the Leipzig-Jakarta list item; 'ant' – meaning; (3. 817) – index number of World loanword database available online at [http://wold.clld.org/meaning].

In this survey the comparative method is used as the main method for linguistics to reveal the homogeneity of comparable objects.

We also use an interdisciplinary approach and different data in studying the classification status of the Teleut and Siberian Tatar idioms.

3 Discussions

As analysis shows, the Leipzig-Jakarta list of Teleut differs from that of Altai in 1 item:

1 'ant' (3. 817) tel. *qüzürüm* (< tel. **qusurum* < **qumurs* < **qïmïrs* (cf. turk. *qïmïrsa-* (< **qïmïrs* + a-) 'to creep, to swarm about insects' (< tu. [ESTJ, p. 141, 2000]) > yak. **qïmïrsakač* (< *qïmïrsa-* + -*kač*) > yak. *qïmïrdaɣas*); alt. *čimali* (< mo.: *čubali, čumali* 'ant'), cf. kyrg. *qumursqa* (< tu.), *čimeli* [Zhumaev] (< mo.);

Teleut and Altai differ also in synonyms in 21 items: 4, 15, 18, 24, 26, 27, 34, 41, 42, 44, 50, 51, 61, 63, 67, 68, 77, 78, 87, 96, and 97. It should be noted that one of the synonyms does not match in Teleut or Altai, e.g.:

4 'back' (4.19) tel. *arqa* (< tu. [ESTJ, p. 174, 1974]), *kögüs* (< tu. [ESTJ, p. 54, 1980]); alt. *uča* (< tu. [ESTJ, p. 613, 1974]), *sïrt* (< tu. [ESTJ, p. 418, 2003]), *bel* (< tu. [ESTJ, p. 136, 1978]), *arqa* (< tu. [ESTJ, p. 174, 1974]) etc.

Other Teleut and Altai items are similar. However, in some cases the items under study (16, 30, 51, 52, 64, 5, 25, 35, 59, and 62) vary phonetically. For example, initial consonants are voiceless in Teleut but voiced in Altai:

16 'child (kin term)' (2.43) tel. *pala* [p-] (< tu. [ESTJ, p. 47, 1978] < i.e. [8, p. 376]); alt. *bala* [b-] (< tu. < i.-e. [ESTJ, p. 47, 1978]) < i.e.), cf. kyrg. *bala* [b-] (< tu. < i.e.);

5 'big' (12.55) tel. *yaan* [t'ān] (< tu. [ESTJ, p. 60, 1989; Räs, pp. 177b-178a, 1969] < chin. *čoŋ* [Räs, p. 116a, 1969]); alt. *jaan* [d'ān] (< tu. < chin.), cf. kyrg. *čoŋ* 'big, large; great' < chin. *čoŋ* [Räs, p. 116a, 1969], *zor* 'big, huge' (< pers. *zūr, zor* 'strength, power' [Räs, p. 532b, 1969]) etc.

Thus, the Teleut idiom has its own phonetic features that make it different from Altai.

The comparative study of the Leipzig-Jakarta list has revealed phonetic and lexical features of the Teleut and Altai and demonstrated that actually the Teleut and Altai idioms are the separate Turkic languages, having a close relationship.

It should be also mentioned that according to the Russian Federation Government Resolution (March 24, 2000, No. 255) the Teleuts were recognized as the separate Turkic people.

These facts justify inclusion of the Teleut language to the peripheral group along with Kyrgyz and Altai.

Thus, we consider that the Teleut language is one of the peripheral Kipchak languages according to scientific, linguistic (the Leipzig-Jakarta list, phonetic features) and political (the Russian Federation Government Resolution) data because despite having close relations with the Altai language, it still differs from it.

Analysis also shows that the Leipzig-Jakarta list of Siberian Tatar has full correspondence with the Bashkir and Tatar ones, e.g.:

1 'ant' (3. 817) tat. *kïrmïska* (< tu. [ESTJ, p. 140, 2000]), stat. *qïmïrïsqa*; bash. *kïrmïðka* (< tu.);

2 'arm' (4.31), 'hand' (4.33) tat. *kul* (< tu. [ESTJ, p. 37, 2000]), stat. *qul*; bash. *kul* (< tu.);

3 'ash' (1.84) tat. *köl* (< tu. [ESTJ, p. 137, 1997]), stat. *köl*; bash. *köl* (< tu.) etc.

Some units of the Siberian Tatar Leipzig-Jakarta list coincide either with Tatar or Bashkir, for example:

4 'back' (4.19) tat. *arka* (< tu. [ESTJ, p. 174, 1974]), dial. tat. *jilkä* (< tu. [ESTJ, p. 181, 1989]), stat. *yältkä* (< tu.), *sïrt* (< tu. [ESTJ, p. 418, 2003]); bash. *arka* (< tu.), *hïrt* (< tu.);

7 'to bite' (4.58) tat. *tešläü* (< tu. [ESTJ, p. 242, 1980]), dial. tat. *čäynäü* (< tu. [Räs, p. 95b, 1969]), stat. *cäynägälä* (< tu.); bash. *tešläü* (< tu.) etc.

The analyzed Siberian Tatar items of the Leipzig-Jakarta list are considerably different phonetically from the Tatar and Bashkir ones. So the initial consonants are voiceless in Siberian Tatar and voiced in Tatar and Bashkir:

16 'child (kin term)' (2.43) tat. *bala* [b-] (< tu. [ESTJ, p. 47, 1978] < i.e.), stat. *pala* [p-]; bash. *bala* [b-] (< tu. < i.e.);

30 'fish' (3.65) tat. *balïk* [b–] (< tu. [ESTJ, p. 59, 1978]), stat. *palïk* [p-]; bash. *balïk* [b-] (< tu.);

58 'neck' (4.28) tat. *muyen* [m-] (< tu. [ESTJ, p. 180, 1978]), stat. *puyïn* [p-]; bash. *muyïn* [m-] (< tu.), *üŋäs* (< tu. [ESTJ, p. 536, 1974]) etc.

Based on the performed analysis, we can conclude that the Leipzig-Jakarta list of Siberian Tatar fully corresponds to the Uralian group of the Kipchak languages. However, the idiom of the Siberian Tatars displays linguistic (phonetic) phenomena that distinguishes them from the respective Kazan Tatar and Bashkir ones. This circumstance allows turning to historical and genetic data for establishing the classification status of the Siberian Tatars since modern studies of the peoples are interdisciplinary and complex.

As it is known, historically the ancestors of modern Siberian and Kazan Tatars were represented in different medieval feudal states that formed after the collapse of the Golden Horde. The development of the Siberian Tatars took place within the Siberian Khanate (till 1598), while the Kazan Tatars were settled in the Kazan Khanate (till 1552).

Tomilov claims that the total number of all Siberian Tatars increased from 16.500 people in the late 17th century up to 28.500 at the end of the 18th century, to more than 47.000 at the end of the 19th century, to 170.000 in the late of 1980s and 180.000 in 2002 [15]. However, there are other data that significantly change the number of all modern Siberian Tatars and distinguish them from the Volga Tatars. Bakiyeva indicates 323.052 Tatars in Western Siberia according to 2010 Census, 239.995 in Tyumen region, and only 41.870 in Omsk region, 24.158 in Novosibirsk region, and 17.029 in Tomsk region. The ratio of all Siberian to the Volga Tatars in the four regions taken together is calculated as 242.289 to 80.763 people [1]. According to the results of the 2002 All-Russian Census, the number of Siberian Tatars was 9.611 people [3], according to the 2010 All-Russian Census, there were 6.779 of the Siberian Tatar people [4]. Despite the varying data on the number of all modern Siberian Tatars, they still remain unrecognized as the separate people. L.A. Shamsutdinova reveals the administrative reasons (starting from the 27th of November, 2008) for the Siberian Tatars not to be recognized as the indigenous people [12].

To further address the issue of classification status of the indigenous Siberian Tatars, it is necessary to turn to genetic research. Currently it is an important tool for establishing the indigenous status of the community. Genetically the indigenous Tobol-Irtysh Siberian Tatars are not connected by common origin with the Kazan and Crimean Tatars [11].

Therefore, the Siberian Tatars differ from the Kazan Tatars both historically and genetically.

The account of the interdisciplinary complex data (linguistic, historical and genetic) allows concluding that the Siberian Tatars are the separate Turkic people whose language is a special one, belonging to the Uralian group of the Kipchak languages along with Tatar and Bashkir.

The analysis of the Leipzig-Jakarta list shows that Chulym Turkic differs from the languages of the peripheral group in 5 items (5, 7, 11, 16, 27), for example:

5 'big' chul. *uluu* (< tu. [ESTJ, p. 593, 1974]); kyrg. *čoŋ* 'big, large; great' (< chin. *čoŋ* [Räs, p. 116a, 1969]), *zor* 'big, huge' (< pers. *zūr, zor* 'strength, power' [Räs, p. 532b, 1969]); alt. *jaan* (< tu. < chin.); tel. *yaan* (< tu. < chin.) etc.

The comparison of the Chulym Turkic Leipzig-Jakarta list with the Uralian one has indicated difference in 6 items (7, 14, 16, 27, 50, and 95) like:

95 'wide' (12.61) chul. *jalbak* (< tu. [ESTJ, p. 100, 1989]); tat. *kiŋ* (< tu. [ESTJ, p. 46, 1980]); stat. *kiŋ* (< tu.); bash. *kiŋ* (< tu.), *iŋle* (< tu. [ESTJ, p. 352, 1974]), *yaðï* (< tu. [ESTJ, p. 155, 1989]) etc.

In order to study the relation of Chulym Turkic to Cuman or Ponto-Caspian (Karachay-Balkar, Kumyk, Karaim) and Kangly or Aralo-Caspian (Kazakh, Karakalpak, Nogai) groups of the Kipchak languages, a further research is required.

4   Conclusion

According to the Russian Federation Government Resolution (March 24, 2000, No. 255), the Teleuts were recognized as the separate Turkic people. The Teleut language has its own phonetic and lexical features that differ it from the Altai one. And these facts give evidence in favour of including the Teleut language to the peripheral group as the separate language along with the Kyrgyz and Altai ones.

As a result of the interdisciplinary complex research, it is established that the Siberian Tatars are the separate Turkic people whose language is a special one, belonging to the Uralian group of the Kipchak languages along with Tatar and Bashkir.

The comparison of the Leipzig-Jakarta list of Chulym Turkic with the rosters of the peripheral and Uralian Kipchak languages indicates that Chulym Turkic is closer to the peripheral rather than Uralian Kipchak Turkic languages.

Acknowledgments. We sincerely thank the Russian Federation Government for the grant "Yazykovoe i etnokulturnoe raznoobrazie Yuzhnoi Sibiri v sinkhronii i diakhronii: vzaimodeistvie yazykov i kultur" [The linguistic and ethno-cultural diversity of Southern Siberia in synchrony and diachrony: the interaction of languages and cultures] (Project 2016-220-05-150); this publication was prepared within the framework of this research project.

We also sincerely thank Y.E. Zhumaev who checked and added items to the Kyrgyz materials.

Abbreviations.

alt.	– Altai
bash.	– Bashkir
chin.	– Chinese
chul.	– Chulym
dial.	– dialectological
i.e.	– Indo-European
kyrg.	– Kyrgyz

mo.   – Written Mongolian
pers.   – Persian
stat.   – SiberianTatar
tat.   – Tatar
tel.   – Teleut
tu.   – Turkic
turk.   – Turkmenian
yak.   – Yakut

ESTJ 1974   – Sevortyan, E. Etimologicheskiy slovar' tyurkskikh yazykov. Obshchetyurkskie i mezhtyurkskie osnovy na glasnye (in Russian) [Etymological dictionary of the Turkic languages. All-Turkic and cross-Turkic stems starting with vowels]. Nauka, Moscow (1974)

ESTJ 1978   – Sevortyan, E., Gadzhieva, N. (ed). Etimologicheskiy slovar tyurkskikh yazykov. Obshchetyurkskie i mezhtyurkskie osnovy na bukvu "B" (in Russian) [Etymological dictionary of the Turkic languages. All-Turkic and cross-Turkic stems starting with letter "B"]. Nauka, Moscow (1978)

ESTJ 1980   – Sevortyan, E., Gadzhieva, N. (ed). Etimologicheskiy slovar tyurkskikh yazykov. Obshchetyurkskie i mezhtyurkskie osnovy na bukvy "V", "G", "D", (in Russian) [Etymological dictionary of the Turkic languages. All-Turkic and cross-Turkic stems starting with letters "V", "G", and "D"]. Nauka, Moscow (1980)

ESTJ 1989   – Levitskaya, L. (ed) Etimologicheskiy slovar tyurkskikh yazykov. Obshchetyurkskie i mezhtyurkskie osnovy na bukvy "J", "ZH", "Y" (in Russian) [Etymological dictionary of the Turkic languages. All-Turkic and cross-Turkic stems starting with letters "J", "ZH", and "Y"]. Nauka, Moscow (1989)

ESTJ 1997   – Etimologicheskiy slovar tyurkskikh yazykov. Obshchetyurkskie i mezhtyurkskie leksicheskie osnovy na bukvy "K", "Q" (in Russian) [Etymological dictionary of the Turkic languages. All-Turkic and cross-Turkic lexical stems starting with letters "K" and "Q"]. Yazyki russkoy kultury, Moscow (1997)

ESTJ 2000   – Etimologicheskiy slovar tyurkskikh yazykov. Obshchetyurkskie i mezhtyurkskie leksicheskie osnovy na bukvy "K" (in Russian) [Etymological dictionary of the Turkic languages. All-Turkic and cross-Turkic lexical stems starting with letter "K"]. Yazyki russkoy kultury, Moscow (2000)

ESTJ 2003   – Etimologicheskiy slovar tyurkskikh yazykov. Obshchetyurkskie i mezhtyurkskie leksicheskie osnovy na bukvy "L", "M", "N", "P", "S" (in Russian) [Etymological dictionary of the Turkic languages. All-Turkic and cross-Turkic lexical stems starting with letter "L", "M", "N", "P", and "S"]. Vostochnaya literatura RAN, Moscow (2003)

Räs 1969   – Räsänen, M. Versuch einen etymologischen Wörterbuchs der Türksprachen. Suomalais-Ugrilainen Seura, Helsinki: (1969).

References

1. Bakiyeva, G., Kvashnin, Yu.: Povolzhskie tatary v zapadnoi Sibiri: osobennosti rasseleniya i etnokulturnogo razvitiya (in Russian) [The Volga Tatars in Western Siberia: features of resettlement and ethno-cultural development]. Vestnik arkheologii, antropologii i etnografii 3(22), 157–158 (2013)
2. Baskakov, N.: Altayskiy yazyk (in Russian) [The Altai language]. In: Yartseva, V. (eds.) Yazyki mira. Tyurkskie yazyki, pp. 179–187. Indrik, Moscow (1997)
3. Census – 2002. Vserossiyskaya perepis naseleniya (in Russian) [All-Russian Census]. http://www.perepis2002.ru/index.html?id=11. Accessed 05 Jan 2017
4. Census – 2010. Vserossiyskaya perepis naseleniya (in Russian) [All-Russian Census]. http://www.gks.ru/free_doc/new_site/perepis2010/croc/perepis_itogi1612.htm. Accessed 05 Jan 2017
5. Erdal, M.: The Turkic-Mongolic relationship in view of the Leipzig-Jakarta list. In: Novgorodov, I. (eds.) Unpublished Proceedings of 2016 International Symposium "The Leipzig-Jakarta list of the Turkic languages as a source of the interdisciplinary comprehensive studies" (2016)
6. Janhunen, J.: The altaic hypothesis: what is it about? In: Novgorodov, I. (eds.) Unpublished proceedings of 2016 International Symposium "The Leipzig-Jakarta list of the Turkic languages as a source of the interdisciplinary comprehensive studies" (2016)
7. Mudrak, O.: Ob utochnenii klassifikatsii tyurkskikh yazykov s pomoshch'yu morfologicheskoy lingvostatistiki (in Russian) [On the specifying of the Turkic languages classification using morphological linguostatistics]. In: Tenishev, E. (eds.) Sravnitelno-istoricheskaya grammatika tyurkskikh yazykov. Regionalnyye rekonstruktsii, pp. 732–737. Nauka, Moscow (2002)
8. Munkasci, B.: Beiträge zu den alten Lehnwörtern im Türkischen. Keleti-Szemple VI **2–3**, 376–379 (1905)
9. Novgorodov, I., Lemskaya, V., Gainutdinova, A., Ishkildina, L.: The Chulym Turkic language is of the Kipchak Turkic language origin according to the Leipzig-Jakarta list. Türkbilig **29**, 1–18 (2015)
10. Novgorodov, I.: Ustojchivyi slovarnyi fond Tyurkskikh yazykov (in Russian) [The Leipzig-Jakarta list of the Turkic languages]. SMIK, Yakutsk (2016)
11. Padyukova, A., Agdzhoian, A., Dolinina, D., Lavriashina, M., Kuznetsova, M., Skhaliakho, R., Balaganskaia, O., Ulianova, M., Tychinskikh, Z., Balanovskii, O., Balanovskaia, E.: Mnogoobrazie i svoeobrazie genofonda sibirskikh tatar (in Russian) [Variety and originality of the gene pool of the Siberian Tatars]. In: Alishina, Kh. (eds.) Conference Proceedings of Scientific-Practical Conference Suleimanov Readings (19): Preservation of the Tatar language, folklore, traditions and customs in modern conditions, pp. 208–210. Pechatnik, Tyumen (2016)
12. Shamsutdinova, L.: K istorii problemy statusa korennogo naroda sibirskikh tatar (in Russian) [To the problem history of the indigenous people status for the Siberian Tatars]. Istoriia i sovremennost **1**(17), 158–162 (2013)
13. Shcherbak, A.: Vvedeniye v sravnitelnoye izucheniye tyurkskikh yazykov (in Russian) [Introduction to the Comparative Study of Turkic Languages]. In: Pokrovskaya, L. (eds.) Nauka, St. Petersburg (1994)
14. Tadmor, U.: The Leipzig-Jakarta list of basic vocabulary. In: Haspelmath, M., Tadmor, U. (eds.) Loanwords in the World's Languages: A Comparative Handbook, pp. 68–75. Mouton de Gruyter, Berlin (2009)

15. Tomilov, N.: Sibirskiye tatary (in Russian) [The Siberian Tatars]. In: Lotkin, I., Tomilov, N. (eds.) Narody Zapadnoi i Srednei Sibiri: kultura i etnicheskie protcessy, pp. 174–229. Nauka, Novosibirsk (2002)
16. Vovin, A.: Koreo-Japonica: A Re-evaluation of a Common Genetic Origin. Hawai'i Studies on Korea. University of Hawai'i Press, Honolulu (2010)
17. Vovin, A.: Lexical and paradigmatic morphological criteria for the establishing language genetic relationships. In: Novgorodov, I. (eds.) Unpublished Proceedings of 2016 International Symposium "The Leipzig-Jakarta list of the Turkic languages as a source of the interdisciplinary comprehensive studies"
18. Zakiyev, M.: Tatarksiy yazyk (in Russian) [The Tatar language]. In: Yartseva, V. (eds.) Yazyki mira. Tyurkskie yazyki, pp. 357–371. Indrik, Moscow (1997)

The Teleut Language is of the Kipchak Turkic Language Origin According to the Leipzig–Jakarta List

Innokentiy N. Novgorodov[1]([⊠]) (iD), Albina F. Gainutdinova[2] (iD),
Linara K. Ishkildina[3] (iD), and Denis M. Tokmashev[4] (iD)

[1] North-Eastern Federal University, Yakutsk 677000, Russian Federation
i.n.novgorodov@mail.ru
[2] Academy of Sciences of Tatarstan Republic, Kazan 420111,
Russian Federation
albina_gain@mail.ru
[3] Ufa Scientific Center of Russian Academy of Sciences, Ufa 450054,
Russian Federation
lina86_08@mail.ru
[4] Tomsk Polytechnic University, Tomsk 634050, Russian Federation
kogutei@yandex.ru

Abstract. This article is about a classification status of the Teleut language according to the Leipzig–Jakarta list. The Teleuts are the Turkic people located in southwestern Siberia, Kemerovo Oblast, Russia. According to the 2010 census, there were 2,643 Teleuts in Russia.

Materials and methods. Research materials are words of the Leipzig–Jakarta list of the Turkic languages. These most resistant words were written out from the dictionaries and publications. The Leipzig–Jakarta list is a 100 word list to test the degree of a relationship of the languages. In this survey the comparative method is used as the main one. A quantitative method is applied to count the similarities and discrepancies in the Leipzig-Jakarta list of the Teleut idiom in comparison with the Oghuz and Kipchak languages. Also different data to study the classification status of the Teleut idiom are employed.

Discussions. Teleut as the Altai language dialect was previously considered to belong to the Kipchak languages according to Baskakov's conception. On the other hand, according to Mudrak's opinion Teleut was included into the Siberian group of the Turkic languages together with Khakass, Shor, Saryg-Yughur, Yakut, Tuvinian, Altai and others.

Conclusions. Authors come to conclusion that the Teleut language is of the Kipchak Turkic language origin according to the Leipzig–Jakarta list. Also the Teleut language has its own linguistic (phonetic and lexical) features that differ it from the Altai language and this fact proves including the Teleut language to peripheral group along with the Kyrghyz and Altai languages.

Keywords: Language · Leipzig-Jakarta list · Teleut · Oghuz · Kipchak

© Springer International Publishing AG 2018
A. Filchenko and Z. Anikina (eds.), *Linguistic and Cultural Studies:*
Traditions and Innovations, Advances in Intelligent Systems and Computing 677,
DOI 10.1007/978-3-319-67843-6_35

1 Introduction

As it is known, the modern Turkic languages are classified into different groups: the Oghuz, Kipchak, Karluk and others. Each group has its own members. For example, the Oghuz languages include the Turkish, Azerbaijani, Turkmen and others.

Previously we came to the conclusion that Teleut is more similar to the Kipchak languages [5]. In this article we are discussing the classification status of the Teleut language according to the Leipzig-Jakarta list. Before discussing, a few words should be said about its speakers. The Teleuts are a Turkic people located in southwestern Siberia. Majority of the people live in Kemerovo Oblast, Russia. According to 2010 Census, there were 2.643 Teleuts in Russia [10].

2 Materials and Methods

To study the relationship of the Teleut language, the Leipzig–Jakarta list was taken into consideration. For the convenience of analysis, the Leipzig–Jakarta list was taken from open electronic sources rather than [9] due to material arrangement (e.g., the alphabetical order of vocabulary and a single lexeme for identifying each vocabulary sample).

The Leipzig–Jakarta list is a 100 word list to test the degree of languages relationship by comparing words that are resistant to borrowing [4, 9]. The indicated 100 most resistant words are used to establish the relationship of Teleut among the Kipchak and Oghuz languages.

The Leipzig–Jakarta list has already been published on the modern Turkic languages [4, 6].

It should be mentioned that previously we came to the preliminary conclusion that the Turkic languages are divided into two main groups [4]. The first one is the Yakut and the Kipchak languages, and the second one – the Chuvash and Oghuz languages.

In order to establish the Teleut language relationship among the Kipchak and Oghuz ones, we take into consideration the Turkish language (which belongs to the Oghuz group) and the Tatar and Bashkir languages (which belong to the Kipchak group).

In this study, publications about Teleut are used.

In this survey, the comparative method is used as the main method.

Before presenting materials of the Leipzig–Jakarta list, it should be noted that 1 is a number of the Leipzig–Jakarta list item; 'ant' – meaning; (3. 817) – index number of World loanword database, available online at http://wold.clld.org/meaning.

3 Discussions

First of all, it should be mentioned that synonyms (18 items: 4, 15, 18, 20, 24, 26, 27, 42, 44, 50, 51, 61, 78, 82, 87, 95, 96, and 97) are traced in the Teleut language, e.g.:

4 'back' (4.19) tel. *arqa* (< tu. [ESTJ, p. 174, 1974]), *kögüs* (< tu. [ESTJ, p. 54 1980]); tur. *arka* (< tu.), dial. *dal* 'back, shoulder'(< tu. [ESTJ, p.131, 1980]), *sırt*

(< tu. [ESTJ, p. 418, 2003]); tat. *arka* (< tu.), dial. *jilkä* (< tu. [ESTJ, p. 181, 1989]); bash. *arka* (< tu.) etc.

It should be noted that Teleut reflects the Kipchak (Bashkir) synonyms in 4 items (15, 24, 42 and 78), e.g.:

15 'to carry' (10.61) tel. *taži-* (< tu. [ESTJ, p. 170, 1980]), *akel-* (< tel. *alip kel-* < tu. [ESTJ, p. 127, 1974; ESTJ, p. 14, 1980]), *apar-* (< tel. *alip bar-* < tu. [ESTJ, p. 127, 1974; ESTJ, p. 64, 1978]); tur. *iletmek* (< tu. [ESTJ, p. 267, 1974]), *taşımak* (< tu.), *götürmek* (< tu. [ESTJ, p. 86, 1980]); tat. *küterep yötü* (*yörtü*) (< tat. < tu. [ESTJ, p. 86, 1980; ESTJ, p. 229, 1989]); bash. *tashïu* (< tu.), *aparïu-* (< tu.) etc.

The synonyms of the Teleut, Turkish, Tatar and Bashkir languages (items 27, 44, and 82) can indicate the Turkic protolanguage synonymy as these words are traced in the Oghuz and Kipchak languages:

27 'to fall' (10.23) tel. *tüš-* (< tu. [ESTJ, p. 330, 1980]), *yïgïl-* (< tu. [ESTJ, p. 273, 1989]); tur. *düşmek* (< tu.), *yıkılmak* (< tu.); tat. *tösü* (< tu.), *egïlu* (< tu. [ESTJ, p. 69, 1974]); bash. *yïγïlïu* (< tu.), *auïu-* (< tu.), *töšü* (< tu.) etc.

The majority of forms (75 items: 2, 3, 4, 6, 7, 8, 9, 10, 17, 19, 20, 21, 22, 23, 24, 25, 26, 28, 29, 30, 31, 32, 33, 34, 35, 36, 37, 38, 40, 43, 44, 45, 46, 47, 48, 50, 51, 53, 54, 55, 56, 58, 59, 60, 62, 63, 64, 65, 66, 68, 70, 71, 72, 73, 74, 78, 79, 80, 81, 82, 83, 84, 85, 86, 88, 89, 90, 91, 92, 93, 94, 95, 96, 97, 98, and 100) of the Teleut language are similar to those of the Turkish, Tatar and Bashkir languages and these forms are of the Turkic origin, e.g.:

2 'arm' (4.31), 'hand' (4.33) tel. *qol* (< tu. [ESTJ, 2000: 37]); tur. *kol* (< tu.), *el* (< tu. [ESTJ, p. 260, 1974]); tat. *kul* (< tu.); bash. *kul* (< tu.) etc.

Early Indo Europian borrowings (items 16, 75) are exposed among the words of the Turkic Leipzig-Jakarta list:

16 'child (kin term)' (2.43) tel. *pala* (< tu. [ESTJ, p. 47, 1978] < i.e. [3, p. 376]); tur. *evlât* (< ar. [Räs, p. 52b, 1969]), dial. *bala* (< tu. < i.e.); tat. *bala* (< tu. < i.e.); bash. *bala* (< tu. < i.e.);

75 'skin/hide' (4.12) tel. *tere* (< tu. [ESTJ, p. 207, 1980] < i.e. [7, p. 561]); tur. *deri* (< tu. < i.e.); tat. *tire* (< tu. < i.e.); bash. *tire* (< tu. < i.e.);

Two Mongolian loanwords (items 18 and 96) are revealed in the Leipzig–Jakarta list of the Teleut language, e.g.:

18 'to crush/to grind' (5.56) tel. *tart-* 'to crush in a mill' (< tu. [ESTJ, p. 154, 1980]), *soq-* 'to crush something hard' (< tu. [ESTJ, p. 286, 2003]), *nïqï-* 'to crush something soft', cf. khak. *nïxï-* 'to press with force', 'to ram' (< mo. *nixu-* 'to rub; to press, to knead');

96 'wind' (1.72) tel. *salqïn* (< mo. [Räs, p. 398b, 1969]), *yel* (< tu. [ESTJ, p. 174, 1989]);

Forms of several words (18 items: 4, 5, 11, 14, 15, 18, 20, 26, 39, 49, 50, 51, 61, 67, 87, 95, 96, and 97) of the Teleut language are not found in the same meaning in the Turkish, Tatar and Bashkir languages, e.g.:

4 'back' (4.19) tel. *kögüs* (< tu. [ESTJ, p. 54, 1980]); tur. *arka* (< tu.), dial. *dal* 'back, shoulder'(< tu. [ESTJ, p. 131, 1980]), *sırt* (< tu. [ESTJ, p. 418, 2003]); tat. *arka* (< tu.), dial. *jilkä* [< tu. [ESTJ, p. 181, 1989]); bash. *arka* (< tu.) etc.

Words of Turkic origin, which are not traced in the Turkish, Tatar and Bashkir languages in the same form and meaning, reveal specifics of the Teleut language.

Teleut as the Altai language dialect was previously considered to belong to the Kipchak languages, according to Baskakov's conception [1], and Siberian group of the Turkic languages that also included the Khakas, Shor, Saryg-Yughur, Yakut, Tuvinian, Altai and other languages according to Mudrak's opinion [2]. We disagree with Mudrak's statement and support Baskakov's conception in part of Teleut relationship.

According to the Russian Federation Government Resolution (March 24, 2000, No. 255), the Teleuts were recognized as the separate Turkic people.

The Leipzig-Jakarta list of the Teleuts differed from the Altai one in 1 item:

1 'ant' (3. 817) tel. *qüzürüm* (< tel. **qusurum* < **qumurs* < **qïmïrs* (cf. turk. *qïmïrsa-* (< **qïmïrs* + *a-*) 'to creep, to swarm about insects' (< tu. [ESTJ, p. 141, 2000]) > yak. **qïmïrsakač* (< *qïmïrsa-* + *-kač*) > yak. *qïmïrdaγas*); alt. *čïmalï* (< mo. *čubali, čumali* 'ant');

Teleut and Altai are also different in synonyms in 21 items: 4, 15, 18, 24, 26, 27, 34, 41, 42, 44, 50, 51, 61, 63, 67, 68, 77, 78, 87, 96 and 97. It should be noted that one of the synonyms does not match in Teleut or Altai, e.g.:

4 'back' (4.19) tel. *arqa* (< tu. [ESTJ, p. 174, 1974]), *kögüs* (< tu. [ESTJ, p. 54, 1980]); alt. *uča* (< tu. [ESTJ, p. 613, 1974]), *sïrt* (< tu. [ESTJ, p. 418, 2003]), *bel* (< tu. [ESTJ, p. 136, 1978]), *arqa* (< tu. [ESTJ, p. 174, 1974]) etc.

Other Teleut and Altai items are similar. And in some cases the studied items (16, 30, 51, 52, 64, 5, 25, 35, 59, and 62) differed phonetically. For example, voiceless initial consonants are present in Teleut, whereas voice ones are revealed in Altai, e.g.:

16 'child (kin term)' (2.43) tel. *pala* [p-] (< tu. [ESTJ, p. 47, 1978] < i.e. [3, p. 376]); alt. *bala* [b-] (< tu. < i.-e.);

5 'big' (12.55) tel. *yaan* [t'ān] (< tu. [ESTJ, p. 60, 1989; Räs, pp. 177b-178a, 1969]); alt. *jaan* [d'ān] (< tu.) etc.

Comparative study of the Leipzig-Jakarta list and revealed phonetic features of Teleut and Altai demonstrate that the Teleut and Altai languages are separate Turkic languages.

As it is known, the modern Kipchak languages are classified into different groups: peripheral (Kyrghyz and Altai), Cuman or Ponto-Caspian (Karachay-Balkar, Kumyk, and Karaim), Tatar-Bashkir or Uralian (Tatar and Bashkir) and Kangly or Aralo-Caspian (Kazakh, Karakalpak and Nogai) [8]. And it is reasonable to include the Teleut language to peripheral group of the modern Kipchak languages.

Analysis of the Leipzig–Jakarta list shows that 75 items out of 100 items: (2, 3, 4, 6, 7, 8, 9, 10, 17, 19, 20, 21, 22, 23, 24, 25, 26, 28, 29, 30, 31, 32, 33, 34, 35, 36, 37, 38, 40, 43, 44, 45, 46, 47, 48, 50, 51, 53, 54, 55, 56, 58, 59, 60, 62, 63, 64, 65, 66, 68, 70, 71, 72, 73, 74, 78, 79, 80, 81, 82, 83, 84, 85, 86, 88, 89, 90, 91, 92, 93, 94, 95, 96, 97, 98, and 100) are found in the Teleut, Turkish, Tatar and Bashkir languages simultaneously and these items are similar in form and meaning. This fact demonstrates that these languages have originated from the Proto-Turkic source.

13 items (1, 12, 13, 18, 24, 41, 42, 52, 57, 69, 77, 87, and 99) reveal that the Teleut list in form and meaning is more similar to the Kipchak (Tatar, Bashkir) Turkic languages than the Oghuz (Turkish) ones, e.g.:

1 'ant' (3. 817) tel. *qüzürüm* (< tel. **qusurum* < **qumurs* < **qïmïrs* (cf. turk. *qïmïrsa-* (< **qïmïrs* + *a-*) 'to creep, to swarm about insects' (< tu. [ESTJ, p. 141,

2000]) > yak. *qïmïrsakač (< qïmïrsa- + -kač) > yak. qïmïrdaɣas); tur. karınca (< tu. [ESTJ, p. 323, 1997]); tat. kïrmïska (< tu. [ESTJ, p. 140, 2000]); bash. kïrmïðka (< tu.) etc.

4 Conclusion

So, in total 88 items out of 100 of the Teleut language match the Kipchak (Tatar and Bashkir) Turkic items in form and meaning, and 75 items are cognate to the Oghuz (Turkish) ones in the same way. Facts of similarity of Teleut and Kipchak (Bashkir) synonymy (4 items: 15, 24, 42, and 78) are also taken into consideration.

Thus, we consider that the Teleut language is more similar to the Kipchak languages than the Oghuz Turkic ones. This circumstance allows us to consider that the Teleut language is of the Kipchak Turkic language origin. Also, the Teleut language has its own linguistic (phonetic and lexical) features that differ it from the Altai language and this fact proves including the Teleut language to the peripheral group along with the Kyrghyz and Altai languages.

Acknowledgements. We sincerely thank the Russian Federation Government for the grant "Yazykovoe i etnokulturnoe raznoobrazie Yuzhnoi Sibiri v sinkhronii i diakhronii: vzaimodeistvie yazykov i kultur" [The linguistic and ethno-cultural diversity of Southern Siberia in synchrony and diachrony: the interaction of languages and cultures] (Project 2016-220-05-150); this publication was prepared within the framework of this research project.

Abbreviations.

alt. –   Altai
ar. –   Arabian
bash. –   Bashkir
dial. –   Dialectological
i.e. –   Indo-European
mo. –   Written Mongolian
tat. –   Tatar
tel. –   Teleut
tur. –   Turkish
tu. –   Turkic
yak. –   Yakut

ESTJ 1974 –   Sevortyan, E. Etimologicheskiy slovar' tyurkskikh yazykov. Obshchetyurkskie i mezhtyurkskie osnovy na glasnye (in Russian) [Etymological dictionary of Turkic languages. All-Turkic and cross-Turkic stems starting with vowels]. Nauka, Moscow (1974)

ESTJ 1978 –   Sevortyan, E., Gadzhieva, N (ed). Etimologicheskiy slovar tyurkskikh yazykov. Obshchetyurkskie i mezhtyurkskie osnovy na bukvu "B" (in Russian) [Etymological dictionary of Turkic languages. All-Turkic and cross-Turkic stems starting with letter "B"]. Nauka, Moscow (1978)

ESTJ 1980 –   Sevortyan, E., Gadzhieva, N (ed). Etimologicheskiy slovar tyurkskikh yazykov. Obshchetyurkskie i mezhtyurkskie osnovy na bukvy "V", "G", "D", (in Russian) [Etymological dictionary of Turkic languages. All-Turkic and cross-Turkic stems starting with letters "V", "G" and "D"]. Nauka, Moscow (1980)

ESTJ 1989 –   Levitskaya, L. (ed) Etimologicheskiy slovar tyurkskikh yazykov. Obshchetyurkskie i mezhtyurkskie osnovy na bukvy "J", "ZH", "Y" (in Russian) [Etymological dictionary of Turkic languages. All-Turkic and cross-Turkic stems starting with letters "J", "ZH", and "Y"]. Nauka, Moscow (1989)

ESTJ 1997 –   Etimologicheskiy slovar tyurkskikh yazykov. Obshchetyurkskie i mezhtyurkskie leksicheskie osnovy na bukvy "K", "Q" (in Russian) [Etymological dictionary of Turkic languages. All-Turkic and cross-Turkic lexical stems starting with letters "K" and "Q"]. Yazyki russkoy kultury, Moscow (1997)

ESTJ 2000 –   Etimologicheskiy slovar tyurkskikh yazykov. Obshchetyurkskie i mezhtyurkskie leksicheskie osnovy na bukvy "K" (in Russian) [Etymological dictionary of Turkic languages. All-Turkic and cross-Turkic lexical stems starting with letter "K"]. Yazyki russkoy kultury, Moscow (2000). Yazyki russkoy kultury, Moscow (2000)

ESTJ 2003 –   Etimologicheskiy slovar tyurkskikh yazykov. Obshchetyurkskie i mezhtyurkskie leksicheskie osnovy na bukvy "L", "M", "N", "P", "S" (in Russian) [Etymological dictionary of Turkic languages. All-Turkic and cross-Turkic lexical stems starting with letters "L", "M", "N", "P", and "S"]. Vostochnaya literatura RAN, Moscow (2003)

Räs 1969 –   Räsänen, M. Versuch einen etymologischen Wörterbuchs der Türksprachen. Suomalais-Ugrilainen Seura, Helsinki: (1969)

References

1. Baskakov, N.: Altayskiy yazyk [The Altai language]. In: Yartseva, V. (ed.) Yazyki mira. Tyurkskie yazyki, pp. 179–187. Indrik, Moscow (1997). (in Russian)
2. Mudrak, O.: Ob utochnenii klassifikatsii tyurkskikh yazykov s pomoshch'yu morfologicheskoy lingvostatistiki [On the specifying of Turkic languages classification using morphological linguostatistics]. In: Tenishev, E. (ed.) Sravnitelno-istoricheskaya grammatika tyurkskikh yazykov. Regionalnyye rekonstruktsii, pp. 733–737. Nauka, Moscow (2002). (in Russian)
3. Munkasci, B.: Beiträge zu den alten Lehnwörtern im Türkischen. Keleti-Szemple VI(2–3), 376–379 (1905)
4. Novgorodov, I., Lemskaya, V., Gainutdinova, A., Ishkildina, L.: The Chulym Turkic language is of the Kipchak Turkic language origin according to the Leipzig-Jakarta list. Türkbilig 29, 1–18 (2015)

5. Novgorodov, I., Tokmashev, D., Gainutdinova, A., Ishkildina, L.: Remarks on the Leipzig-Jakarta list of the Teleut language. Crede Expero: transport, obshhestvo, obrazovanie, jazyk **4**, 48–58 (2016)
6. Novgorodov, I.: Ustojchivyj slovarnyj fond Tyurkskikh yazykov [The Leipzig-Jakarta list of the Turkic languages]. SMIK, Yakutsk (2016). (in Russian)
7. Pedersen, H.: Türkische Lautgesetze. Zeitschrift der Deutschen Morgenländlischen Gesellschaft LVII, pp. 535–561 (1903)
8. Shcherbak, A.: Vvedeniye v sravnitelnoye izucheniye tyurkskikh yazykov [Introduction to the Comparative Study of Turkic Languages]. In: Pokrovskaya, L. (eds.) Nauka, St. Petersburg (1994). (in Russian)
9. Tadmor, U.: The Leipzig-Jakarta list of basic vocabulary. In: Haspelmath, M., Tadmor, U. (eds.) Loanwords in the World's Languages: A Comparative Handbook, pp. 68–75. Mouton de Gruyter, Berlin (2009)
10. Census: Vserossiyskaya perepis naseleniya (in Russian) [All-Russian Census] (2010). http://www.gks.ru/free_doc/new_site/perepis2010/croc/perepis_itogi1612.htm. Accessed 05 January 2017

Verbs of Movement in the Selkup Language

Natalia V. Polyakova[1] ⓘ and Anastasia S. Persidskaya[2](✉) ⓘ

[1] Tomsk State Pedagogical University, Tomsk 634061, Russian Federation
nvp@tspu.edu.ru
[2] Municipal Autonomous General Education Institution Tomsk Secondary
School № 47, Tomsk 634006, Russian Federation
yatsan86@bk.ru

Abstract. In the Selkup dialects the lexical-semantic group of movement verbs is represented by a large amount of the verbs describing various ways of movement, its nature, general orientation against the speaker, environment in which this movement occurs, and some other additional components. Most Selkup verbs of movement (60%) are monosemantic, i.e. about 60% of meanings does not repeat, they are peculiar to one verb of movement. In the Selkup lexico-semantic group of verbs of movement the synonymic relations are poorly developed in comparison with Russian in spite of the fact that among the Selkup verbs of movement the synonymy is more developed than in other lexico-semantic groups of the studied language.

Selkup verbs of movement can be divided into two groups: verbs of undirected movement and verbs of directed movement. The group of verbs of directed movement includes two subgroups: verbs with absolute orientation (verbs of vector movement in horizontal direction, verbs of vector movement in vertical direction) and verbs with relative orientation (verbs denoting orientation of the movement to an object, verbs denoting orientation of the movement from an object).

The verb *qənqo* is the central element of the lexico-semantic group of verbs of movement in the Taz dialect of the Selkup language. The semantics of this verb does not include the description of conditions in which the movement occurs, its direction, way and mode; however, some meanings of this verb are oriented in space. The verbs *čāǯịgu* and *kwangu* are the central elements of the lexico-semantic group of verbs of movement in the southern and central dialects of the Selkup language. Description of the general direction of the movement (approach or removal) and orientation in space is the most essential feature of the Selkup verbs of movement. There are several specifics of the lexico-semantic group of verbs of movement in the Selkup dialects: existence of the space-oriented verbs; verbs denoting movements in space made by means of certain devices/vehicles; almost full incompatibility of verbs of movement with abstract nouns and nouns denoting objects, incapable of the independent movement.

Keywords: Selkup · Verbs of directed movement · Verbs of undirected movement

© Springer International Publishing AG 2018
A. Filchenko and Z. Anikina (eds.), *Linguistic and Cultural Studies:*
Traditions and Innovations, Advances in Intelligent Systems and Computing 677,
DOI 10.1007/978-3-319-67843-6_36

1 Introduction

In all languages the group of verbs of movement denotes the vital concept – the movement – the most amazing phenomenon in the world. The movement is truly considered to *be the main property* and *the main sign of life* in the modern world. Today *the movement* is understood not only as the movement of some bodies against others, but also as the change of body temperature, electromagnetic radiation, it is connected with any changes and currents in society and nature. Thus, the concept *movement* has such width of interpretation which allows considering the concept *movement* as one of fundamental categories in different sciences, first of all, in physics, philosophy and linguistics. The practical, symbolic and axiological role of the movement also defines interest of linguistics in this concept. At the same time verbs of movement in linguistics are understood as any lexemes denoting change of position of a subject or its parts against any reference point.

Verbs of movement are united on the basis of presence of the general categorial and lexical sema 'movement in space' in their meaning. Further these verbs are differentiated according to the abstract and semantic category 'general orientation of the movement', 'environment of the movement', 'way of the movement', 'speed of the movement', 'figurative characteristic of the movement'. As well as in other Samoyed languages, an organic sign of verbs of movement in the Selkup dialects is the movement orientation which is strictly differentiated in semantic structure of verbs. Differentiation is carried out according to the following abstract semantic categories: the movement in vertical direction – the movement in horizontal direction, the movement towards the subject of observation – the movement from the subject of observation.

In this article the verbs of the directed and undirected movement in the Selkup dialects are analyzed.

2 Verbs of Undirected Movement

The vocabulary of the Selkup language which has recently acquired the system of writing is a source of historical, cultural, ethnographic data on material and spiritual culture of the Selkup ethnos. This factor encouraged researchers to study various lexico-semantic groups of the Selkup dialects [2–5].

The lexico-semantic group of verbs of movement is one of the groups of verbs in the Selkup dialects which is quite difficult to analyze. It is caused not only by their considerable amount, but also by their semantic complexity and heterogeneity.

Selkup verbs of movement are subdivided into two groups: (1) verbs of directed (focused) movement; (2) verbs of undirected movement, based on presence of the significant of focused (in one direction) movement or undirected (without exact indication of the direction) movement in their meanings.

2.1 Core of Lexico-Semantic Group of Undirected Movement Verbs

The verb *qənqo,* the semantics of which does not contain the significant of the direction of the movement, environment in which the movement occurs, its way and mode, is one of the central elements in the lexico-semantic group of verbs of movement in the Taz dialect of the Selkup language [3].

The sememe of the verb *qənqo* includes an archiseme 'to move', a differential seme 'a means of movement (by means of legs)', a differential seme 'environment of the movement (land)', a potential seme 'mode of the movement':

- to set out, to go somewhere: Taz. *təp qənpa takky* 'he has gone to the north'; *mat qəntak qäĺtyryqo* 'I will go (I will set out) to walk';
- to leave (to go away, to depart, to fly away, to sail away): Taz. *tō qənäš!* 'Go away' – this meaning can be considered as focused in space;
- to move on the ground by means of legs, to go, to walk: Taz. *topysä qəntak* 'I will go on foot' [3, p. 78].

In the Taz dialect of the Selkup language the verb *qənqo* is used to denote the movement made by means of special devices or by vehicles:

- to move, by means of any type of transport (to float, to drive): Taz. *antysä qəntak* 'I will float by a branch (I will float on a branch)'
- to move by means of special devices (about transport): Taz. *alako qəssa* 'the boat has floated' [3, p. 78].

The verbs *čāǯįgu*, *čāǯygu* and *kwangu* are the most frequent verbs in the central and southern dialects of the Selkup language that denote independent movement on a firm surface [1]. Examples showing the use of verbs of movement recorded in Selkup folklore texts and the texts of everyday speech are worth overviewing:

Okkyr bar čāǯa, qonǯyrnyt, pēge āmda 'one day he goes and sees a hazel grouse sitting' [6, p. 138]; Ob. Sh. *azįt maďond kwanba sūrəm kwatku aza tömba* 'the father went to hunt to the taiga and did not return' [7, p. 170], Ob. Sh. *onǯ šöt kwanba* 'he went to the forest' [8, p. 164].

Semantics of the above-stated verbs is identical to semantics of the verb *qənqo*: they denote the movement on a firm surface which is made by means of legs; however these verbs can also be used to denote the movement carried out by means of special devices or with use of any vehicles:

- Ob. Sh. *mat kwanǯak kyba andse čāǯygu* 'I will go to sail the oblasok for a while' [8, p. 163]; *mat kįba ande čandzak kįyįt, šabįrgak, labe tüak* 'I am going down the river by oblasok, I am fishing, I am rowing with an oar' [9, p. 134].

Comparing concordance of verbs of movement in the Taz dialect of the Selkup language with the same aspect of Russian verbs of movement, it should be noted that the Russian verb *идти* 'to go' is used both with animate and inanimate objects, compare: *дети идут* 'children are going', *снег идет* 'it is snowing', *дождь идет* 'it is raining', *часы идут* 'time goes', etc. while in the Taz dialect verbs of movement are combined only with animate objects. However cases showing the use of verbs of movement with names of atmospheric precipitation are recorded in the southern and central dialects of the Selkup language which can be explained by the influence of Russian: Ket. *sįrį čāǯan* 'it is snowing' [9, p. 72]. It should be noted that cases when the verbs of movement are used with abstract nouns are not the rule, but the exclusion.

2.2 Periphery of Lexico-Semantic Group of the Undirected Movement Verbs

Selkup verbs of undirected movement include the following semantic components in their meanings:

- 'environment in which the movement occurs': *pɛntyqo* 'to go down the stream, without rowing'; *ūrqo* 'to swim'; *tiĺtyrqo*, *tīmpyqo* 'to fly (about birds, planes, the objects scattered by the wind)': UO *tep ūrnə ūdan ēlɣan* 'he swims under water'; Kel. *nun inne piĺaqət śipa timbotin* 'the duck is flying in the sky';
- 'way of the movement': *toĺcysä qäĺtyryqo* 'go skiing'; *tōtqo (tōtyrqo) toĺcysä* 'to ski'; *qaqlyttyqo* 'to move by sledge' [1, p. 79, 81]. These verbs are in the inclusion relation with the verb *qənqo* which can obtain the meanings that these verbs transfer, as well as many other. The verbs *tōtqo, qaqlyttyqo* have a single meaning 'to ski', 'to move by sledge';
- 'speed of the movement': *qoptyryqo* 'to run fast'; *soqqyqo* 'to crawl': Nap. *tuk soqqa nüt pudoyot* 'a snake is slithering in grass';
- 'the figurative characteristic of the movement': *ńeśyrqo* 'to slide'.

Verbs of undirected movement can acquire any postpositions with spatial meaning the choice of which is defined by a speech situation. Compare: Nap. *ēdat qanqit qoryo qwejamba* 'a bear roamed near the village'; Fark. *mett't kimit qelimpisak* '(I) walked in the middle of the road'; Nap. *ija kananə uyout čaža* 'the boy is walking ahead of a dog'.

3 Verbs of Directed Movement

The verbs focused in space accounts for 20% of total number of the Selkup verbs of movement.

The orientation sign of verbs of directed movement receives a specification. It is lexically expressed in specification of the direction by serial postpositions; grammatically it is specified by the formants coinciding with suffixes of the inflectional system.

3.1 Verbs with Absolute Orientation

Verbs of Vector Movement in Vertical Direction. *Verbs with the meaning 'the downward movement'.* Verbs of movement of this subgroup are united by a general meaning 'to go down, to fall down'.

In the Taz dialect of the Selkup language the verbs *panišqo, paišqo, pańcyrqo, quptyqo* denote the downward movement. As a rule, the verbs *panišqo, paišqo* have the meanings 'to go down (about various subjects)', and the verb *quptyqo* also means 'to fall down': *pōmyn illä panišäš* 'go down the tree!' [3, p. 82].

In the southern and central dialects of the Selkup language the downward movement is denoted by the verbs *āĺdigu, p'ngəlgu* with the meaning 'to fall down somewhere or from somewhere': Lask. *qibače elle qət paroyondo aĺča* 'the child has fallen down from the mountain'; Iv. *tap aĺd'ä perenát pārónt* 'she fell onto the bed';

Fark. *šibá tot kinti al'čisa* 'the duck fell into the middle of the lake'; UO *ütče qən* 'the child fell downhill'.

Verbs with the meaning 'the upward movement'. The upward movement is expressed by the verbs *βašegu* 'to fly up', *siγilgu* 'to climb up': *ńab mat č'ad βaǯeǯ'a* 'the duck flew up in front of me'; *qāqe qən pāryndə siγəlgin assə nuniḳkịn, qāqe üttə patkun, assə ḳịγịn pregu* 'a riddle: when it goes uphill, it does not get tired, when it goes into water, it does not want to drink' [10, p. 34].

Verbs of Vector Movement in Horizontal Direction. Verbs of this subgroup differ on the basis of an orientation of the movement 'inside' and 'from inside'.

The meaning 'to come into' is delivered by the verb *šērqo*: *mat šērlāk, pūt tan šērāšyk* 'I come into, then you come into'; *təp tūsy man mōnny tətty, mōtty cäŋka šēräty* 'he reached my house, (but) he did not come into the house' [3, p. 81]; *tab mādap nömbad, korend šerba* 'she opened the door and came into the booth [6, p. 85]; *matkịt wargak, qal'da äral'dziga čünde töγa mekka, māt šerna, mekka čenča: tabel eja ili n'etuγak?* 'I lived in the house, an old man arrived to me on a horse, he came into the house and told me: 'Do you have squirrels, don't you?'" [9, p. 115]; Lask. *nälγup madət šüńńenǯ šērna* 'the woman came into the house'; Iv. *pajaga mat puǯond šerna* 'the old woman came into the house'.

The verbs *patqo, patqylqo* have the meaning 'to get into something (so that it was not visible)': *cēlyty patty* 'the sun declined'; *kβịnnat, üdịmba, čeldị patpa* 'we left, it got dark, the sun declined' [9, p. 128]; *mat īm eγarmit, tetti pot mödečḳimba, šịdị bar čačemba, nagurumǯel čut puǯot patpinda* 'my son was in the army, four years he was at war (there was a war on), he was wounded two times, with the third wound he went under the earth'.

The antonymous meaning 'to go out, to leave' is denoted by the verbs *taryqo, tantyqo*, 'to go out quickly from somewhere, to jump out' – by the verb *putylmōtqo*: *lōsyt näl'a ponä tary* "the devil's daughter went out"; *mōtqyny ponä tantysa* 'I went out of the house'; *mōtqyny putylmōtqo* 'to jump out of the house' [3, p. 82].

The verbs denoting the movement from inside *čarigu* 'to jump out, to go out'; *qončel'd'igu* 'to jump out'; *paktirgu* 'to jump out' acquire postpositions containing suffixes of the initial case: Fark. *noma pot toapqene paqtisa* 'the hare jumped out from under the tree'; Kel. *ńöma paqt'itol'sit pot put'inoń* 'the hare jumped out from under the tree'; Lask. *šepka tīdət sünǯeut čara* 'the chipmunk jumped out from the osier-bed'; Iv. *mad't mogogindo qončel'd'imba kanak* 'the dog jumped out from behind the house'.

3.2 Verbs with Relative Orientation

Verbs Indicating Orientation of the Movement to an Object. In the Taz dialect of the Selkup language the verbs *tüqo* and *tül'cyqo* 'to come, to arrive' are used in the meaning 'to reach some limit': *təp tūsy tammyt tətty, tymty utyrysyty* 'he reached this place, there he stopped (the sledge)'; *šittymtäl aj na tünty* 'he came again the second time' [3, p. 81]; in the central and southern dialects of the Selkup language the verbs *tögu* and *tüγu* are used in the same meaning: *nädek pajan tömba* 'the girl came to an old woman' [7, p. 166]; *hɛr čušpeliγa, mi kare tüγaj* 'snow began to melt, we came home' [9, p. 108]; *ambam timn'ad tömbadɟt* 'the mother and the father arrived' [9, p. 15].

The verbs *tögu* 'to come, to approach', *medigu* 'to reach, to approach' contain an orientation of the movement forward to a subject in the semantics. At the same time they denote the achievement of any limit, level: *okkɪr čel medak māt* 'one day arrived home'; *aza öndeďist, qanduk pajat medymba* 'I did not hear how the wife approached'.

Verbs Indicating Orientation of the Movement From an Object. This subgroup includes the verbs indicating removal from a reference point: *kunyqo* 'to run away from somewhere': *təp kunnɛja, topysä qənny* 'he ran away, he went on foot', *tab kunymba* 'she ran away'; *kwanəšpugu* 'to leave, to get ready for the trip'; *kuralgo* 'to escape': *newa loγanando kuralba* 'the hare escaped from the fox' and also the verbs *qənqo, paktyqo, kwangu* among others and the meaning 'to leave, to escape, to run off': *azɪt maďond kwanba sūrəm kwatku aza tömba* 'the father went to hunt to the taiga (and) did not return'; *onʒ šöt kwanba* 'he went to the forest'.

4   Conclusion

Designation of the general orientation of the movement (approach or removal) and orientations in space are the most essential for the Selkup verbs of movement. Specifics of the lexico-semantic group of verbs of movement in the Selkup dialects are the existence of the verbs for orientation in space; the verbs denoting movements in space with use of certain devices/vehicles; poor development of the synonymic relations among verbs of movement; almost full incompatibility of these verbs with abstract nouns and nouns denoting objects, incapable of the independent movement. The movement and the ability to the move is a marker of animateness, on the basis of this marker representatives of the Selkup ethnos differentiate animate and inanimate, i.e. alive and dead. Everything that moves is animated.

Verbs of motion in the Selkup dialects are divided into verbs of undirected movement and verbs of directed movement. In the lexico-semantic group of verbs of undirected movement we can distinguish the core which is formed by the most frequent polysemantic verbs *kwangu, čāʒɪgu* and *qənqo*, and the periphery which is formed by the verbs of undirected movement containing the lexico-semantic components 'conditions in which the movement occurs', 'way of the movement', 'speed of the movement', 'the figurative characteristic of the movement' in the meanings.

Verbs of directed movement are subdivided into verbs with absolute orientation (verbs of vector movement in horizontal direction, verbs of vector movement in vertical direction) and verbs with relative orientation (verbs denoting orientation of the movement to an object, verbs denoting orientation of the movement from an object).

Abbreviations.

Iv. –	Ivankino
Kel. –	Kellog
Ket. –	The Ket dialect

Lask. –   Laskino
Nap. –   Napas
Ob. Sh. –   The Ob Sheshkups' dialect
Taz –   The Taz dialect
UO –   Ust-Ozernoje
Fark. –   Farkovo

References

1. Bykonya, V.V., Kuznetsova, N.G., Maksimova, N.P.: Sel'kupsko-russkij dialektnyj slovar' [Selkup and Russian dialect dictionary]. TSPU publishing house, Tomsk (2005). (in Russian)
2. Bykonya, V.V., Karmanova, Y.A.: Lingvokul'turologicheskij analiz naimenovanij nasekomyh v sel'kupskom yazyke [Cultural and linguistic analysis of insects names in the Selkup language]. Sib. Phil. J. **2**, 163–174 (2012). (in Russian)
3. Kuznetsova, A.I., Helimskij, E.A., Grushkina, E.V.: Ocherki po sel'kupskomy yazyky. Tazovsky dialekt [Essays on the Selkup language. The Taz dialect], vol. 1. Moscow University publishing house, Moscow (1980). (in Russian)
4. Normanskaya, J., Krasikova, N.: Nekotorye prichiny semanticheskih izmenenij (na material genezisa i razvitiya sistemy glagolov plavaniya v sel'kupskom yazyke) [Some reasons of the semantic changes (on the material of genesis and devolopment of the verbs of swimming in the Selkup language]. Ural-Altaic Stud. **1**, 49–60 (2009). (in Russian)
5. Persidskaya, A., Polyakova, N.: Linguocultural Analysis of Nomination of Fingers in the Selkup Language. Mediterr. J. Soc. Sci. **6**(3), 448–452 (2015)
6. Filchenko, A., Potanina, O., Kryukova, E., Baydak, A., et al.: Annotirovannye folklornyke teksty obsko-eniseyskogo yazykovogo areala [Annotated Folklore Prose Texts of Ob-Yenissei Linguistic Area], vol. 1. Veter publishing house, Tomsk (2010). (in Russian)
7. Filchenko, A., Potanina, O., Kryukova, E., Baydak, A., et al.: Sbornik annotirovannykh folklornykh i bytovykh tekstov obsko-eniseyskogo yazykovogo apeala [Annotated Folklore and Daily Prose Texts of Ob-Yenissei Linguistic Area], vol. 2. Vajar publishing house, Tomsk (2012). (in Russian)
8. Filchenko, A., Potanina, O., Tanoyan, M., Kurganskaya, Y., et al.: Sbornik annotirovannykh folklornykh i bytovykh tekstov obsko-eniseyskogo yazykovogo apeala [Annotated Folklore and Daily Prose Texts of Ob-Yenissei Linguistic Area], vol. 3. Vajar publishing house, Tomsk (2013). (in Russian)
9. Filchenko, A., Potanina, O., Tanoyan, M., Kurganskaya, Y., et al.: Sbornik annotirovannykh folklornykh i bytovykh tekstov obsko-eniseyskogo yazykovogo apeala [Annotated Folklore and Daily Prose Texts of Ob-Yenissei Linguistic Area], vol. 4. TML-Press publishing house, Vajar publishing house, Tomsk (2015). (in Russian)
10. Filchenko, A.Y.: Assimetrichnoe otritsanie v Vostochno-Khantyiskom i Yuzhno-Sel'kupskom [Assymetric Negation in Eastern Khanty and Southers Selkup]. Tomsk J. Linguist. Anthropol. **2**(2), 29–49 (2013). (in Russian)

"The World of Shakespeare" in the Stroganovs' Book Collection: Literary and Graphic Works

Irina A. Poplavskaya[1] , Irina V. Novitskaya[1(✉)] ,
and Victoria V. Vorobeva[2]

[1] Tomsk State University, Tomsk 634050, Russian Federation
poplavskaj@rambler.ru, irno2012@yandex.ru
[2] Tomsk Polytechnic University, Tomsk 634050, Russian Federation
victoriavorobeva@mail.ru

Abstract. The Stroganovs' personal library as an organic whole unit is housed in the Scientific Library at Tomsk State University and contains some Shakespeare's literary works in English which are accompanied by numerous graphic illustrations to his tragedies, comedies and historical chronicles. This part of the personal multilingual book collection is analyzed in the present article in terms of the intermedia studies approach which allows to reveal how "the cult of Shakespeare" was being constructed in the British literature of the XVIII-XIX centuries. This task entails considering the role that Shakespeare's works played in the development of the national self-awareness and national identity concept in the British society at that time as well as the construction of the Imperial myth in the British literature. Besides, a case study of the English editions of Shakespeare's works kept in the Scientific library enables one to trace the evolution of Shakespeare studies in Europe. It can also give an insight into the reception of the Briton's works in the Russian society in the XIX century. The article addresses the issue of the impact that the library owner's personality has on the composition of the book collection and the ensuing interpretation of how the Western European and Russian literatures interact within the Stroganovs' library.

Keywords: Shakespeare · The Stroganovs' book collection · Tomsk State University · Graphic works · National identity · Western European world

1 Introduction

The myth of Shakespeare has been developing in the British culture since the time of the poet's death. An important role in this process has been played by the practice of publishing the poet's biographies, collected works, as well as paintings based on the plots of his plays and chronicles in the XVIII–XIX centuries. Taken from this perspective, a study of the published books and illustrations to them, as well as critiques of the performances of his plays can help one to comprehend the inception of Shakespeare studies as a scientific branch. It can also give an insight into Shakespeare as a cultural phenomenon in the Western European countries and Russia, reveal some features

A. Filchenko and Z. Anikina (eds.), *Linguistic and Cultural Studies:*
Traditions and Innovations, Advances in Intelligent Systems and Computing 677,
DOI 10.1007/978-3-319-67843-6_37

specific to the Russian reception of the poet's heritage while examining his translated works and books from a personal library.

The Scientific Library of Tomsk State University is the place where one of the unique book collections in Siberia is being kept. It is the Stroganovs' family library donated to the Imperial Tomsk University in 1879. The library's last owner was Earl Grigoriy Alexandrovich Stroganov (1770–1857), a famous statesman at the times of Alexander I (1777–1825) and Nicholas I (1796–1855), the Russian envoy in Madrid (1805–1809), the plenipotentiary minister in Stockholm (1812–1816), the head of the diplomatic mission in Constantinople (1816–1821), a member of the State Council, an honorary member of St. Peterburgh Academy of Sciences, a representative of the Russian Empire at the British Queen Victoria's coronation ceremony in London in 1838.

At present the Stroganovs' library contains about 24,000 volumes in French, German, English, Spanish, Russian and other languages, which is indicative of the "Russian European" earl Stroganov's wide range of interests tied to his professional activities, personal literary and artistic preferences and tastes. Among the factors that are believed to have contributed to the formation of the library are the earl's continuous exposure to the influence of the European cultures during his stays abroad and proficiency in some European languages. The fact that the library contains several volumes of Shakespeare's works speaks in favour of Stroganov's interest in the British literature in general and the great poet's works in particular. This interest may have been deepened and maintained by the earl's personal cultural interests as well as contacts with the representatives of the royal families from Britain, Spain, Portugal, Sweden and with Western European diplomats and famous cultural figures at the turn of the XIX century. According to some research [25], when G.A. Stroganov was appointed the envoy in Spain in 1805, he did not go to Madrid at once, but chose to stop in London for some time to be able to meet with the British Prime Minister W. Pitt, the Younger (1759–1806). It can be assumed that the earl might have purchased the English editions of Shakespeare's works during that visit. It is also worth mentioning that Stroganov's great-uncle, Baron Alexander Grigorjevich Stroganov (1699–1754) was the first to translate John Milton's "Paradise Lost" in 1745 [14, pp. 134–136].

The present article aims to study the development process for the cult of Shakespeare based on the materials of one of the internal subsections in the Stroganovs' personal library. This subsection includes Shakespeare's works published in the English language and containing graphic illustrations. The examination of Shakespeare's heritage in English allows to address such issues as the formation of the national culture in the course of nation-building in England, the inception of Shakespeare studies in the British literature critique of the XVIII–XIX centuries, the practice of intermedia construction of the characters in literature and arts.

2   Research Context

The theory underpinning the present study is developed by numerous researchers of the book, the national and personal library, the history of reading and the theory of communication [2, 4, 5, 10, 17, 18, 21].

The study of Shakespeare's collected works in various languages along with illustrations to his stories is based on a variety of methods, including the imagological [12, 22], receptive [7] and intermedia ones [6, 19]. An important role is given to the methods allowing to investigate both the history and culture of certain ethnic groups and the practice of nation-building [3, 23]. The focus on the English section of the library results in a description of the national approach to the presentation of the poet's personality and works within the framework of the intermedia narratology [19, 27]. This creates an opportunity to compare the British tradition with similar practices in other European cultures.

In addition to that, the methods helpful to study family libraries belonging to the Russian gentlefolk of the XVIII–XIX centuries are the historical and cultural [11] and meta-communicative ones [24]. These approaches allow to reveal, how different language and ethnic-cultural worlds come to interact within the Stroganovs' library and how they are refracted in the reader's consciousness of the library owner. According to Tomsk philological school, a library should be holistically considered as a unit of culture and communication. It is, on the one hand, a universal meta-cultural form, and a meta-communicative model of culture, on the other [24]. The communicative model of the personal library is a multilayered structure which is composed of the primary (sender – text – receiver) and secondary (reader – library – owner; reader – bibliographical text (catalogue) – librarian) acts of communication.

3   A Study of Shakespeare's Works in the Stroganovs' Book Collection

3.1   Description of the Internal Collection

The largest section of the Stroganovs' personal library, which amounts to 20,000 volumes, is made of books in French. It contains some handwritten documents dating back to the XII–XVII centuries, four books of which used to be a part of the library belonging to French King Louis XVI (1754–1793). Literature in other European languages (English, German, Spanish, Russian and other) makes up a small section of the library and contains around 4,000 volumes.

There is a substantial number of the western European and Russian works of graphic arts, including some graphic images of the paintings housed in the Uffizi Gallery Museum and the Pitti Palace Museum in Florence, Italy, in the Dresden Gallery and Munich Pinakotheka in Germany, copies of the murals found in the Italian city of Herculanium, a collection of engravings from John Boydell's Shakespeare Gallery which illustrate 27 works of the British bard, William Turner's graphic works and lithographic portraits from the collection of the Military Gallery in the Winter Palace.

3.2   British Shakespeare Through the Eyes of Russian Reader

The earliest and most valuable part of the English section of the Stroganovs' library is the 15-volume edition titled "The Plays of Shakespeare" [29]. It came out in 1793 in London and was edited by Samuel Johnson (1709–1784), a poet and literary critic, and

George Steevens (1736–1800), a researcher. It was the fourth edition of Shakespeare's collected works prepared by the same editors in collaboration (previous editions were published in 1773, 1778 and 1785). All volumes of this edition are available in the library, each volume has G.A. Stroganov's ex libris.

The genre of the publication can be determined as a scientific, literary and artistic work since it includes the poet's biography, a comprehensive description of the years of his creative work, some textological analyses, other authors' works devoted to Shakespeare and 34 plays of the English playwright with comments to them. What makes this edition a special one is the fact that it contains many engraved portraits of the famous researchers who studies the poet's oeuvre as well as numerous engraved illustrations to his dramatic works. Along with the earlier editions of Shakespeare's collected works (published over the period from 1709 to 1790) this most scientifically and artistically complete edition opens an era of Shakespeare studies in England, France, Germany and Russia.

What draws first attention is the fact that the layout of the edition differs from what is common in the present day publications. The contents section is found in the third volume, while the plays are presented in volumes 3–15. The first two volumes introduce the works devoted to the genius writer.

In the first volume of the edition there is a Shakespeare's biography "Some account of the life of William Shakespeare" written by an English playwright, Nicholas Rowe (1674–1718), who was the first to regard Shakespeare as a speaker of the national tradition in the English dramatic art. It was N. Rowe who presented Shakespeare's dramatic works in the form compliant with the rules established in the XVIII century: he divided the text into scenes and acts, attached a dramatis personae to each play, regularized the entrances and exits of characters during the performance, and specified the scene for each act.

Another type of materials that the editors placed in the first volume is the prefaces to all previously published works by Shakespeare. These prefaces are considered as important milestones in the development of Shakespeare studies which, in its turn, is believed to have had an impact on the creation of the national "myth of Shakespeare" in the English culture of the XVIII century. The list of authors is impressive and includes such names as a famous poet Alexander Pope (1688–1744), a writer and translator Theobald Lewis (1688–1744), an editor Thomas Hanmer (1677–1746), a writer, literary critic and churchman William Warburton (1698–1779), editors Samuel Johnson and George Steevens, a critic and publisher Edward Capell (1713–1781), an English philologist and editor Isaak Reed (1742–1807), and a researcher Edmond Malone (1741–1812). To make it more personalized, each preface is accompanied by a lithographic portrait of its author. It is in these prefaces that we find attempts to critically assess the innovative and original creative approach that Shakespeare chose to pursue in his works. According to Johnson, "Shakespeare's plays are not in the rigorous and critical sense either tragedies or comedies, but compositions of a distinct kind; exhibiting the real state of the sublunary nature, which partakes of good and evil, joy and sorrows, mingled with endless variety of proportion and innumerable modes of combination…" [29, vol. 1, p. 188]. Malon adds, that "Perhaps he was as learned as his dramatic province required; for whatever other learning he wanted, he was master of two books unknown to many of the profoundly read <...> the book of nature and that of man" [29, vol. 1, p. 419].

One of the factors that contributes to the value of the edition is that it contains a list of the plays and poems that are considered to be the earliest works of the poet first published when he was alive. According to the edition, the historical chronicle "Richard III" was successively published seven times (in 1597, 1598, 1602, 1612, 1622, 1629, 1634), "Romeo and Juliet" came out five times, the earliest edition was produced by a notorious pirate printer John Danter in 1597.

An important part of the edition under analysis is given to the poems dedicated to Shakespeare. Among the authors who contributed their poems are William Basse (1583–1653), Ben Jonson (1572–1637), John Milton (1608–1674), James Thomson (1700–1748), Thomas Gray (1716–1771), Samuel Johnson, and others. Their evaluation raises Shakespeare to the rank of the greatest playwright and poet – a "rare tragedian" (W. Basse), "a son of memory, an heir of fame" who manages to create a "live-long monument" (John Milton) in the hearts of the readers.

The artistic charm to the edition is being created by 141 graphic illustrations that accompany Shakespeare's dramatic works and commentaries to them. An interested reader will find lithographic portraits of the English bard, Queen Elisabeth (1533–1603), book publishers, reviewers, as well as illustrations to the plays. For example, there is a lithograph depicting the Rialto Bridge (ital. Ponte di Rialto) mentioned three times in Act 1, Scene 3 of "The Merchant of Venice". The effect produced by a concurrent verbal and visual representation of the bridge is that of emphasis put on the symbolic background of the comedy. The bridge is assumed to be a symbol of wealth and loss, unpredictability of business outcomes and changeability of people's fates, it contributes to the portrayal of the heroes' complicated and ambiguous personalities.

All in all, this abundance of illustrations in the edition helps to create a number of intermedia texts (resulting from an interplay of literature and paintings) which construct the "visual field", i.e. a sort of visual background, to Shakespeare's literary works while simultaneously revealing how the principles of verbal arts are used in visual arts. The visual-verbal images in this edition are associated with an aesthetic portrayal of history and assignment of the mythologized status to some events and personas. Such synaesthetic complex texts are based on the reduplication or multiplication of certain images and motifs (e.g. images of a castle, cathedral, bridge, sea, ship, etc.) which help to convey the national peculiarities of the English literature.

In summation, this edition of Shakespeare's works can be considered as a handbook of his oeuvre, because it provides an overview of his life and dramatic pieces over the course of the XVIII century. The materials included in the edition contribute to the formation of "the myth of Shakespeare" as a founder of the national theatre and the national language which was closely tied to the nurturing of the national concepts of "englishness" and "britishness" as well as with the practices of national and cultural building in the XVIII–XIX centuries [1, 13, 20].

It can also be inferred that the library owner, G.A. Stroganov, tended to acquire the most valuable, in terms of content and aesthetic impact, books that met his personal tastes and preferences. Stroganov's undeniable interest in the English literature and culture may also have been inspired by Karamzin's "Letters of the Russian Traveller" who translated Shakespeare's "Julius Caesar" in 1786 and visited England in 1790.

Repetitive references to the English authors and their works testify for the wide introduction of the English literature to the Russian readership. Viewed from this perspective, Shakespeare's texts are assumed to have functioned as text-mediators between the two cultures.

Another type of evidence of the ongoing process of the English culture acquisition by the Russian readers is the fact that the Stroganovs's library possesses one of the unique world's graphic collections [30]. It consists of 67 prints included in "The Collection of Prints, From Pictures Painted for the Purpose of Illustrating the Dramatic Works of Shakspeare, by the Artists of Great-Britain" published in folio in 1791–1803 in London [28, 30]. 65 prints from the Stroganovs' library reproduce the scenes from Shakespeare's 27 plays: historical chronicles "Henry IV", "Henry VI", "Henry VIII", "Richard III", tragedies "Romeo and Juliet", "King Lear", "Macbeth", "Hamlet", and comedies "Merry Wives of Windsor" and "Much Ado about Nothing". There is also an engraving of the sculpture by Thomas Banks depicting Shakespeare standing between the Dramatic Muse and the Genius of Painting. One more engraving depicts the great English actor, playwright, theatre manager and producer David Garrick (1717–1779).

It is a well-known fact that the person who initiated and carried through the project on establishing a permanent exhibition of paintings based on Shakespeare's works was John Boydell (1720–1804). He managed to persuade a great number of renowned artists of that period to contribute their paintings to the Boydell Shakespeare Gallery that was opened in the center of London in 1789 [8, 16]. Later he published engravings of the commissioned paintings as illustrations for his edition of Shakespeare's plays [26]. In the Boydell Shakespeare Gallery each painting was accompanied by an excerpt from Shakespeare's plays which helped to convey the intricate verbal-visual nature of his artistic imagery. Thematically, all graphic works form 5 groups: engravings depicting the main characters who have transformed into the archetypal figures of the world literature (e.g. King Lear, Hamlet, Romeo, Juliet); engravings featuring mythological characters (e.g. Troilus and Cressida); engravings devoted to the plots of Shakespeare's historical chronicles (e.g. Richard III, Henry IV) as well as battle scenes (e.g. Henry VI); engravings depicting daily life (e.g. Merry Wives of Windsor, Much Ado about Nothing, Henry IV); engravings presenting imaginary (e.g. A midsummer Night's Dream, The Tempest) or landscape scenes (e.g. The Merchant of Venice, As You Like it).

Among the artists presented in the Stroganovs' print collection are the first President of the Royal Academy of Arts Joshua Reynolds (1723–1792), his disciple, portraitist James Northcote (1746–1831), a Swiss-born artist Henry Fuseli (1741–1825), an American portraitist Benjamin West (1738–1820), a Swiss neoclassical painter Angelica Kauffmann (1741–1807), artists John Opie (1761–1807), Robert Smirke (1752–1845), William Hamilton (1751–1801), printer, painter and engraver Josiah Boydell (1752–1817) and others. The famous engravers whose plates are found in the Stroganovs' library are Italians Francesco Bartolozzi (1727–1815) and Luigi Schiavonetti (1765–1810), Jan Pierre Simon (1750–1810), an appointed historical engraver to the Prince of Wales Robert Thew (1758–1802), a woman-engraver Caroline Watson (1761–1814) and others.

According to some researchers, one of the reasons that accounts for the appearance of the Boydell Gallery and the Collection of Prints is the process known as

"Shakespeare appropriation" affecting all arts: new editions of Shakespeare's collected works, their translation into German, French, and other languages, stage performances of his plays in the countries of Europe and in the USA led to an increased demand for the illustrated Shakespeare [15]. "Mood rather than motive now appears the primary quality by which character and action may be judged" [9]. The Historic Gallery opened by Robert Bowyer before 1806 in London, Thomas Macklin's Gallery of the Poets in 1790–1795 in London, James Woodmason's Shakespeare Gallery in 1792 in Dublin – these are just a few examples of the efforts to promote Shakespeare's works and foster a school of British history painting.

Another illustrated edition kept in the Stroganovs' library is "The Shakspeare Gallery, containing the principal female characters in the plays of the great poet" issued by an engraver, illustrator and publisher Charles Heath (1785–1848) in 1836 in London [31]. It contains 45 black-and-white engravings depicting main female characters of Shakespeare's plays that were made by such English engravers as Henry Richard Cooke (1800–1845), William Henry Mote (1803–1871), John Henry Robinson (1796–1871) on the basis of the paintings by an illustrator Joseph Kenny Meadows (1790–1874), a portraitist John Hayter (1880–1891), a painter and sculptor Edward Henry Corbould (1815–1905) and others.

The most striking and memorable images in this portrait gallery are those of Cordelia (King Lear), Ophelia (Hamlet), Titania (A Midsummer Night's Dream), Jessica (The Merchant of Venice), Miranda (The Tempest), and Cleopatra (Antony and Cleopatra). Owing to these artistic portraits, the literary characters acquire a visual embodiment and, as such, provide some associative ties between the text of a play and its "visualization" on the stage. The genre of portrait enables a painter to convey the emotional state of the heroines by means of their appearance and posture, expressive household scenes or landscape images. Viewed from this perspective, portraits are believed to be contributing to the intermedia representation of Shakespeare's artistic imagery meaning that the verbal text is associated with its theatrical interpretation via painting. For example, engraved by W.H. Mote after J.K. Meadow Cordelia is captured at the moment she is reading a letter informing her of her father's mistreatment. The woman is depicted sitting on the throne holding the letter in one hand and a kerchief – in the other. Such details as the letter and kerchief are assumed to reveal an association among the outer, plot-based and inner, psychological lines in the tragedy.

4 Conclusion

The article presents a case study aiming at the reconstruction of the "world of Shakespeare" as it is rendered by the three English editions of Shakespeare's works which are a part of the Stroganovs' personal book collection in the Scientific Library of Tomsk State University. These editions are the collected works "The Plays of Shakespeare" (1793), 67 prints from "The Collection of Prints, From Pictures Painted for the Purpose of Illustrating the Dramatic Works of Shakspeare, by the Artists of Great-Britain" (1791–1803) and "The Shakspeare Gallery, containing the principal female characters in the plays of the great poet" (1836) [29–31].

The 15-volume edition of Shakespeare's works (1793) summarizes the research outcomes of Shakespeare's oeuvre in England over the XVII and XVIII centuries and to a large extent encourages the formation of Shakespeare studies and Shakespeare appropriation as a distinct branch in the modern industry of culture. The major impact of the edition is believed to be connected with the inception of the myth of Shakespeare introducing him as the greatest playwright and poet in England, as well as the founder of the English theatre and the standardized English language.

It is argued that Shakespeare's historical chronicles are associated with the development of the British Imperial myth in literature. Both reading and watching his plays at the theatre are believed to be inseparable from the formation of the feeling of national identity which was based on the "englishness" and "britishness" - the two key factors underpinning the ideology and practice of the national and cultural building in Great Britain.

The abovementioned graphic works based on the poet's plays contribute to the formation of the cult of Shakespeare in the Western European and world culture. The prints from the Boydell Shakespeare Gallery (1791–1803) and the gallery of female characters (1836) help to reveal a process in which the English history and culture were acquiring some aesthetic characteristics and mythological value. It is argued that these graphic works enable one to identify the general principles of creating artistic imagery brought about by the interplay of literature, painting and theatre. Owing to the inter-media approach it is possible to consider these verbal and visual works as a synaes-thetic type of the text.

In Russia the reception of Shakespeare's "English text" was promoted by N.M. Karamzin's oeuvre who visited the Boydell Shakespeare Gallery in London in 1790. The factors that encouraged interest in Shakespeare's works at that time in Russia were their translations into Russian and a common practice of collecting books for personal libraries amongst the aristocratic families in the country. Shakespeare's works were by all means the most significant part of those libraries.

Being one of the brightest representatives of the "Russian European" type, earl G. A. Stroganov showed his keen cultural interest in the oeuvre of the British playwright and in purchasing the most valuable literary, artistic and graphic editions of the bard's works. Viewed from this perspective, a study of Shakespeare's works in German and French that are a part of the Stroganovs' library is seen as a continuation of the present paper.

Acknowledgements. This research is supported by "Tomsk State University D.I. Mendeleev's Scientific Fund" (Project №. 8.1.25.2017).

References

1. Barczewski, S.: Myth and National in XIXth Century Britain: the Legends of King Arthur and Robin Hood. Oxford University Press, Oxford (2000)
2. Belovitskaja, A.A.: Knigovedenije. Obshchee Knigovedenije [The General Theory of the Book]. MGUP, Moscow (2007). (in Russian)
3. Bhabha, H.J.: The Location of Culture. Routledge, London (1994)

4. Carrière, J.-C., Eco, U.: Ne nadeites' izbavitsa ot knig! [Do not try to get rid of books!]. In: Simposium, St. Petersburg (2010). (in Russian)
5. Chartier, R.: Lectures et lecteurs dans la France de l'Ancien Régime. Seule, Paris (1987)
6. Cluver, C.: Intermediality and interarts studies. In: Arvidson, J., Askander, M., Bruhn, J., Fuhrer, H. (eds.) Changing Borders: Contemporary Positions in Intermediality, pp. 19–37. Intermedia Studies Press, Lund (2007)
7. Fish, S.: Is There a Text in This Class? The Authority of Interpretive Communities. Harvard UP, Cambridge (1980)
8. Friedman, W.H.: Boydell's Shakespeare Gallery. Garland Publishing Inc., New York (1976)
9. Gidal, E., Sillars, S.: The illustrated Shakespeare, 1709-1875. Nineteenth-Century Lit. **65**(1), 93–96 (2010)
10. Iljina, O.N.: Izuchenije lichnykh bibliotek v Rossii: materialy k ukazatelju literatury na russkom jazyke za 1934-2006 gg [Studies of Personal Libraries in Russia: Data for the Literature Reference List Over 1934-2006]. Sudarynja, St. Petersburg (2008). (in Russian)
11. Kolosova, G.I.: Russian fiction of the first half of XIX century in the book collection of count G.A. Stroganov. Tomsk State Univ. J. Cult. Stud. Art Hist. **2**(14), 45–53 (2014)
12. Leersseen, J.: Imagology: history and method. In: Beller, M., Leerseen, J. (eds.) Imagology: the Cultural Construction and Literary Representation of National Characters. A Critical Survey, pp. 17–32. Rodopi, Amsterdam (2007)
13. Lenman, B.: England's Colonial Wars 1550-1688: Conflicts Empire and National Identity. Longman, New York (2001)
14. Levin, YuD: Vosprijatije anglijskoi literatury v Rossii [Reception of the English Literature in Russia]. Nauka, Leningrad (1990). (in Russian)
15. Lukov, V.A.: Predromantizm [Pre-Romanticism]. Nauka, Moscow (2006). (in Russian)
16. Martineau, J., Shawe-Taylor, D.: Shakespeare in Art. Merrell, London (2003)
17. McLuhan, M.: The Gutenberg Galaxy: The Making of Typographic Man. 2nd edn. Akademicheskiy Project, Gaudeamus, Moscow (2013)
18. Migon, K.: Nauka o knige: Ocherk problematiki [The Theory of the Book: An Overview of the Issues]. Kniga, Moscow (1991). (in Russian)
19. Muller, J.E.: Intermediality and Media Historiography in the Digital Era. Acta Univ. Sapientiae, Film and Media Studies **2**, 15–38 (2010)
20. Murdock, A.: British History 1660-1832: National Identity and Local Culture. MacMillan Press/St. Martin's Press, London/New York (1998)
21. Nemirovskiy, E.L.: Bolshaja kniga o knige: spravochno-entsiklopedicheskoje izdanije [A Big Book About the Book: A Handbook]. Vremja, Moscow (2010). (in Russian)
22. Pageaux, D.-H.: De l'imagerie culturelle à l'imaginaire. Précis de littérature comparée. PUF, Paris (1989)
23. Plakhy, S.: The Origins of the Slavic Nations Premodern Identities in Russia Ukraine, and Belarus. Cambridge University Press, Cambridge (2006)
24. Poplavskaya, I.A.: Problemy izuchenija biblioteki Stroganovykh v Tomske: knigi frantsuzskikh pisatelej XIX v. [Research on the Stroganovs' collection in Tomsk: books of French writers of the XIX century]. Tomsk State Univ. J. Philol. **4**(20), 87–97 (2012). (in Russian). doi:10.17223/19986645/20/9
25. Saplin, A.I.: Rossijskij posol v Ispanii (1805-1809) [The Russian envoy in Spain (1805-1809)]. Voprosy istorii **3**, 178–184 (1987). (in Russian)
26. Sillars, S.: The Illustrated Shakespeare, 1709-1875. Cambridge University Press, Cambridge (2008)
27. Schröter, J.: Discourses and models of intermediality. CLCWeb: Comp. Lit. Cult. **13**(3) (2011). doi:10.7771/1481-4374.1790

28. Vasen'kin, N.V.: Shakespeare in the English engravings of the XVIII century. Collection of the Scientific Library of Tomsk State University. Catalogue. Tomsk State University Publishers, Tomsk (2012). (in Russian). http://www.lib.tsu.ru/win/vystavki/ork_1/sh.html. Accessed 18 May 2017
29. Shakespeare, W.: The Plays of Shakespeare: With the Corrections and Illustrations of Various Commentators. Notes by Samuel Johnson and George Steevens, 4th edn., vol. 15. Longman, London (1793)
30. The Collection of Prints, From Pictures Painted for the Purpose of Illustrating the Dramatic Works of Shakspeare, by the Artists of Great-Britain (1791-1803). Shakspeare Gallery, London (1805)
31. Heath, Ch.: The Shakespeare Gallery: Containing the Principal Female Characters in the Plays of the Great Poet, 4th edn. C. Tilt, London (1836)

Author Index

Printed in the United States
By Bookmasters